T0286277

Advances in Biochemistry

Advances in Biochemistry

Edited by **Oliver Stone**

New York

Published by Callisto Reference,
106 Park Avenue, Suite 200,
New York, NY 10016, USA
www.callistoreference.com

Advances in Biochemistry
Edited by Oliver Stone

International Standard Book Number: 978-1-63239-037-0 (Hardback)

Contents

Preface

This book was inspired by the evolution of our times; to answer the curiosity of inquisitive minds. Many developments have occurred across the globe in the recent past which has transformed the progress in the field.

Over the years, biochemistry has become significant in classifying living processes so much so that many scientists in the field of life sciences are involved in biochemical research. This book presents an analysis of the research area of proteins, enzymes, cellular mechanisms and chemical compounds that are used in relevant methods. It includes the basic issues and some of the current advancements in biochemistry. This book caters to students, researchers, biologists, chemists, chemical engineers and professionals who are keen to know more about biochemistry, molecular biology and other related fields. The chapters within the book have been contributed by renowned international scientists with expertise in protein biochemistry, enzymology, molecular biology and genetics; many of whom are active in biochemical and biomedical research. It will provide information for scientists about the complexities of some biochemical procedures; and will stimulate both professionals and students to devote a part of their future research in understanding related mechanisms and methods of biochemistry.

This book was developed from a mere concept to drafts to chapters and finally compiled together as a complete text to benefit the readers across all nations. To ensure the quality of the content we instilled two significant steps in our procedure. The first was to appoint an editorial team that would verify the data and statistics provided in the book and also select the most appropriate and valuable contributions from the plentiful contributions we received from authors worldwide. The next step was to appoint an expert of the topic as the Editor-in-Chief, who would head the project and finally make the necessary amendments and modifications to make the text reader-friendly. I was then commissioned to examine all the material to present the topics in the most comprehensible and productive format.

I would like to take this opportunity to thank all the contributing authors who were supportive enough to contribute their time and knowledge to this project. I also wish to convey my regards to my family who have been extremely supportive during the entire project.

Editor

Part 1

Metabolism and Mechanism

Enzyme-Mediated Preparation of Flavonoid Esters and Their Applications

Jana Viskupicova[1,2], Miroslav Ondrejovic[3,4] and Tibor Maliar[3]
[1]Institute of Experimental Pharmacology and Toxicology, Slovak Academy of Sciences
[2]Department of Biochemistry and Microbiology
Slovak University of Technology in Bratislava
[3]Department of Biotechnology, University of SS. Cyril and Methodius in Trnava
[4]Department of Biocentrum, Food Research Institute in Bratislava
Slovakia

1. Introduction

Flavonoids comprise a group of plant polyphenols with a broad spectrum of biological activities. They have been shown to exert beneficial effects on human health and play an important role in prevention and/or treatment of several serious diseases, such as cancer, inflammation and cardiovascular disease (Middleton et al., 2000; Rice-Evans, 2001). Flavonoids are important beneficial components of food, pharmaceuticals, cosmetics and various commodity preparations due to their antimutagenic, hepatoprotective (Stefani et al., 1999), antiallergic (Berg & Daniel, 1988), antiviral (Middleton & Chithan, 1993) and antibacterial activity (Tarle & Dvorzak, 1990; Tereschuk et al., 1997; Singh & Nath, 1999; Quarenghi et al., 2000; Rauha et al., 2000). They are known to inhibit nucleic acid synthesis (Plaper et al., 2003; Cushnie & Lamb, 2006), cause disturbance in membranes (Stepanovic et al., 2003; Stapleton et al., 2004; Cushnie & Lamb, 2005) and affect energy metabolism (Haraguchi et al., 1998). But the most studied activity is their antioxidant action since they can readily eliminate reactive oxygen and nitrogen species or degradation products of lipid peroxidation and are thus effective inhibitors of oxidation (Ross & Kasum, 2002).

However, their commercial applications are limited due to low solubility in lipophilic environment and low availability for a living organism. Although aglycons, prenylated and methoxylated flavonoid derivatives may be implemented into such systems, they are rarely found in nature and are often unstable. In some plant species, the last step in the flavonoid biosynthesis is terminated by acylation which is known to increase solubility and stability of glycosylated flavonoids in lipophilic systems. Selectively acylated flavonoids with different aliphatic or aromatic acids may not only improve physicochemical properties of these molecules (Ishihara & Nakajima, 2003) but also introduce various beneficial properqties to the maternal compound. These include penetration through the cell membrane (Suda et al., 2002; Kodelia et al., 1994) enhanced antioxidant activity (Viskupicova et al., 2010; Katsoura et al., 2006; Mellou et al., 2005), antimicrobial (Mellou et al., 2005), anti-proliferative (Mellou et al., 2006) and cytogenic (Kodelia et al., 1994) effect and improvement of thermostability and light-resistivity of certain flavonoids.

In nature, flavonoid acylation is catalyzed by various acyltransferases which are responsible for the transfer of aromatic or aliphatic acyl groups from a CoA-donor molecule to hydroxyl residues of flavonoid sugar moieties (Davies & Schwinn, 2006). Acylation is widespread especially among anthocyanins; more than 65% are reported to be acylated (Andersen & Jordheim, 2006). While the exact role of plant acylation is not yet fully understood, it is known that these modifications modulate the physiological activity of the resulting flavonoid ester by altering solubility, stability, reactivity and interaction with cellular targets (Ferrer et al., 2008). Acylation might be a prerequisite molecular tag for efficient vacuolar uptake of flavonoids (Kitamura, 2006; Nakayama et al., 2003). Some acylated flavonoids have been found to be involved in plant-insect interactions; they act as phytoalexins, oviposition stimulants, pollinator attractants (Iwashina, 2003), and insect antifeedants (Harborne & Williams, 1998). With respect to novel biological activities, acylation of flavonoids can result in changes in pigmentation (Bloor, 2001), insect antifeedant activity (Harborne & Williams, 1998) and antioxidant properties (Alluis & Dangles, 1999).

Over the past 15 years, there has been a substantial effort to take advantage of this naturally occurring phenomenon and to implement acylation methods into laboratories. However, the use of acyltransferases as modifying agents is rather inconvenient, as they require corresponding acylcoenzyme A, which must be either in stoichiometric amounts or regenerated *in situ*. Natural acyltransferases and cell extracts from *Ipomoea batatas* and *Perilla frutescens* containing acyltransferases were applied for selective flavonoid modification with aromatic acids (Tab.1) (Nakajima et al., 2000; Fujiwara et al., 1998).

Acyltransferase	Plant source	References
hydroxycinnamoyl-CoA:anthocyanin 3-O-glucosid-6''-O-acyltransferase	*Perilla frutescens*	Yonekura-Sakakibara et al., 2000
malonyl-CoA:anthocyanin 3-O-glucosid-6''-O-malonyltransferase	*Dahlia variabilis*	Wimmer et al., 1998
hydroxycinnamoyl-CoA:anthocyanin 5-O-glucosid-6''-O-acyltransferase	*Gentiana triflora*	Tanaka et al., 1996
hydroxycinnamoyl-CoA:anthocyanidin 3-rutinosid acyltransferase	*Petunia hybrida*	Brugliera & Koes, 2003
malonyl-CoA:anthocyanidin 5-O-glucosid-6''-O-malonyltransferase	*Salvia splendens*	Suzuki et al., 2001

Table 1. Acyltransferase catalysis of flavonoid acylation and their nature sources.

To solve this problem, the chemical approach was first investigated. It possessed a low degree of regioselectivity of esterification and drastic reaction conditions had to be applied (Patti et al., 2000). Later on, hydrolytic enzymes (lipases, esterases and proteases) have been recognized as useful agents due to their large availability, low cost, chemo-, regio- and enantioselectivity, mild condition processing and no need of cofactors (Collins & Kennedy, 1999; Nagasawa & Yamada, 1995).

Since the enzymatic preparation of flavonoid derivatives is a matter of several years, commercial applications have just been emerging. There are several patented inventions available to date, oriented on the flavonoid ester production and their use for the manufacture of pharmaceutical, dermopharmaceutical, cosmetic, nutritional or agri-foodstuff compositions

(Fukami et al., 2007; Moussou et al., 2004, 2007; Ghoul et al., 2006; Bok et al., 2001; Perrier et al., 2001; Otto et al., 2001; Nicolosi et al., 1999; Sakai et al., 1994).

This review presents available information on enzyme-mediated flavonoid acylation *in vitro*, emphasizing reaction parameters which influence performance and regioselectivity of the enzymatic reaction. In the second part, the paper focuses on biological effects of synthesized flavonoid esters as well as of those isolated from nature. Finally, the paper ends with application prospects of acylated flavonoids in the food, pharmaceutical and cosmetic industry.

2. Flavonoid esterification

Presently, the enzyme-catalyzed flavonoid esterification in organic media is a well-mastered technique for synthesis of selectively modified flavonoids. Results in this field suggest that a high degree of conversion to desired esters can be achieved when optimal reaction conditions are applied. The key factors, which influence regioselectivity and the performance of the enzymatic acylation of flavonoids, include type and concentration of enzyme, structure and concentration of the substrates (acyl donor, acyl acceptor and their ratio), nature of the reaction media, water content in the media, reaction temperature and nature of the reaction as reviewed in Chebil et al., 2006, 2007.

2.1 Enzymes

To date, the use of proteases, esterases, acyltransferases and lipases has been investigated in order to find the most potent biocatalyst for selective flavonoid acylation. These enzymes are often in the immobilized form which improves enzyme stability, facilitates product isolation, and enables enzyme reuse (Adamczak & Krishna, 2004).

2.1.1 Proteases

Proteases represent a class of enzymes which occupy a pivotal position with respect to their physiological roles as well as their commercial applications. They represent the first group of hydrolytic enzymes used for flavonoid modification. They perform both hydrolytic and synthetic functions. Since they are physiologically necessary for living organisms, proteases occur ubiquitously in diverse sources, such as plants, animals, and microorganisms. They are also classified as serine proteases, aspartic proteases, cysteine proteases, threonine proteases and metalloproteases, depending on the nature of the functional group at the active site.

Proteases have a large variety of applications, mainly in the detergent and food industries. In view of the recent trend of developing environmentally friendly technologies, proteases are envisaged to have extensive applications in leather treatment and in several bioremediation processes. Proteases are also extensively used in the pharmaceutical industry (Rao et al., 1998). Protease subtilisin was the first enzyme used for flavonoid ester synthesis conducted by Danieli et al. (1989, 1990). Later on, subtilisin was used for selective rutin acylation in organic solvents (Xiao et al., 2005; Kodelia et al., 1994). However, it has been reported that reactions catalyzed by subtilisin led to low conversion yields and a low degree of regioselectivity was observed (Danieli et al., 1990). These authors reported that the structure of the sugar moiety affected the regioselectivity. For flavonoid acylation, especially serine proteases (subtilisin) have been used in ester synthesis (Danieli et al., 1989, 1990; Kodelia et al., 1994).

2.1.2 Esterases

Esterases (carboxyl esterases, EC 3.1.1.1) represent a diverse group of hydrolases catalyzing the cleavage and formation of ester bonds with wide distribution in animals, plants and microorganisms. A classification scheme for esterases is based on the specificity of the enzymes for the acid moiety of the substrate, such as the carboxyl esterases, aryl esterases, acetyl esterases, cholin esterases, cholesterol esterases, etc. (Jeager et al., 1999). Esterases show high regio- and stereospecificity, which makes them attractive biocatalysts for the production of optically pure compounds in fine-chemicals synthesis (reviewed in Bornscheuer, 2002).

They have the same reaction mechanism as lipases, but differ from them by their substrate specificity, since they prefer short-chain fatty acids, whereas lipases usually prefer long-chain fatty acids. Another difference lies in the interfacial activation (Hidalgo & Bornscheuer, 2006). In contrast to lipases, only a few esterases have commercial applications in organic synthesis because lipases are generally more entantioselective and resistant to organic solvents. The most widely used esterase is the preparation isolated from pig liver (Hidalgo & Bornscheuer, 2006). The practical applications of esterases in enzymatic transformation of flavonoids are not very attractive as it enables the implementation only of the molecule of a short aliphatic chain length, such as acetate, propionate and butyrate (Sakai et al., 1994).

2.1.3 Lipases

Today lipases stand amongst the most important biocatalysts in industry. Among them, microbial lipases find the biggest application use. They can be classified according to sequence alignment into three major groups: mammalian lipases (e.g. porcine pancreatic lipase), fungal lipases (*Candida rugosa* and *Rhizomucor* family) and bacterial lipases (*Staphylococcus* and *Pseudomonas* family) (Hidalgo & Bornscheuer, 2006). More than 50% of the reported lipases are produced by yeast in the forms of various isozymes (Vakhlu & Kour, 2006).

Lipases (triacylglycerol acylhydrolases, EC 3.1.1.3) belong to the class of serine hydrolases. They catalyze a wide range of reactions, including hydrolysis, inter-esterification, alcoholysis, acidolysis, esterification and aminolysis (Vakhlu & Kour, 2006). Under natural conditions, they catalyze the hydrolysis of ester bonds at the hydrophilic-hydrophobic interface. At this interface, lipases exhibit a phenomenon termed interfacial activation, which causes a remarkable increase in activity upon contact with a hydrophobic surface. The catalytic process involves a series of differentiated stages: contact with the interface, conformational change, penetration in the interface, and finally the catalysis itself (Hidalgo & Bornscheuer, 2006). Under certain experimental conditions, such as in the absence of water, they are capable of reversing the reaction. The reverse reaction leads to esterification and formation of glycerides from fatty acids and glycerol (Saxena et al., 1999). This synthetic activity of lipases is being successfully utilized also in flavonoid ester production.

Candida antarctica lipase B (CALB) is one of the most widely used biocatalysts in organic synthesis on both the laboratory and the commercial scale (Anderson et al., 1998; Uppenberg et al., 1995) due to its ability to accept a wide range of substrates, its non-aqueous medium tolerance and thermal deactivation resistance (Degn et al., 1999; Anderson et al., 1998; Cordova et al., 1998; Drouin et al., 1997). CALB belongs to the α/β hydrolase-fold superfamily with a conserved catalytic triad consisting of Ser105-His224-Asp187 (Uppenberg et al., 1995). It comprises 317 amino acid residues. The active site contains an oxyanion hole which stabilizes the transition state and the oxyanion in the reaction intermediate (Haeffner et al., 1998). Reaction mechanism of CALB follows the bi-bi ping-pong mechanism, illustrated in Fig.1 (Kwon et al., 2007). The substrate molecule reacts with serine of the active site forming a tetrahedral intermediate which is stabilized by catalytic residues of His and Asp. In the next step alcohol is released and the acyl-enzyme complex is created. A nucleophilic attack (water in hydrolysis, alcohol in transesterification) causes another tetrahydral intermediate formation. In the last step, the intermediate is split into product and enzyme and is recovered for the next catalytic cycle (Patel, 2006).

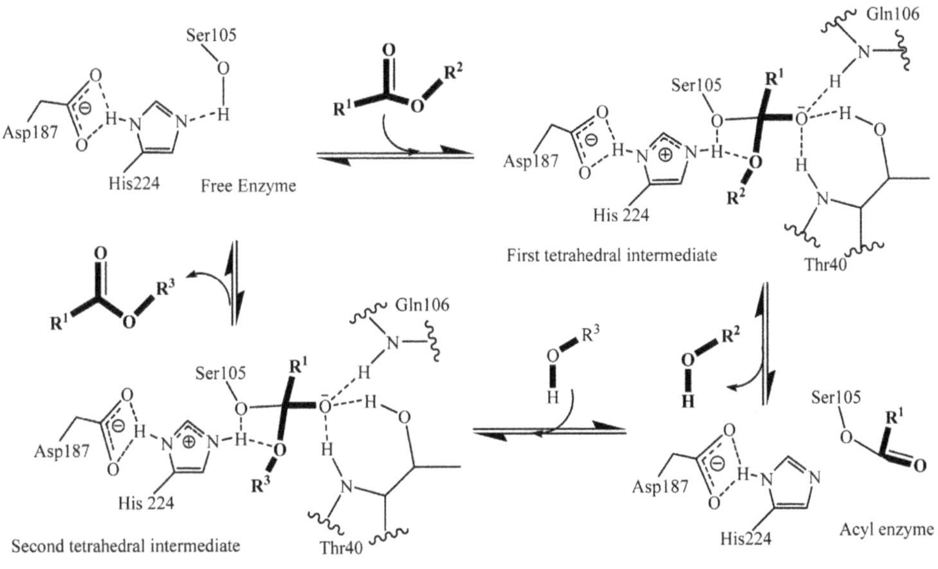

Fig. 1. Reaction mechanism catalyzed by *Candida antarctica* lipase (Kwon et al., 2007).

The active site of CALB consists of a substrate-nonspecific acyl-binding site and a substrate specific alcohol-binding site (Cygler & Schrag, 1997; Uppenberg et al., 1995). It is selective for secondary alcohols (Uppenberg et al., 1995), as reflected by the geometry of the alcohol-binding site (Lutz, 2004). In contrast to most lipases, CALB has no lid covering the entrance to the active site and shows no interfacial activation (Martinelle et al., 1995). CALB is being frequently used in acylation of various natural compounds such as saccharides, steroids and natural glycosides, including flavonoids (Riva, 2002; Davis & Boyer, 2001). The proper enzyme selection plays multiple roles in flavonoid acylation. The biocatalyst significantly influences the regioselectivity of the reaction. Information is available mainly on the use of lipases for flavonoid ester synthesis; especially the use of lipase B from *Candida antarctica*, which is preferred due to its acceptance of a wide range of substrates, good catalytic activity and a high degree of regioselectivity (Viskupicova et al., 2010; Katsoura et al., 2006, 2007; Ghoul et al., 2006; Mellou et al., 2005, 2006; Stevenson et al., 2006; Ardhaoui et al., 2004a, 2004b, 2004c; Passicos et al., 2004; Moussou et al., 2004; Gayot et al., 2003; Ishihara & Nakajima, 2003; Ishihara et al., 2002; Kontogianni et al., 2001, 2003; Nakajima et al., 1999, 2003; Gao et al., 2001; Otto et al., 2001; Danieli et al., 1997).

As for flavonoid aglycons, only two enzymes have been reported to be capable of acylating this skeleton – lipase from *Pseudomonas cepacia* and carboxyl esterase. Lambusta et al. (1993) investigated the use of *P. cepacia* lipase for catechin modification. They discovered that the acylation took place on the C5 and C7 hydroxyls. Sakai et al. (1994) observed that carboxyl esterase showed regioselectivity towards C3-OH of catechin. Sakai et al. (1994) explored the use of carboxyl esterase from *Streptomyces rochei* and *Aspergillus niger* for the 3-O-acylated catechin production.

2.2 Reaction conditions

The performance and regioselectivity of the enzyme-catalyzed flavonoid transformation is affected by several factors, including the type of enzyme, the nature of medium, reaction conditions, water content in the media, structure and concentration of substrates and their molar ratio. By varying these factors, significant changes in ester production and regioselectivity can be achieved.

2.2.1 Reaction media

Reaction media play an important role in enzymatic transformations. Methodologies for enzymatic flavonoid acylation have focused on searching a reaction medium which allows appropriate solubility of polar acyl acceptor (flavonoid glycoside) and nonpolar acyl donor as well as the highest possible enzymatic activity. Moreover, the medium has often been required to be nontoxic and harmless to biocatalyst. In order to meet the above-mentioned requirements, several scientific teams have dealt with proper medium selection (Viskupicova et al., 2006; Mellou et al., 2005; Kontogianni et al., 2001, 2003; Gao et al., 2001; Nakajima et al., 1999; Danieli et al., 1997).

Non-aqueous biocatalysis has several advantages over conventional aqueous catalysis: the suppression of hydrolytic activity of the biocatalyst which is carried out in water (Fossati & Riva, 2006), the enhanced solubility of hydrophobic substrates, the improvement of enzyme enantioselectivity, the exclusion of unwanted side reactions, the easy removal of some products, the enhanced enzyme thermostability and the elimination of microbial contamination (Rubin-Pitel & Zhao, 2006; Torres & Castro, 2004). Laane (1987) pointed out that log P, as a solvent parameter, correlated best with enzyme activity. Zaks & Klibanov (1988) reported that the activity of lipases was higher in hydrophobic solvents than in hydrophilic ones. Narayan & Klibanov (1993) claimed that it was hydrophobicity and not polarity or water miscibility which was important, whereas the log P parameter could be called a measure of solvent hydrophobicity. Trodler & Pleiss (2008), using multiple molecular dynamics simulations, showed that the structure of CALB possessed a high stability in solvents. In contrast to structure, flexibility is solvent-dependent; a lower dielectric constant led to decreased protein flexibility. This reduced flexibility of CALB in non-polar solvents is not only a consequence of the interaction between organic solvent molecules and the protein, but it is also due to the interaction with the enzyme-bound water and its exchange on the surface (Trodler & Pleiss, 2008). In organic solvents, the surface area has been suggested to be reduced, leading to improved packing and increased stability of the enzyme (Toba & Merz, 1997).

Polar aprotic solvents such as dimethyl sulfoxid (DMSO), dimethylformamide (DMF), tetrahydrofuran (THF) and pyridine were first investigated (Nakajima et al., 1999; Danieli et al., 1997). However, it was observed that enzyme activity was readily deactivated in these solvents. To date enzymatic acylation of flavonoids has been successfully carried out in various organic solvents (Tab.2), while the most frequently used are 2-methylbutan-2-ol and acetone because of their low toxicity, their polarity allowing proper solubilization of substrates and high conversion yields.

Solvent	Reference
2-Methylbutan-2-ol	Ghoul et al., 2006; Ardhaoui et al., 2004a, 2004b, 2004c; Passicos et al., 2004; Gayot et al., 2003
Acetone	Ghoul et al., 2006; Mellou et al., 2005, 2006; Kontogianni et al., 2001, 2003; Ishihara et al., 2002, Ishihara & Nakajima, 2003; Nakajima et al., 1999, 2003; Danieli et al., 1997
Acetonitrile	Ghoul et al., 2006; Ishihara & Nakajima, 2003; Nakajima et al., 1997, 1999
2-Methylpropan-2-ol	Ghoul et al., 2006; Stevenson et al., 2006; Mellou et al., 2005; Moussou et al., 2004; Kontogianni et al., 2001, 2003; Otto et al., 2001
Dioxane	Ghoul et al., 2006; Danieli et al., 1997
Pyridine	Danieli et al., 1990, 1997
THF, DMSO, DMF	Kontogianni et al., 2001, 2003; Danieli et al., 1997
Binaric mixtures of solvents	Ghoul et al., 2006; Gao et al., 2001; Nakajima et al., 1999; Danieli et al., 1997

Table 2. Organic solvents used in flavonoid acylation.

The effect of the solvent on conversion yield depends on the nature of both the acyl donor and the flavonoid (Chebil et al., 2006). Although much has been done in this area, it is quite difficult to deduce any general conclusion on solvent choice because the available data are controversial and sometimes even contrary.

Recently, ionic liquids have received growing attention as an alternative to organic solvents used for the enzymatic transformation of various compounds (Katsoura et al., 2006; Kragl et al., 2006; Jain et al., 2005; Lozano et al., 2004; Reetz et al., 2003; Van Rantwick et al., 2003). The potential of these "green solvents" lies in their unique physicochemical properties, such as non-volatility, nonflammability, thermal stability and good solubility for many polar and less polar organic compounds (Jain et al., 2005; Wilkes, 2004; Itoh et al., 2003; Van Rantwick et al., 2003). Probably the most promising advantage of the use of ionic liquids is their potential application in food, pharmaceutical and cosmetic preparations due to their reduced toxicity (Jarstoff et al., 2003). Due to the many above-mentioned advantages of ionic liquids for enzyme-mediated transformations, several flavonoid esters have been recently prepared in such media (Katsoura et al., 2006, 2007; Kragl et al., 2006). The biocatalytic process showed significantly higher reaction rates, regioselectivity and yield conversions compared to those achieved in organic solvents. Thus ionic liquid use seems to be a challenging approach to conventional solvent catalysis.

The solvent-free approach for elimination of the co-solvent of the reaction has been recently introduced as an alternative for conventional solvents (Enaud et al., 2004; Kontogianni et al., 2001, 2003). It is based on the use of one reactant in the role of the solvent. The authors reported rapid reaction rates; however, the conversion yields were slightly decreased. In spite of the attractiveness, the use of solvent-free systems is characterized by a serious drawback due to the necessity to eliminate the excess of the acyl donor for the recovery of the synthesized products (Chebil et al., 2006).

2.2.2 Water content

Water content in reaction media is a crucial parameter in lipase-catalyzed synthesis as it alters the thermodynamic equilibrium of the reaction towards hydrolysis or synthesis. Moreover, it is involved in noncovalent interactions which keep the right conformation of an enzyme catalytic site (Foresti et al., 2007). The amount of water required for the catalytic process depends on the enzyme, its form (native or immobilized), the enzyme support, and on the solvent nature (Arroyo et al., 1999; Zaks & Klibanov, 1988). The influence of water content in the reaction system on enzyme activity is variable with various enzymes (lipase from *Rhizomucor miehei*, *Rhizomucor niveus*, *Humicola lanuginosa*, *Candida rugosa*, *Pseudomonas cepacia*).

In general, the water amount which is considered to be optimal for esterifications in organic solvents is 0.2 – 3% (Rocha et al., 1999; Yadav & Piyush, 2003; Iso et al., 2001). The enzymatic esterification of flavonoids in non-aqueous media is greatly influenced by the water content of the reaction system (Ardhaoui et al., 2004b; Gayot et al., 2003; Kontogianni et al., 2003). Ardhaoui et al. (2004b) observed the best enzyme activity when water content was maintained at 200 ppm. Gayot et al. (2003) found that the optimal value of water in an organic reaction medium equaled 0.05% (v/v). Kontogianni et al. (2003) reported that highest flavonoid conversion was reached when initial water activity was 0.11 or less.

2.2.3 Temperature

Temperature represents a significant physical factor in enzyme-catalyzed reactions. It affects viscosity of the reaction medium, enzyme stability, and substrate and product solubility.

Since lipase from *C. antarctica* belongs to thermostable enzymes, improved catalytic activity was observed at higher temperatures (Arroyo et al., 1999). To date, flavonoid transformation has been carried out in the temperature range 30 – 100°C (Ghoul et al., 2006; Katsoura et al., 2006; Stevenson et al., 2006; Mellou et al., 2005; Ardhaoui et al., 2004a, 2004b, 2004c; Moussou et al., 2004; Passicos et al., 2004; Enaud et al., 2004; Gayot et al., 2003; Kontogianni et al., 2003; Ishihara et al., 2002; Gao et al., 2001; Otto et al., 2001; Nakajima et al., 1999; Danieli et al., 1990). The choice of temperature depends on the enzyme and solvent used. The majority of authors performed flavonoid acylation at 60°C due to the best enzyme activity, good solubility of substrates and highest yields of resulting esters reached (Viskupicova et al., 2006, 2010; Ghoul et al., 2006; Katsoura et al., 2006; Stevenson et al., 2006;

Ardhaoui et al., 2004a, 2004b, 2004c; Moussou et al., 2004; Passicos et al., 2004; Enaud et al., 2004; Gayot et al., 2003; Otto et al., 2001). Our results on the effect of temperature on naringin conversion are presented in Fig.2 and are in accordance with other authors (Viskupicova et al., 2006).

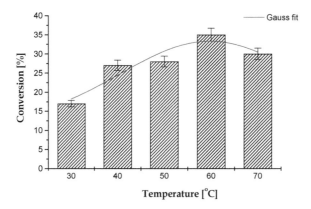

Fig. 2. Effect of temperature on naringin conversion to naringinpalmitate in 2-methylbutan-2-ol catalyzed by *C. antarctica* lipase after 24 h (Viskupicova et al., 2006).

2.3 Acyl donors and acceptors

2.3.1 Acyl donor

Since lipase-catalyzed acylation takes place through the formation of an acyl-enzyme intermediate, the nature of the acyl donor has a notable effect on reactivity. The ideal acyl donor should be inexpensive, fast acylating, and completely non-reactive in the absence of the enzyme (Ballesteros et al., 2006). Many acylating agents have been tested in flavonoid esterification, such as aromatic or aliphatic organic acids, substituted or not (Tab.3). Special attention was attributed to fatty acid ester production (Katsoura et al., 2006; Mellou et al., 2005, 2006; Ardhaoui et al., 2004a, 2004b, 2004c; Enaud et al., 2004; Gayot et al., 2003; Kontogianni et al., 2003). This approach enables to improve flavonoid solubility and stability in lipophilic systems. The proper acyl donor selection may significantly influence not only the physicochemical but also biological properties of the resulting esters.

A simple way to increase the reaction rate and conversion yield in acylation is to use an excess of acyl donor (Patti et al., 2000). Many authors have tried to determine the optimal molar ratio of flavonoid/acyl donor in order to achieve the highest possible yields. The molar ratios 1:1 to 1:15 (acyl acceptor/acyl donor) have been investigated, whereas the majority agreed on the ratio 1:5 to be the most suitable for the best reaction performance (Mellou et al., 2006; Gayot et al., 2003; Ishihara & Nakajima, 2003; Ishihara et al., 2002; Kontogianni et al., 2001). A better solution is offered by the use of special acyl donors which ensure a more or less irreversible reaction. This can be achieved by the introduction of electron-withdrawing substituents (esters), resulting in higher conversion yields and reaction rates. The use of vinyl esters allows a several times faster reaction progress than do other activated esters (Ballesteros et al., 2006). Enzymatic synthesis of flavonoid esters can be realized by two basic approaches, i.e. esterification and transesterification (Fig.3).

Aliphatic acids	Aromatic acids
Acetic*	Caffeic*
Malic*	p-Coumaric*
Malonic*	Ferulic*
Succinic*	Gallic*
Tartaric*	p-Hydroxybenzoic*
Butyric	Sinapic*
Crotonic	Benzoic
n-Butanoic	Cinnamic
Isobutyric	Isoferulic
Isovaleric	Methylsinapic
Lactic	
3-methylbutyric	
Quinic	
Vinylpropionic	
Tiglic	

*acyl donors found in anthocyanins

Table 3. Acyl donors found in flavonols, flavones (Williams, 2006) and anthocyanins (Andersen & Jordheim, 2006).

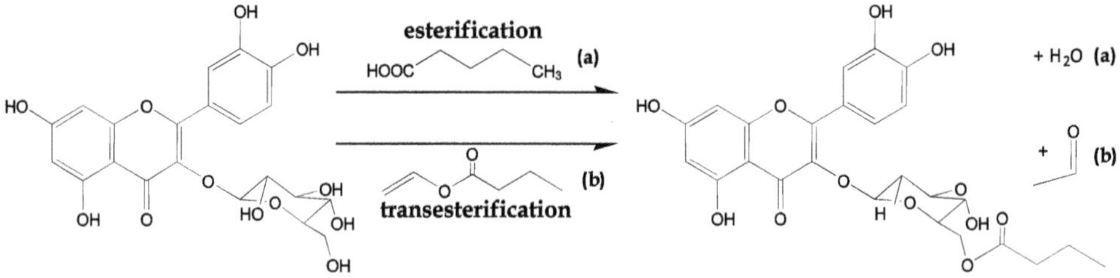

Fig. 3. Mechanism of isoquercitrin esterification and transesterification (Chebil et al., 2006).

Pleiss et al. (1998) studied the acyl binding site of CALB and found the enzyme to be selective for short and medium fatty acid chain length. This fact may be attributed to the structure of the lipase acyl binding pocket, which is an elliptical, narrow cleft of 9.5 × 4.5 Å. With increasing carbon number of a fatty acid or molecule size, the steric hindrance is involved resulting in low efficiency of the enzymatic reaction (Riva et al., 1988; Wang et al., 1988; Carrea et al., 1989). This fact was experimentally confirmed by Katsoura et al. (2006) and by Viskupicova & Ondrejovic (2007) whose results showed higher performance of the naringin and rutin esterification when fatty acids up to C10 were introduced. On the other hand, Ardhaoui et al. (2004b) and Kontogianni et al. (2003) reported that the fatty acid chain length had no significant effect on conversion yield when fatty acids of a medium and high chain length were used.

Thus, the effect of fatty acid chain length on flavonoid acylation still remains a matter of discussion. Our team conducted a series of experiments with both saturated and unsaturated fatty acids and found a correlation between log P of the acids tested and conversion yields (Viskupicova et al., 2010). It would be interesting to take this parameter into consideration when assessing the influence of an acyl donor on the reaction progress.

Only little progress has been achieved in flavonoid esterification with aromatic acids (Stevenson et al., 2006; Enaud et al., 2004; Gao et al., 2001; Nakajima et al., 2000). It has been observed that the performance of the process depends mainly on the nature of the substitutions, the position of the hydroxyls and the length of the spacers.

2.3.2 Acyl acceptor

The structure of acyl acceptor (flavonoid), especially stereochemistry of glycosidic bonds, plays an important role in flavonoid acylation. The structural differences, such as the number and position of hydroxyl groups, the nature of saccharidic moiety, as well as the position of glycosidic bonds, influence the flavonoid solubility, and thus affect the overall conversion yield.

Available studies are concerned mainly with acylation on flavonoid glycosides. Among polyphenolic compounds, naringin and rutin are the most widely used substrates. For the naringin molecule, which possesses a primary hydroxyl group on glucose, the acylation takes place on the 6''-OH (Katsoura et al., 2006; Konntogianni et al., 2001, 2003; Ishihara et al., 2002; Gao et al., 2001; Otto et al., 2001; Danieli et al., 1990) since the primary hydroxyl is favored by CALB (Fig.4). However, in rutin, which has no primary hydroxyl available, either the 3''-OH of glucose (Ishihara et al., 2002; Danieli & Riva, 1994) or the 4'''-OH of rhamnose (Fig.4) (Viskupicova et al., 2010; Mellou et al., 2006; Ardhaoui et al., 2004a, 2004b, 2004c) can be acylated. Danieli et al. (1997) observed the rutin-3'',4'''-O-diester formation. When subtilisin was used as biocatalyst, naringin-3''-O-ester and rutin-3''-O-ester were synthesized (Danieli et al., 1990).

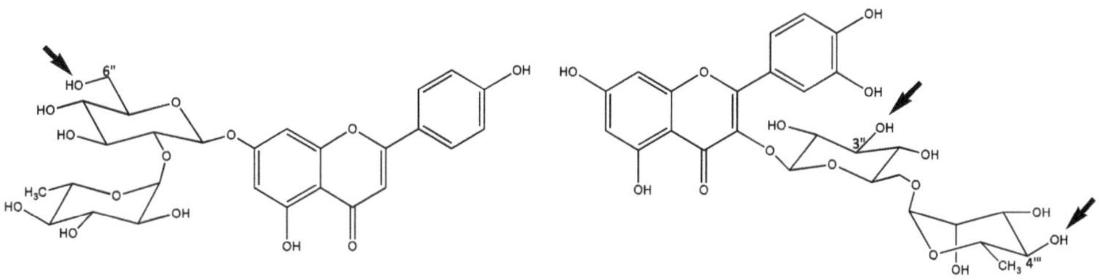

Fig. 4. Acylation sites of naringin (left) and rutin (right) molecule.

The concentration of the flavonoid also affects the performance of the acylation reaction. The conversion yield and the initial rate rise with increasing flavonoid concentration. However, the amount of flavonoid is limited by its solubility in a reaction medium (Chebil et al., 2006, 2007).

3. Influence of flavonoid derivatization on biological activities

3.1 Esters with aromatic acids

Aromatic acids, along with flavonoids, belong to the group of phenols of secondary metabolism of living organisms. The described secondary metabolites represent a store of biologically active compounds, displaying various biological activities. We can therefore assume that physicochemical and biological properties of the initial flavonoids may be improved by acylation of flavonoids with aromatic acids. However, by this reaction a new compound can also gain novel activities provided by the aromatic acids.

Flavonoid acylation with aromatic acids was reported to improve physiological activities, such as UV-absorbing capacity, radical scavenging ability (Delazar et al., 2005; Ishihara & Nakajima, 2003; Harborne & Williams, 2000; Alluis & Dangles 1999; Jungblut et al., 1995) pigment stabilization (especially anthocyanins) (Ishihara & Nakajima 2003), and interaction with cellular targets (Ferrer et al., 2008).

Flavonoid esters acylated with *p*-coumaric acid were found to increase antioxidant (Pajero et al., 2005) and anti-inflammatory activities (Harborne & Williams, 2000), as well as antiproliferative and cytotoxic effects on various cancer cell lines (Mitrokotsa et al., 1993). Moreover, *p*-coumaroyl esters of quercetin and kaempferol were reported to have positive effects on cerebrovascular disorders (Calis et al., 1995). Similarly, flavonoid esters esterified with cinnamic acid were shown to exhibit antiproliferative activity against several human cancer cell lines (Duarte-Almeida et al., 2007). Flavonoid acylation with caffeic acid contributes to the enhancement of antioxidant properties (Pajero et al., 2005). Flavonolignans acylated with truxinic acid were shown to possess hepatoprotective as well as anticancer activity (Sharma et al., 2003).

3.2 Esters with aliphatic acids

Biological activities of aliphatic acids are not of a big importance in comparison with aromatic acids. These compounds are mainly accepted as energy storage and components of several compartments of cells, such as membranes, enzymes, surfactants, etc. In the literature, more studies can be found describing changes in biological activities of flavonoids after their acylation with aliphatic acids.

The aliphatic acylation of anthocyanins with malonic acid is important for enhancing the pigment solubility in water, protecting glycosides from enzymatic degradation and stabilizing anthocyanin structures (Nakayama et al., 2003). Several *in vitro* observations suggest that acylation with malonic acid or sinapic acid is crucial for efficient flavonoid accumulation in plants.

Fatty acid esters of catechins were reported to display antitumor, antibacterial and 5-α reductase inhibiting activity (Fukami et al., 2007) as well as antioxidant properties (Sakai et al., 1994). Lee et al. (2003) reported anti-atherogenic activity of two naringenin derivatives, 7-*O*-oleic ester and 7-*O*-cetyl ether.

Acylation of the flavonoid molecule with polyunsaturated fatty acids introduces potential antitumor and antiangiogenic properties (Mellou et al., 2006). Anticarcinogenic effects were observed also in silybin esters acylated with butyric and lauric acid (Xanthakis et al., 2010). Recently, we found that acylation of rutin with unsaturated fatty acids, such as oleic, α-linoleic and linolenic, increased the antioxidant potential of the initial compound (Viskupicova et al., 2010). This observation is in accordance with the results of Mellou et al. (2006) and Katsoura et al. (2006).

In the field of fatty acid ester synthesis, information on the photoprotective effectiveness of new quercetin derivatives acylated with acetic, propionic and palmitic acids, has been reported. The authors found that esterification with a short side-chain (such as acetate or propionate) may improve migration through the aqueous environment and interaction with or penetration into phospholipid membranes (Saija et al., 2003).

Recent experimental findings indicate that acylation of flavonoid may increase enzyme inhibitory activity. Lin et al. (2010) observed increased 5α-reductase inhibition after acylation of (-)-epigallocatechins. Salem et al. (2011) showed that the acylation of isorhamnetin-3-O-glucoside with different aliphatic acids enhanced its capacity to inhibit xanthine oxidase. Our recent investigations showed that lipophilic rutin and naringin esters were strong inhibitors of transport enzymes such as sarcoplasmic reticulum Ca^{2+}-ATPase and plasma membrane Ca^{2+}-ATPase (Augustyniak et al., 2010; Viskupicova et al., 2009), and thus might be useful in calcium regulation. We presume that there might be a general mechanism involved in the enhanced inhibitory activity of the acylated flavonoids on structurally diverse classes of enzymes which seems to be donated by the medium to long fatty acid chains.

4. Application perspectives

The following section provides a summary of patented inventions available in the commercial sphere. These include practical applications in food, pharmaceuticals and cosmetics.

4.1 Food

The major contribution of acylated flavonoids in the food industry lies in the improvement of stability and solubility of initial molecules, e.g. by reducing lipid oxidation in oil/fat based food systems, desirable modification of unwanted sensory properties of certain flavonoids, taking advantage of pigment stabilization by the means of flavonoid acylation, or other food characteristics. Furthermore, selectively acylated flavonoids may cause significant changes in their bioavailability and bioactivity, and when consumed, may thus play a role in preventing diseases.

Flavonoid acylation is a useful tool for modification of sensory properties of food. While flavonoids provide a variety of health benefits, flavonoid-containing food often suffers from bitter and astringent taste. Degenhardt et al. (2007) found that certain glycosylation and acylation patterns can effectively modulate these negative taste factors in edible preparations, pharmaceutical preparations and cosmetics with mouth contact (i.e. tooth paste, mouth wash). Both the taste intensity and the taste profile perception are improved by the novel compounds. Ghoul et al. (2006) introduced a process for the selective preparation of acylated flavonoid glycosides with improved stability and solubility in various preparations with their antioxidant effect remaining intact or being improved.

Another particular advantage obtained by these modified flavonoids is the bifunctional character of their molecule with higher biological activity. Free unsaturated fatty acids represent a potential risk because they are highly reactive and by creating free radicals they cause undesirable damage in food. Enzymatic synthesis of flavonoids with unsaturated fatty acids was found to be a useful solution for the stabilization of these highly oxidizable acids (Viskupicova et al., 2010; Mellou et al., 2006).

Another important benefit of acylated anthocyanins lies in the use as food colorants which can serve as a useful alternative to synthetic additives (Giusti & Wrolstad, 2003; Fox, 2000; Asen et al., 1979). The discovery of acylated anthocyanins with increased stability has shown that these pigments may provide food products with the desirable color and stability at a wide pH range. Examples of suitable acylated anthocyanin sources may be radishes, red potatoes, red cabbage, black carrots, and purple sweet potatoes (reviewed in Giusti & Wrolstad, 2003). The invention of Asen et al. (1979) refers to a stable food colorant from a natural source. It relates to an anthocyanin isolated from the Heavenly Blue Morning Glory (*Ipomoea tricolor* Cav cv), peonidin 3-(dicaffeylsophoroside)-5-glucoside, which is characterized by the stability of colors ranging from purplish-red to blue produced in food and beverage products at pH values from about 2.0 to about 8.0. Fox (2000) reported the invention referring to a stable, ruby red natural colorant (anthocyanins acylated with chlorogenic acid) derived from purple sunflower hulls, useful as a coloring agent in food products, cosmetics, pharmaceuticals and other materials.

4.2 Pharmaceuticals

In recent years, coronary artery diseases, such as atherosclerosis and hypercholesterolemia, represent a major cause of death, exceeding even oncological causes or infectious diseases. Novel acylated flavanone derivatives are effective in the treatment or prevention of elevated blood lipid level-related diseases, e.g. hyperlipidemia, arteriosclerosis, angina pectoris, stroke and hepatic diseases since they exert inhibitory effects on acylcolicholesterol acyl transferase activity and HMG-CoA reductase activity. In spite of their potent efficacies, the flavanone derivatives exhibited no toxicity or mitogenicity in tests using mice (Bok et al., 2001).

Mellou et al. (2005) carried out enzymatic acylation on Greek endemic plants and reported that this modification increased both their antioxidant activity towards isolated low-density lipoproteins (LDL) and serum model and antimicrobial activity against two Gram-positive bacteria, *Staphylococcus aureus* and *Bacillus cereus*. Katsoura et al. (2006) also found that biocatalytic acylation of rutin with various acyl donors affected its antioxidant potential towards both isolated LDL and total serum model *in vitro*. A significant increase in antioxidant activity was observed for rutin-4'''-oleate.

The 6''-O-esterification of kaempferol-3-O-glucoside (astragalin) with p-coumaric acid was found to increase its anti-inflammatory activity eight times compared to the initial flavonoid, while addition of another p-coumaroyl group at 2'' position gave an activity 30 times greater than that of astragalin (Harborne & Williams, 2000). Another kaempferol derivative, kaempferol 3-(2'',3''-di-E-p-coumaroylrhamnoside), was found to possess a cytotoxic effect. It significantly modulated the proliferation of promyelocytic cell line HL60 and MOLT3 (a T-ALL with phenotypic characteristics of cortical thymocytes) (Mitrokotsa et al., 1993). Also Demetzos et al. (1997) synthesized novel flavonoid esters with cytotoxic activity. These acetylated esters of tiliroside exhibited a strong cytotoxic effect against four leukemic cell lines (HL60, DAUDI, HUT78 and MOLT3), whilst the maternal compound had no effect (Demetzos et al., 1997). Tricin-7-O-β-(6''-methoxycinnamic)-glucoside, a flavone from sugarcane, was found to exhibit antiproliferative activity against several human cancer cell lines, with higher selectivity toward cells of the breast resistant NIC/ADR line (Duarte-Almeida et al., 2007). Mellou et al. (2006) provided evidence that flavonoid derivatives esterified with polyunsaturated fatty acids were able to decrease the production of vascular endothelial growth factor by K562 human leukemia cells unlike the initial flavonoids, indicating that these novel compounds might possess improved anti-angiogenic and anti-tumor properties. Anticancer acitivity was established also in two O-acylated flavonoids, daglesiosides I and II, which were isolated from the leaves of *Pseudotsuga menziesii* (Sharma et al., 2003).

Parejo et al. (2005) examined quercetagetin glycosides acylated with caffeic and *p*-coumaric acid for antioxidant activity. They found that these compounds exhibited a high radical scavenging activity in comparison with reference compounds. Fatty acid derivatives of catechins are described as having antitumorigenesis promoting activity or 5-α reductase inhibiting activity, as well as antibacterial activity (Fukami et al., 2007). Since these acylated catechin compounds have a greatly superior solubility in fats and oils than any catechins previously known, they may be used as a highly effective antioxidative agents (Sakai et al., 1994).

A different catechin derivative, 3-*O*-octanoyl-(+)-catechin, was synthesized by Aoshima et al. (2005) by incorporation of an octanoyl chain into (+)-catechin. This ester was found to be more efficient than catechin in inhibiting the response of ionotropic gamma-aminobutyric acid receptors and Na+/glucose cotransporters expressed in *Xenopus oocytes* in a noncompetitive manner. Moreover, it induced a nonspecific membrane current and decreased the membrane potential of the oocyte. This newly synthesized catechin derivative possibly binds to the lipid membrane more strongly than do catechin, (-)-epicatechin gallate, or (-)-epigallocatechin-3-gallate, and as a result it perturbs the membrane structure (Aoshima et al., 2005).

4.3 Cosmetics

The majority of cosmetic or dermopharmaceutical compositions consist of a fatty phase, the oily products of which have a certain tendency to oxidize, even at room temperature. The consequence of this oxidation is to profoundly modify the properties, which makes them unusable after a variable time period. In order to protect the compositions with respect to these oxidation phenomena, it is common practice to incorporate protective agents which act as antioxidizing agents (N'guyen, 1995). By virtue of the skin-protecting and skin-cleansing properties of flavonoids and their effects against aging, against skin discoloration and on the appearance of the skin, they have been used as constituents of cosmetic or dermopharmaceutical compositions. They also act on the mechanical properties of hair (Ghoul et al., 2006).

Moussou et al. (2007) found that the esters of flavonoids with omega-substituted C6 to C22 fatty acids have the property to protect the skin cells against damage caused by UV radiation. According to the invention, these esters of flavonoids protect skin cells against UVA and UVB radiation in a more effective manner than flavonoids alone. Moreover, these esters demonstrated their property to stimulate glutathione metabolism of human skin cells after UVA irradiation, i.e. to stimulate their cellular defenses. They have also anti-inflammatory and soothing properties, as demonstrated by the inhibition of released protein kinase PGE2 after UVB irradiation. Thus these flavonoid esters may be used to protect the skin and scalp and/or to fight against UV and sun damage, erythema, sunburn, mitochondrial or nuclear DNA damage, to prevent or fight photo-aging, providing improvement for signs of aging as skin wrinkles, elasticity lost and decrease in skin thickness (Moussou et al., 2007).

Perrier et al. (2001) discovered that specific flavonoid esters can be stabilized while preserving their initial properties, particularly free radical inhibition and enzyme inhibition, and for applications associated with these properties: venous tonics, agents for increasing the strength of blood capillaries, inhibitors of blotchiness, inhibitors of chemical, physical or actinic erythema, agents for treating sensitive skin, decongestants, draining agents, slimming agents, anti-wrinkle agents, stimulators of the synthesis of the components of the extracellular matrix, toners for making the skin more elastic and anti-ageing agents (Perrier et al., 2001).

5. Conclusions

Flavonoids, having a wide spectrum of health-beneficial activities, seem to be applicable in various areas of national management from food additivization to pharmaceutical preparations with the purpose of prevention and/or treatment of important civilization diseases. Their chemical structure determines not only biological effects on human health but also their solubility, stability and bioavailability. Recently, selective enzyme-mediated acylation of flavonoids has been introduced to confer improved biological properties to the novel compounds including both biological activity of initial flavonoid and other parameters determined by the chemical structure of an acyl donor. In the past, proteases, esterases and acyltransferases were used for the preparation of acylated flavonoids. In light of our review, immobilized lipases, especially *Candida antarctica* B lipase, are suitable for this purpose. Not only the given enzyme but also the reaction conditions have a distinct influence on the performance of acylation. This aspect must be considered when producing acylated flavonoids in technology scale for potential uses in the food, pharmaceutical and cosmetic industry.

6. Acknowledgment

The work was supported by The Slovak Research and Development Agency in the frame of the Project APVV-VMSP-II-0021-09 and by The Agency of the Ministry of Education, Science, Research and Sport of the Slovak Republic for the Structural Funds of EU, OP R&D of ERDF in the frame of the Project „Evaluation of natural substances and their selection for prevention and treatment of lifestyle diseases (ITMS 26240220040).

7. References

Adamczak, M. & Krishna, S.H. (2004). Strategies for improving enzymes for efficient biocatalysis. *Food Technology and Biotechnology*, Vol.42, pp. 251-264.

Alluis, B. & Dangles, O. (1999). Acylated flavone glucosides: synthesis, conformational investigation, and complexation properties. *Helvetica Chimica Acta*, Vol.82, pp. 2201-2212.

Andersen, Ø.M. & Jordheim, M. (2006). The anthocyanins. In: Andersen, Ø.M.; Markham K.R.: Flavonoids: chemistry, biochemistry and applications. New York: Taylor & Francis Group, CRC Press, 471 – 551. ISBN: 0-8493-2021-6.

Anderson, E.M.; Larsson, K.M. & Kirk, O. (1998). One biocatalyst – many applications: the use of *Candida antarctica* B-lipase in organic synthesis. *Biocatalysis and Biotransformation*, Vol.16, pp. 181-204.

Aoshima, H.; Okita, Y.; Hossain, S.J.; Fukue, K.; Mito, M.; Orihara, Y.; Yokoyama, T.; Yamada, M.; Kumagai, A.; Nagaoka, Y.; Uesatos, S. & Hara, Y. (2005). Effect of 3-O-octanoyl-(+)-catechin on the responses of GABAa receptors and Na+/glucose cotransporters expressed in *Xenopus oocytes* and on the oocyte membrane potential. *Journal of Agricultural and Food Chemistry*, Vol.53, pp. 1955-1959.

Ardhaoui, M.; Falcimaigne, A.; Engasser, J.-M.; Moussou, P.; Pauly, G. & Ghoul, M. (2004a). Acylation of natural flavonoids using lipase of *Candida antarctica* as biocatalyst. *Journal of Molecular Catalysis B: Enzymatic*, Vol.29, pp. 63-67.

Ardhaoui, M.; Falcimaigne, A.; Ognier, S.; Engasser, J.-M.; Moussou, P.; Pauly, G. & Ghoul, M. (2004b). Effect of acyl donor chain length and substitutions pattern on the enzymatic acylation of flavonoids. *Journal of Biotechnology*, Vol.110, pp. 265-272.

Ardhaoui, M.; Falcimaigne, A.; Engasser, J.-M.; Moussou, P.; Pauly, G. & Ghoul, M. (2004c). Enzymatic synthesis of new aromatic and aliphatic esters of flavonoids using *Candida antarctica* as biocatalyst. *Biocatalysis and Biotransformation*, Vol.22, pp. 253-259.

Arroyo, M.; Sánchez-Montero, J.M. & Sinisterra, J.V. (1999). Thermal stabilization of immobilized lipase B from *Candida antarctica* on different supports: Effect of water activity on enzymatic activity in organic media. *Enzyme & Microbial Technology*, Vol.24, pp. 3-12.

Asen, S.; Stewart, R.N. & Norris, K.H. (1979). Stable foods and beverages containing the anthocyanin, peonidin-3-(dicaffeylsophoroside)-5-glucosid. Patent US 4172902.

Augustyniak, A.; Bartosz, G.; Cipak, A.; Duburs, G.; Horakova, L.; Luczaj, W.; Majekova, M.; Odysseos, A.D.; Rackova, L.; Skrzydlewska, E.; Stefek, M.; Strosova, M.; Tirzitis, G.; Viskupicova, J.; Vraka, P.S. & Zarkovic, N. (2010). Natural and synthetic antioxidants: an updated overview. *Free Radical Research*, Vol.44, pp. 1216-1262.

Ballesteros, A.; Plou, F.J.; Alcade, M.; Ferrer, M.; Garcia-Arellano, H.; Reyes-Duarte, D. & Ghazi, I. (2006). Enzymatic synthesis of sugar esters and oligosaccharides from renewable resources. In: Patel, R.N.: Biocatalysis in the pharmaceutical and biotechnology industries. New York: CRC Press, Taylor & Francis Group, 463-488. ISBN: 0-8493-3732-1.

Berg, P.A. & Daniel, P.T. (1988). Plant flavonoids in biology and medicine II. *Progress in Clinical and Biological Research*, Vol.280, pp. 157-171.

Bloor, S.J. (2001). Deep blue anthocyanins from blue *Dianella* berries. *Phytochemistry*, Vol.58, pp. 923-927.

Bok, S.-H.; Jeong, T.-S.; Lee, S.-K.; Kim, J.-R.; Moon, S.-S.; Choi, M.-S.; Hyun, B.-H.; Lee, C.-H. & Choi, Y.-K. (2001). Flavanone derivatives and composition for preventing or treating blood lipid level-related diseases comprising same. Patent US 6455577.

Bornscheuer, U.T. (2002). Microbial carboxyl esterases: classification, properties and application in biocatalysis. *FEMS Microbiology Reviews*, Vol.26, pp. 73-81.

Calis, I.; Ozipek, M. & Ruedi, P. (1995). Enzyme-mediated regioselective acylation of flavonoid glycosides. FABAD Journal of Pharmaceutical Sciences, Vol.20, pp. 55-59.

Carrea, G.; Riva, S.; Secundo, F. & Danieli, B. (1989). Enzymatic synthesis of various 1-O-sucrose and 1-O-fructose esters. *Journal of the Chemical Society Perkin Transactions*, Vol.1, pp. 1057-1061.

Chebil, L.; Humeau, C.; Falcimaigne, A.; Engasser, J.-M. & Ghoul, M. (2006). Enzymatic acylation of flavonoids. *Process Biochemistry*, Vol.41, pp. 2237-2251.

Chebil, L.; Anthoni, J.; Humeau, C.; Gerardin, C.; Engasser, J.-M. & Ghoul, M. (2007). Enzymatic acylation of flavonoids: Effect of the nature of the substrate, origin of lipase, and operating conditions on conversion yield and regioselectivity. *Journal of Agricultural and Food Chemistry*, Vol.55, pp. 9496-9502.

Collins, A.M. & Kennedy, M.J. (1999). Biotransformations and bioconversions in New Zealand: Past endeavours and future potential. *Australasian Biotechnology*, Vol.9, pp. 86-94.

Cordova A.; Iverson T. & Hult K. (1998). Lipase catalysed formation of macrocycles by the ring opening polymerisation of ε-caprolactone. *Polymer*, Vol.39, pp. 6519-6524.

Cushnie, T.P.T. & Lamb, A.J. (2005). Detection of galangin-induced cytoplasmic membrane damage in *Staphylococcus aureus* by measuring potassium loss. *Journal of Ethnopharmacology*, Vol.101, pp. 243-248.

Cushnie, T.P.T. & Lamb, A.J. (2006). Assessment of the antibacterial activity of galangin against 4-quinolone resistant strains of *Staphylococcus aureus*. *Phytomedicine*, Vol.13, pp. 191-197.

Cygler, M., & Schrag, J. (1997). Structure as a basis for understanding interfacial properties of lipases. *Methods in Enzymology*, Vol.284, pp. 3-27.

Danieli, B.; De Bellis; P.; Carrea, G. & Riva, S. (1989). Enzyme-mediated acylation of flavonoid monoglycosides. *Heterocycles*, Vol.29, pp. 2061-2064.

Danieli, B.; De Bellis, P.; Carrea, G. & Riva, S. (1990). Enzyme-mediated regioselective acylations of flavonoid disaccharide monoglycosides. *Helvetica Chimica Acta*, Vol.73, pp. 1837-1844.

Danieli, B.; Luisetti, M.; Sampognaro, G.; Carrea, G. & Riva, S. (1997). Regioselective acylation of polyhydroxylated natural compounds catalyzed by *Candida antarctica* lipase B (Novozym 435) in organic solvents. *Journal of Molecular Catalysis B: Enzymatic*, Vol.3, pp. 193-201.

Davis, B. G. & Boyer, V. (2001). Biocatalysis and enzymes in organic synthesis. *Natural Product Reports*, Vol.18, pp. 618-640.

Davies, K.M. & Schwinn, K.E. (2006). Molecular biology and biotechnology of flavonoid biosynthesis. In: Andersen, Ø.M.; Markham, K.R.: Flavonoids: chemistry, biochemistry and applications. New York: Taylor & Francis Group, CRC Press, 143-218. ISBN: 0-8493-2021-6.

Degenhardt, A.; Ullrich, F.; Hofmann, T. & Stark, T. (2007). Flavonoid sugar addition products, method for manufacture and use thereof. Patent US 20070269570.

Degn P.; Pedersen L.H.; Duus J. & Zimmerman, W. (1999). Lipase catalysed synthesis of glucose fatty acid esters in *t*-butanol. *Biotechnology Letters*, Vol.21, pp. 275-280.

Delazar, A.; Celik, S.; Gokturk, R.S.; Nahar, L. & Sarker, S.D. (2005). Two acylated flavonoid glycosides from *Stachys bombycina*, and their free radical scavenging activity. *Pharmazie*, Vol.60, pp. 878-880.

Demetzos, C.; Magiatis, P.; Typas, M.A.; Dimas, K.; Sotiriadou, R.; Perez, S. & Kokkinopoulos, D. (1997). Biotransformation of the flavonoid tiliroside to 7-methylether tiliroside: bioactivity of this metabolite and of its acetylated derivative. *Cellular and Molecular Life Sciences*, Vol.53, pp. 587-592.

Drouin, J.; Costante, J. & Guibé-Jampel, E. (1997). A thermostable microbial enzyme for fast preparative organic chemistry: Preparation of R-(+)-1-phenylethanol from (±)-1-phenylethyl pentanoateand *n*-butanol. *Journal of Chemical Education*, Vol.74, pp. 992-995.

Duarte-Almeida, J.M.; Negri, G.; Salatino, A.; De Carvalho, J.E. & Lajolo, F.M. (2007). Antiproliferative and antioxidant activities of a tricin acylated glycoside from sugarcane (*Saccharum officinarum*) juice. *Phytochemistry*, Vol.68, pp. 1165-1171.

Enaud, E.; Humeau, C.; Piffaut, B. & Girardin, M. (2004). Enzymatic synthesis of new aromatic esters of phloridzin. *Journal of Molecular Catalysis B: Enzymatic*, Vol.27, pp. 1-6.

Ferrer, J.-L.; Austin, M.B.; Stewart, C. & Noel, J.P. (2008). Structure and function of enzymes involved in the biosynthesis of phenylpropanoids. *Plant Physiology and Biochemistry*, Vol.46, pp. 356-370.

Foresti, M.L.; Pedernera, M.; Bucalá, V. & Ferreira, M.L. (2007). Multiple effects of water on solvent-free enzymatic esterifications. *Enzyme & Microbial Technology*, Vol.41, pp. 62-70.

Fossati, E. & Riva, S. (2006). Stereoselective modifications of polyhydroxylated steroids. In: Patel, R.N.: Biocatalysis in the pharmaceutical and biotechnology industries. New York: CRC Press, Taylor & Francis Group, 591-604. ISBN: 0-8493-3732-1.

Fox, G.J. (2000). Natural red sunflower anthocyanin colorant with naturally stabilized color qualities, and the process of making. Patent US 6132791.

Fujiwara, H.; Tanaka, Y.; Yonekura-Sakakibara, K.; Fukuchi-Mizutani, M.; Nakao, M.; Fukui, Y.; Yamaguchi, M.; Ashikari, T. & Kusumi, T. (1998). cDNA cloning, gene expression and subcellular localization of anthocyanin 5-aromatic acyltransferase from *Gentiana triflora*. *The Plant Journal*, Vol.16, pp. 421-431.

Fukami, H.; Nakao, M.; Namikawa, K. & Maeda, M. (2007). Esterified catechin, process for producing the same, food and drink or cosmetic containing the same. Patent EP 1849779.

Gao, C.; Mayon, P.; Macmanus, D.A. & Vulfson, E. N. (2001). Novel enzymatic approach to the synthesis of flavonoid glycosides and their esters. *Biotechnology & Bioengineering*, Vol.71, pp. 235-243.

Gayot, S.; Santarelli, X. & Coulon, D. (2003). Modification of flavonoid using lipase in non-conventional media: effect of the water content. *Journal of Biotechnology*, Vol.101, pp. 29-36.

Ghoul, M.; Engasser, J.-M.; Moussou, P.; Pauly, G.; Ardhaoui, M. & Falcimaigne, A. (2006). Enzymatic production of acyl flavonoid derivatives. Patent US 20060115880.

Giusti, M.M. & Wrolstad, R.E. (2003). Acylated anthocyanins from edible sources and their applications in food systems. *Biochemical Engineering Journal*, Vol.14, pp. 217-225.

Haeffner, F.; Norin, T. & Hult, K. (1998). Molecular modeling of the enantioselectivity in lipase-catalyzed transesterification reactions. *Biophysical Journal*, Vol.74, pp. 1251-1262.

Haraguchi, H.; Tanimoto, K.; Tamura, Y.; Mizutani, K. & Kinoshita, T. (1998). Mode of antibacterial action of retrochalcones from *Glycyrrhiza inflata*. *Phytochemistry*, Vol.48, pp. 125-129.

Harborne, J.B. & Williams, C.A. (1998). Anthocyanins and other flavonoids. *Natural Product Reports*, Vol.15, 631-652.

Harborne, J.B. & Williams, C.A. (2000). Advances in flavonoid research since 1992. *Phytochemistry*, Vol.55, pp. 481-504.

Hidalgo, A. & Bornscheuer, U.T. (2006). Direct evolution of lipases and esterases for organic synthesis. In: Patel, R.N.: Biocatalysis in the pharmaceutical and biotechnology industries. New York: CRC Press, Taylor & Francis Group, 159-179. ISBN: 0-8493-3732-1.

Ishihara, K. & Nakajima, N. (2003). Structural aspects of acylated plant pigments: stabilization of flavonoid glucosides and interpretation of their functions. *Journal of Molecular Catalysis B: Enzymatic*, Vol.23, No.2-6, pp. 411-417.

Ishihara, K.; Nishimura, Y.; Kubo, T.; Okada, C.; Hamada, H. & Nakajima, N. (2002). Enzyme-catalyzed acylation of plant polyphenols for interpretation of their functions. *Plant Biotechnology*, Vol.19, pp. 211-214.

Iso, M.; Chen, B.; Eguchi, M.; Kudo, T. & Shrestha, S. (2001). Production of biodiesel fuel from triglycerides and alcohol using immobilized lipase. *Journal of Molecular Catalysis B: Enzymatic*, Vol.16, pp. 53-58.

Itoh, T.; Nishimura, Y.; Ouchi, N. & Hayase, S. (2003). 1-Butyl-2,3-dimethylimidazolium tetrafluoroborate: the most desirable ionic liquid solvent for recycling use of enzyme in lipase-catalyzed transesterification using vinyl acetate as acyl donor. *Journal of Molecular Catalysis B: Enzymatic*, Vol.26, pp. 41-45.

Iwashina, T. (2003). Flavonoid function and activity to plants and other organisms. *Biological Science in Space*, Vol.17, pp. 24-44.

Jain, N.; Kumar, A.; Chauhan, S. & Chauhan, S.M.S. (2005). Chemical and biochemical transformations in ionic liquids. *Tetrahedron*, Vol.61, pp. 1015-1060.

Jarstoff, B.; Störmann, R.; Ranke, J.; Mölter, K.; Stock, F.; Oberheitmann, B.; Hoffmann, J.; Nüchter, M.; Ondruschka, B. & Filser, J. (2003). How hazardous are ionic liquids? Structure–activity relationships and biological testing as important elements for sustainability evaluation. *Green Chemistry*, Vol.5, pp. 136-142.

Jeager, K.E.; Dijkstra, B.W. & Reetz, M.T. (1999). Bacterial biocatalysts: molecular biology, three-dimensional structures, and biotechnological applications. *Annual Review of Microbiology*, Vol.53, pp. 315-351.

Jungblut, T.P.; Schnitzler, J.P.; Hertkorn, N.; Metzger, J.W.; Heller, W. & Sandermann, H. (1995). Acylated flavonoids as plant defense compounds against environmentally relevant ultraviolet-B radiation in *Scots pine* seedlings. Phytochemicals and health. Proceedings, tenth annual Penn State Symposium in Plant Physiology, Rockville, Md.: American Society of Plant Physiologists, 266 – 267.

Katsoura, M.H.; Polydera, A.C.; Tsironis, L.; Tselepis, A.D. & Stamatis, H. (2006). Use of ionic liquids as media for the biocatalytic preparation of flavonoid derivatives with antioxidant potency. *Journal of Biotechnology*, Vol.123, pp. 491-503.

Katsoura, M.H.; Polydera, A.C.; Katapodis, P.; Kolisis, F.N. & Stamatis, H. (2007). Effect of different reaction parameters on the lipase-catalyzed selective acylation of polyhydroxylated natural compounds in ionic liquids. *Process Biochemistry*, Vol.42, pp. 1326-1334.

Kitamura, S. (2006). Transport of flavonoids: from cytosolic synthesis to vacuolar accumulation. In Grotewold, E.: The science of flavonoids. Springer Science Business Media, New York, 123-146. ISBN: 0-387-28821-X

Kodelia, G.; Athanasiou, K. & Kolisis, F.N. (1994). Enzymatic synthesis of butyryl-rutin ester in organic solvents and its cytogenetic effects in mammalian cells in culture. *Applied Biochemistry and Biotechnology*, Vol.44, pp. 205-212.

Kontogianni, A.; Skouridou, V.; Sereti, V.; Stamatis, H. & Kolisis, F.N. (2003). Lipase-catalyzed esterification of rutin and naringin with fatty acids of medium carbon chain. *Journal of Molecular Catalysis B: Enzymatic*, Vol.21, pp. 59-62.

Kontogianni, A.; Skouridou, V.; Sereti, V.; Stamatis, H. & Kolisis, F.N. (2001). Regioselective acylation of flavonoids catalyzed by lipase in low toxicity media. *European Journal of Lipid Science and Technology*, Vol.103, pp. 655-660.

Kragl, U.; Kaftzik, N.; Schofer, S. & Wasserscheid, P. (2006). U.S. Patent No. 20060211096.

Kwon, C.H.; Dae, Y.S.; Jong, H.L. & Seung, W.K. (2007). Molecular modeling and its experimental verification for the catalytic mechanism of *Candida antarctica* lipase B. *Journal of Microbiology and Biotechnology*, Vol.17, No.7, pp. 1098-1105.

Laane, C. (1987). Medium engineering for bioorganic synthesis. *Biocatalysis*, Vol.30, pp. 80-87.

Lambusta, D.; Nicolosi, G.; Patti, A. & Piattelli, M. (1993). Enzyme-mediated regioprotection-deprotection of hydroxyl groups in (+)-catechin. *Synthesis*, Vol.11, pp. 1155-1158.

Lee, S.; Lee, C.-H.; Moon, S.-S.; Kim, E.; Kim, C.-T.; Kim, B.-H.; Bok, S.-H.; Jeong, T.-S. (2003). Naringenin derivatives as anti-atherogenic agents. *Bioorganic & Medicinal Chemistry Letters*, Vol.13, pp. 3901-3903.

Lin, S.F.; Lin, Y.-H.; Lin, M.; Kao, Y.-F.; Wang, R.-W.; Teng, L.-W.; Chuang, S.-H.; Chang, J.-M. Yuan, T.-T.; Fu, K.C.; Huang, K.P.; Lee, Y.-S.; Chiang, C.-C.; Yang, S.-C.; Lai, C.-L.; Liao, C.-B.; Chen, P.; Lin, Y.-S.; Lai, K.-T.; Huang, H.-J.; Yang, J.-Y.; Liu, C.-W.; Wei, W.-Y.; Chen, C.-K.; Hiipakka, R.A.; Liao, S. & Huang, J.-J. (2010). Synthesis and structure-activity relationship of 3-*O*-acylated (-)-epigallocatechins as 5α-reductase inhibitors. *European Journal of Medicinal Chemistry*, Vol.45, pp. 6068-6076.

Lozano, P.; De Diego, T.; Carrie, D.; Vaultier, M. & Iborra, J. L. (2004). Synthesis of glycidyl esters catalyzed by lipases in ionic liquids and supercritical carbon dioxide. *Journal of Molecular Catalalysis A: Chemical*, Vol.214, pp. 113-119.

Lutz, S. (2004). Engineering lipase B from *Candida antarctica*. *Tetrahedron: Asymmetry*, Vol.15, pp. 2743-2748.

Martinelle, M.; Holmquist, M. & Hult, K. (1995). On the interfacial activation of *Candida antarctica* lipase A and B as compared with *Humicola lanuginosa* lipase. *Biochimica et Biophysica Acta*, Vol.1258, pp. 272-276.

Mellou, F.; Lazari, D.; Skaltsa, H.; Tselepis, A.D.; Kolisis, F.N. & Stamatis, H. (2005). Biocatalytic preparation of acylated derivatives of flavonoid glycosides enhances their antioxidant and antimicrobial activity. *Journal of Biotechnology*, Vol.116, pp. 295-304.

Mellou, F.; Loutrari, H.; Stamatis, H.; Roussos, C. & Kolisis, F.N. (2006). Enzymatic esterification of flavonoids with unsaturated fatty acids: Effect of novel esters on vascular endothelial growth factor release from K562 cells. *Process Biochemistry*, Vol.41, pp. 2029-2034.

Middleton, E.Jr. & Chithan, K. (1993). The impact of plant flavonoids on mammalian biology: implications for immunity, inflammation and cancer. In.: Harborne, J.B.: The flavonoids: advances in research since 1986. London, Chapman and Hall, 619-652, ISBN 0-412-48070–0.

Middleton, E.Jr.; Kandaswami, C. & Theoharides, T.C. (2000). The effects of plant flavonoids on mammalian cells: Implications for inflammation, heart disease, and cancer. *Pharmacological Reviews*, Vol.52, pp. 673-751.

Mitrokotsa, D.; Mitaku, S.; Demetzos, C.; Harvala, C.; Mentis, A.; Perez, S. & Kokkinopoulos, D. (1993). Bioactive compounds from the buds of *Platinus orientalis* and isolation of a new kaempferol glycoside. *Planta Medica*, Vol.59, pp. 517-520.

Moussou, P.; Falcimaigne, A.; Ghoul, M.; Danoux, L. & Pauly, G. (2007). Esters of flavonoids with ω-substituted C6-C22 fatty acids. Patent US 20070184098.

Moussou, P.; Falcimaigne. A.; Pauly, G.; Ghoul, M.; Engasser, J.-M. & Ardhaoui, M. (2004). Preparation of flavonoid derivatives. Patent EP1426445.

Nagasawa, T. & Yamada, H. (1995). Microbial production of commodity chemicals. *Pure & Applied Chemistry*, Vol.67, pp. 1241-1256.

Nakajima, N.; Ishihara, K.; Hamada, H.; Kawabe, S.-I. & Furuya, T. (2000). Regioselective acylation of flavonoid glucoside with aromatic acid by an enzymatic reaction system from cultured cells of *Ipomoea batatas*. *Journal of Bioscience & Bioengineering*, Vol.90, pp. 347-349.

Nakajima, N.; Ishihara, K.; Itoh, T.; Furuya, T. & Hamada, H. (1999). Lipase catalysed direct and regioselective acylation of flavonoid glucoside for mechanistic investigation of stable plant pigments. *Journal of Bioscience and Bioengineering*, Vol.61, pp. 1926-1928.

Nakajima, N.; Sugimoto, M.; Yokoi, H.; Tsuji, H. & Ishihara, K. (2003). Comparison of acylated plant pigments: Light-resistance and radical-scavenging ability. *Bioscience, Biotechnology, and Biochemistry*, Vol.67, pp. 1828-1831.

Nakayama, T.; Suzuki, H. & Nishino, T. (2003). Anthocyanin acyltransferases: specificities, mechanism, phylogenetics, and applications. *Journal of Molecular Catalysis B: Enzymatic*, Vol.23, pp. 117-132.

Narayan, V.S. & Klibanov, A.M. (1993). Are water-immiscibility and apolarity of the solvent relevant to enzyme efficiency? *Biotechnology & Bioengineering*, Vol.41, pp. 390-393.

N'guyen, Q.L. (1995). Cosmetic or dermopharmaceutical composition containing, in combination, a lauroylmethionate of a basic amino acid and at least one polyphenol. Patent US 5431912.

Nicolosi, G.; Piattelli, M.; Lambusta, D. & Patti, A. (1999). Biocatalytic process for the preparation of 3-*O*-acyl-flavonoids. Patent WO/1999/066062.

Otto, R.; Geers, B.; Weiss, A.; Petersohn, D.; Schlotmann, K. & Schroeder, K.R. (2001). Novel flavone glycoside derivatives for use in cosmetics, pharmaceuticals and nutrition. Patent EP 1274712.

Parejo, I.; Bastida, J.; Viladomat, F. & Codina, C. (2005). Acylated quercetagetin glycosides with antioxidant activity from *Tagetes maxima*. *Phytochemistry*, Vol.66, pp. 2356-2362.

Patel, R.N. (2006). Biocatalysis for synthesis for chiral pharmaceutical intermediates. In: Patel, R.N.: Biocatalysis in the pharmaceutical and biotechnology industries. New York: CRC Press, Taylor & Francis Group, 103-158. ISBN: 0-8493-3732-1.

Passicos, E.; Santarelli, X. & Coulon, D. (2004). Regioselective acylation of flavonoids catalyzed by immobilized *Candida antarctica* lipase under reduced pressure. *Biotechnology Letters*, Vol.26, pp. 1073-1076.

Patti, A.; Piattelli, M. & Nicolosi, G. (2000). Use of *Mucor miehei* lipase in the preparation of long chain 3-*O*-acylcatechins. *Journal of Molecular Catalysis B: Enzymatic*, Vol.10, pp. 577-582.

Perrier, E.; Mariotte, A.-M.; Boumendjel, A. & Bresson-Rival, D. (2001). Flavonoide esters and their use notably in cosmetics. Patent US 6235294.

Plaper, A.; Golob, M.; Hafner, I.; Oblak, M.; Solmajer, T. & Jerala, R. (2003). Characterization of quercetin binding site on DNA gyrase. *Biochemical and Biophysical Research Communications*, Vol.306, pp. 530-536.

Pleiss, J.; Fischer, M. & Schmid, R. D. (1998). Anatomy of lipase binding sites: the scissile fatty acid binding site. *Chemistry and Physics of Lipids*, Vol.93, pp. 67-80.

Quarenghi, M.V.; Tereschuk, M.L.; Baigori, M.D. & Abdala, L.R. (2000). Antimicrobial activity of flowers from *Anthemis cotula*. *Fitoterapia*, Vol.71, pp. 710-712.

Rao, M.B.; Tanksale, A.M.; Ghatge, M.S. & Deshpande, V.V. (1998). Molecular and biotechnological aspects of microbial proteases. *Microbiology and Molecular Biology Reviews*, Vol.62, pp. 597-635.

Rauha, J.P.; Remes, S.; Heinonen, M.; Hopia, A.; Kähkönen, M.; Kujala, T.; Pihlaja, K.; Vuorela, H. & Vuorela, P. (2000). Antimicrobial effect of Finnish plant extracts containing flavonoids and other phenolic compounds. *International Journal of Food Microbiology*, Vol.56, pp. 3-12.

Reetz, M.T.; Wiesenhöffer, W.; Francio, G. & Leitner, W. (2003). Continuous flow enzymatic kinetic resolution and enantiomer separation using ionic liquid/supercritical carbon dioxide media. *Advanced Synthesis & Catalysis*, Vol.345, pp. 01221-01228.

Rice-Evans, C.A. (2001). Flavonoid Antioxidants. *Current Medicinal Chemistry*, Vol.8, pp. 797-807.

Riva, S. (2002). Enzymatic modification of the sugar moieties of natural glycosides. *Journal of Molecular Catalysis B: Enzymatic*, Vol.19-20, pp. 43-54.

Riva, S.; Chopineau, J.; Kieboom, A.P.G. & Klibanov, A.M. (1988). Protease-catalyzed regioselective esterification of sugars and related compounds in anhydrous dimethylformamide. *Journal of the American Chemical Society*, Vol.110, pp. 584-589.

Rocha, J.; Gil, M. & Garcia, F. (1999). Optimisation of the enzymatic synthesis of n-octyl oleate with immobilized lipase in the absence of solvents. *Journal of Chemical Technology & Biotechnology*, Vol.74, pp. 607-612.

Ross, J.A. & Kasum, C.M. (2002). Dietary flavonoids: bioavailability, metabolic effects, and safety. *Annul Reviews in Nutrition*, Vol.22, pp. 19-34.

Rubin-Pitel, S.B. & Zhao, H. (2006). Recent Advances in biocatalysis by directed enzyme evolution. *Combinatorial Chemistry & High Throughput Screening*, Vol.9, pp. 247-257.

Saija, A.; Tomaino, A.; Trombetta, D.; Pellegrino, M.L.; Tita, B.; Messina, C.; Bonina, F.P.; Rocco, C.; Nicolosi, G. & Castelli, F. (2003). 'In vitro' antioxidant and photoprotective properties and interaction with model membranes of three new quercetin esters. *European Journal of Pharmaceutics and Biopharmaceutics*, Vol.56, pp. 167-174.

Sakai, M.; Suzuki, M.; Nanjo, F. & Hara, Y. (1994). 3-O-acylated catechins and methods of producing same. Patent EP 0618203.

Salem, J.H.; Chevalot, I.; Harscoat-Schiavo, C.; Paris, C.; Fick, M. & Humeau, C. (2011). Biological activities of flavonoids from *N. retusa* and their acylated derivatives. *Food Chemistry*, Vol.124, pp. 486-494.

Saxena, R.K.; Ghosh, P.K.; Gupta, R.; Davidson, W.S.; Bradoo, S. & Gulati, R. (1999). Microbial lipases: potential biocatalysts for the future industry. *Current Science*, Vol.77, pp. 101-115.

Sharma, G.; Singh, R.P.; Chan, D.C. & Agarwal, R. (2003) Silibinin induces growth inhibition and apoptotic cell death in human lung carcinoma cells. *Anticancer Research*, Vol.23, pp. 2649-2655.

Singh, R.K. & Nath, G. (1999). Antimicrobial activity of *Elaeocarpus sphaericus*. *Phytotherapy Research*, Vol.13, pp. 448-450.

Stapleton, P.D.; Shah, S.; Anderson, J.C.; Hara, Y.; Hamilton-Miller, J.M.T. & Taylor, P.W. (2004). Modulation of beta-lactam resistance in *Staphylococcus aureus* by catechins and gallates. *International Journal of Antimicrobial Agents*, Vol.23, pp. 462-467.

Stefani, E.D.; Boffetta, P.; Deneo-Pellegrini, H.; Mendilaharsu, M.; Carzoglio, J.C.; Ronco, A. & Olivera, L. (1999). Dietary antioxidants and lung cancer risk: a case-control study in Uruguay. *Nutrition and Cancer*, Vol.34, pp. 100–110.

Stepanovic, S.; Antic, N.; Dakic, I. & Svabic-Vlahovic, M. (2003). *In vitro* antimicrobial activity of propolis and synergism between propolis and antimicrobial drugs. *Research in Microbiology*, Vol.158, pp. 353-357.

Stevenson, D.E.; Wibisono, R.; Jensen, D.J.; Stanley, R.A. & Cooney, J.M. (2006). Direct acylation of flavonoid glycosides with phenolic acids catalyzed by *Candida antarctica* lipase B (Novozym 435®). *Enzyme and Microbial Technology*, Vol.39, pp. 1236-1241.

Suda, I.; Oki, T.; Masuda, M.; Nishiba, Y.; Furuta, S.; Matsugano, K.; Sugita, K. & Terahara, N. (2002). Direct absorption of acylated anthocyanin in purple-fleshed sweet potato into rats. *Journal of Agricultural and Food Chemistry*, Vol.50, pp. 1672-1676.

Tarle, D. & Dvorzak, I. (1990). Antimicrobial activity of the plant *Cirsium oleraceum* (L.). Scop. *Acta Pharmaceutica Jugoslavica*, Vol.40, pp. 567-571.

Tereschuk, M.L.; Riera, M.V.; Castro, G.R. & Abdala, L.R. (1997). Antimicrobial activity of flavonoids from leaves of *Tagetes minuta*. *Journal of Ethnopharmacology*, Vol.56, pp. 227-232.

Toba, S. & Merz, K. M. (1997). The concept of solvent compatibility and its impact on protein stability and activity enhancement in non-aqueous solvents. *Journal of the American Chemical Society*, Vol.119, pp. 9939-9948.

Torres, S. & Castro, G.R. (2004). Non-aqueous biocatalysis in homogeneous solvent systems. *Food Technology & Biotechnology*, Vol.42, pp. 271-277.

Trodler, P. & Pleiss, J. (2008). Modeling structure and flexibility of *Candida antarctica* lipase B in organic solvents. *BMC Structural Biology* [online].

Uppenberg, J.; Öhrner, N.; Norin, M.; Hult, K.; Kleywegt, G.J.; Patkar, S.; Waagen, V.; Anthonsen, T. & Jones, T.A. (1995). Crystallographic and molecular-modelling studies of lipase B from *Candida antarctica* reveal a stereospecificity pocket for secondary alcohols. *Biochemistry*, Vol.33, pp. 16838-16851.

Van Rantwick, F.; Lau, R.M. & Sheldon, R.A. (2003). Biocatalytic transformations in ionic liquids. *Trends in Biotechnology*, Vol.21, pp. 131-138.

Vakhlu, J. & Kour, A. (2006). Yeast lipases: enzyme purification, biochemical properties and gene cloning. *Electronic Journal of Biotechnology* [online].

Viskupicova, J.; Danihelova, M.; Ondrejovic, M.; Liptaj, T. & Sturdik, E. (2010). Lipophilic rutin derivatives for antioxidant protection of oil-based foods. *Food Chemistry*, Vol.123, pp. 45-50.

Viskupicova, J.; Maliar, T.; Psenakova, I. & Sturdik, E. (2006). Eznymatic acylation of naringin. *Nova Biotechnologica*, Vol.6, pp. 149-159 [In Slovak].

Viskupicova, J. & Ondrejovic, M. (2007). Effect of fatty acid chain length on enzymatic esterification of rutin. In: Book of abstracts of the 1st International Conference of Applied Natural Sciences, November 7 – 9, 2007 (pp. 59), Trnava: UCM.

Viskupicova, J.; Strosova, M.; Sturdik, E. & Horakova, L. (2009). Modulating effect of flavonoids and their derivatives on sarcoplasmic reticulum Ca^{2+}-ATPase oxidized by hypochloric acid and peroxynitrite. *Neuroendocrinology Letters*, Vol.30, pp. 148-151.

Wang, Y.F.; Lalonade, J.J.; Momongan, M.; Bergbreiter, D.E. & Wong, C.H. (1988). Lipase-catalyzed irreversible transesterifications using enol esters as acylating reagents: preparative enantio and regioselective synthesis of alcohols, glycerol derivatives, sugars, and organometallics. Journal of the American Chemical Society, Vol.110, pp. 7200-7205.

Williams, C.A. (2006). Flavone and flavonol O-glycosides. In: Andersen, Ø.M.; Markham, K.R.: Flavonoids: chemistry, biochemistry and applications. New York: Taylor & Francis Group, CRC Press, 397-441. ISBN: 0-8493-2021-6.

Wilkes, J.S. (2004). Properties of ionic liquids solvents for catalysis. *Journal of Molecular Catalysis A: Chemical*, Vol.214, pp. 11-17.

Xanthakis, E.; Theodosiou, E.; Magkouta, S.; Stamatis, H. & Loutrari, H. (2010). Enzymatic transformation of flavonoids and terpenoids: structural and functional diversity of the novel derivatives. *Pure & Applied Chemistry*, Vol.82, pp. 1-16.

Xiao, Y.-M.; Wu, Q.; Wu, W.-B.; Zhang, Q.-Y. & Lin, X.-F. (2005). Controllable regioselective acylation of rutin catalyzed by enzymes in non-aqueous solvents. *Biotechnology Letters*, Vol.27, pp. 1591-1595.

Yadav, G.D. & Piyush, S.L. (2003). Kinetics and mechanism of synthesis of butyl isobutyrate over immobilised lipases. *Biochemical Engineering Journal*, Vol.16, pp. 245-252.

Zaks, A. & Klibanov, A.M. (1988). Enzyme catalysis in nonaqueous solvents. *Journal of Biological Chemistry*, Vol.263, pp. 3194-3201.

2

Enzymology of Bacterial Lysine Biosynthesis

Con Dogovski[1*] et al.

[1]*Department of Biochemistry and Molecular Biology, Bio21 Molecular Science and Biotechnology Institute, University of Melbourne, Parkville, Victoria Australia*

1. Introduction

Lysine is an essential amino acid in the mammalian diet, but can be synthesised *de novo* in bacteria, plants and some fungi (Dogovski et al., 2009; Hutton et al., 2007). In bacteria, the lysine biosynthesis pathway, also known as the diaminopimelate (DAP) pathway (Fig. 1), yields the important metabolites *meso*-2,6-diaminopimelate (*meso*-DAP) and lysine. Lysine is utilised for protein synthesis in bacteria and forms part of the peptidoglycan cross-link structure in the cell wall of most Gram-positive species; whilst *meso*-DAP is the peptidoglycan cross-linking moiety in the cell wall of Gram-negative bacteria and also Gram-positive *Bacillus* species (Burgess et al., 2008; Mitsakos et al., 2008; Voss et al., 2010) (Fig. 1).

The synthesis of *meso*-DAP and lysine begins with the condensation of pyruvate (PYR) and L-aspartate-semialdehyde (ASA) by the enzyme *dihydrodipicolinate synthase* (DHDPS, EC 4.2.1.52) (Blickling et al., 1997a; Mirwaldt et al., 1995; Voss et al., 2010; Yugari & Gilvarg, 1965). The product of the DHDPS-catalysed reaction is an unstable heterocycle, 4-hydroxy-2,3,4,5-tetrahydro-L,L-dipicolinic acid (HTPA) (Fig. 1). HTPA is non-enzymatically dehydrated to produce dihydrodipicolinate (DHDP), which is subsequently reduced by the NAD(P)H-dependent enzyme, *dihydrodipicolinate reductase* (DHDPR, EC 1.3.1.26), to form L-2,3,4,5,-tetrahydrodipicolinate (THDP) (Dommaraju et al., 2011; Girish et al., 2011; Reddy et al., 1995, 1996) (Fig. 1). The metabolic pathway then diverges into four sub-pathways depending on the species, namely the succinylase, acetylase, dehydrogenase and aminotransferase pathways (Dogovski et al., 2009; Hutton et al., 2007) (Fig. 1).

The most common of the alternative metabolic routes is the succinylase pathway, which is inherent to many bacterial species including *Escherichia coli*. This sub-pathway begins with the conversion of THDP to N-succinyl-L-2-amino-6-ketopimelate (NSAKP) catalysed by *2,3,4,5-tetrahydropyridine-2-carboxylate N-succinyltransferase* (THPC-NST, EC 2.3.1.117).

* Sarah. C. Atkinson[1], Sudhir R. Dommaraju[1], Matthew Downton[2], Lilian Hor[1], Stephen Moore[2], Jason J. Paxman[1], Martin G. Peverelli[1], Theresa W. Qiu[1], Matthias Reumann[2], Tanzeela Siddiqui[1], Nicole L. Taylor[1], John Wagner[2], Jacinta M. Wubben[1] and Matthew A. Perugini[1,3]

[1]*Department of Biochemistry and Molecular Biology, Bio21 Molecular Science and Biotechnology Institute, University of Melbourne, Parkville, Victoria, Australia*
[2]*IBM Research Collaboratory for Life Sciences-Melbourne, Victorian Life Sciences Computation Initiative, University of Melbourne, Parkville, Victoria, Australia*
[3]*Department of Biochemistry, La Trobe Institute for Molecular Science, La Trobe University, Melbourne, Australia*

NSAKP is then converted to N-succinyl-*L,L*-2,6,-diaminopimelate (NSDAP) by *N-succinyldiaminopimelate aminotransferase* (NSDAP-AT, EC 2.6.1.17), which is subsequently desuccinylated by *succinyldiaminopimelate desuccinylase* (SDAP-DS, EC 3.5.1.18) to form *L,L*-2,6-diaminopimelate (*LL*-DAP) (Kindler & Gilvarg., 1960; Ledwidge & Blanchard., 1999; Simms et al., 1984) (Fig. 1). *LL*-DAP is then converted to *meso*-DAP by the enzyme diaminopimelate epimerase (DAPE, EC 5.1.1.7) (Wiseman, & Nichols, 1984) (Fig. 1).

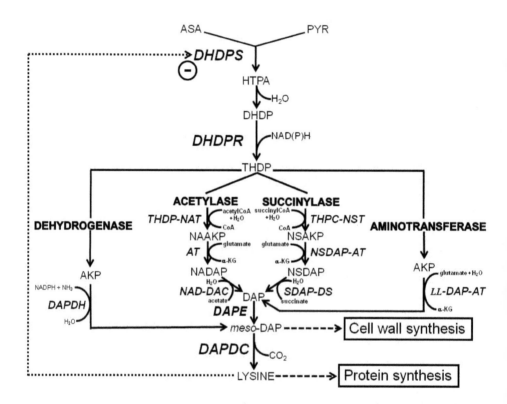

Fig. 1. Diaminopimelate pathway in bacteria.

As for the succinylase pathway, the acetylase pathway involves four enzymatic steps, but incorporates N-acetyl groups rather than N-succinyl moieties. This pathway is common to several *Bacillus* species, including *B. subtilis* and the anthrax-causing pathogen *B. anthracis* (Chatterjee & White., 1982; Peterkofsky & Gilvarg., 1961; Sundharadas & Gilvarg., 1967). The sub-pathway begins with the conversion of THDP to N-acetyl-(*S*)-2-amino-6-ketopimelate (NAAKP) catalysed by *tetrahydrodipicolinate N-acetyltransferase* (THDP-NAT, EC 2.3.1.89), followed by conversion to N-acetyl-(2*S*)-2,6,-diaminopimelate (NADAP) by *aminotransferase A* (ATA, EC 2.6.1). NADAP is subsequently deacetylated to form DAP by the enzyme *N-acetyldiaminopimelate deacetylase* (NAD-DAC, EC 3.5.1.47) (Fig. 1). As in the succinylase pathway, *LL*-DAP is then converted to *meso*-DAP by DAPE (Fig. 1).

There are also two additional sub-pathways that are less common to bacteria. The aminotransferase pathway, catalysed by the enzyme *diaminopimelate aminotransferase* (LL-DAP-AT, EC 2.6.1.83), is found in plant, eubacterial and archaeal species (Hudson et al., 2006). This sub-pathway involves the conversion of the acyclic form of THDP, 1,2-amino-6-ketopimelate (AKP), to *meso*-DAP in a single step. LL-DAP is then converted in the second step of the sub-pathway to *meso*-DAP by DAPE, as for the acetylase and succinylase pathways (Fig. 1). The dehydrogenase pathway, which is common to *Corynebacterium* and some *Bacillus* species, converts THDP to *meso*-DAP, also in a single step (Misono et al., 1976). This sub-pathway employs the NADPH-dependent enzyme, *diaminopimelate dehydrogenase* (DAPDH, EC 1.4.1.16), which also employs AKP as the substrate (Fig. 1).

All four alternative pathways then converge to utilise the same enzyme for the final step of lysine biosynthesis, namely diaminopimelate decarboxylase (DAPDC, EC 4.1.1.20) (Ray et al., 2002). DAPDC catalyses the decarboxylation of *meso*-DAP to yield lysine and carbon dioxide. This step is important for the overall regulation of the lysine biosynthesis pathway since the downstream product, lysine, has been shown to allosterically inhibit DHDPS from plants and Gram-negative bacteria (Section 2.1.1, Fig. 1). DHDPS is therefore considered the rate-limiting enzyme of the pathway.

This book chapter will describe the function, structure, and regulation of the key enzymes functioning in the lysine biosynthesis pathway. Furthermore, given that several of these enzymes are the products of essential bacterial genes that are not expressed in humans, the pathway is of interest to antibiotic discovery research (Dogovski et al., 2009; Hutton et al., 2007). Accordingly, the chapter will also review the current status of rational drug design initiatives targeting essential enzymes of the lysine biosynthesis pathway in pathogenic bacteria.

2. Dihydrodipicolinate synthase

2.1 Function of DHDPS

Dihydrodipicolinate synthase (DHDPS, EC 4.2.1.52) was first purified in 1965 from *E. coli* extracts (Yugari & Gilvarg, 1965). The enzyme is the product of the *dap*A gene, which has been shown to be essential in several bacterial species (Dogovski et al., 2009; Hutton et al., 2007). The *dap*A product, DHDPS, catalyses the condensation of pyruvate (PYR) and aspartate semialdehyde (ASA) to form 4-hydroxy-2,3,4,5-tetrahydro-*L,L*-dipicolinic acid (HTPA) (Fig. 1). It was first suggested that the product released by DHDPS was dihydrodipicolinate (DHDP), but studies using ^{13}C-labelled pyruvate support the view that the product is the unstable heterocycle HTPA (Blickling et al. 1997a). Rapid decomposition of the ^{13}C-NMR signals of HTPA following its production indicate that formation of DHDP occurs via a nonenzymatic step.

In all cases examined, the DHDPS-catalysed reaction proceeds via a ping-pong kinetic mechanism in which pyruvate binds the active site, resulting in the release of a protonated water molecule. ASA then binds and is condensed with pyruvate to form the heterocyclic product, HTPA (Blickling et al. 1997a).

In the first step of the mechanism, the active site lysine, (Lys161 in *E. coli* DHDPS) forms a Schiff base with pyruvate (Laber et al., 1992) (Fig 2). Formation of the Schiff base proceeds

through a tetrahedral intermediate. It is proposed that a catalytic triad of three residues -
Tyr133, Thr44 and Tyr107 (*E. coli* numbering) - act as a proton relay to transfer protons to
and from the active site via a water-filled channel leading to bulk solvent (Dobson et al.,
2004a). The Schiff base (imine) is converted to its enamine form, which then adds to the
aldehyde group of ASA (Blickling et al., 1997a; Dobson et al., 2008). In aqueous solution,
ASA is known to exist in the hydrated form rather than the aldehyde, but the biologically-
relevant form of the substrate remains to be determined. HTPA is then formed by
nucleophilic attack of the amino group of ASA onto the intermediate imine, leading to
cyclisation and detachment of the product from the enzyme, with release of the active site
lysine residue (Fig 2).

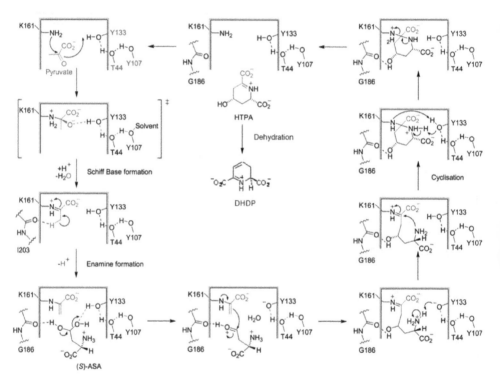

Fig. 2. The catalytic mechanism of DHDPS.

2.1.1 Regulation of DHDPS activity

In some organisms, the activity of DHDPS is regulated allosterically by lysine via a classical
feedback inhibition process. Lysine feedback inhibition of DHDPS has been investigated in
several plant, Gram-negative and Gram-positive bacterial species to date. Studies involving
Daucus carota sativa (Matthews et al., 1979), *Pivus sativum* (Dereppe et al., 1992), *Spinacia
aloeracea* (Wallsgrove et al., 1980), *Triticum aestivium* (Kumpaisal et al., 1987), and *Zea mays*
(Frisch et al., 1991) show that DHDPS from plant species are generally strongly inhibited by

lysine (IC_{50} = 0.01-0.05 mM). In contrast, DHDPS from bacteria are significantly less sensitive to lysine inhibition than their plant counterparts. For example, DHDPS from Gram-negative bacteria, such as *E. coli* (Dobson et al., 2005a; Yugari and Gilvarg, 1965), *Niesseria meningitidis* (Devenish et al., 2009), and *Sinorhizobium. meliloti* (Phenix & Palmer, 2008), display IC_{50} values that range from 0.25 mM to 1.0 mM. Whereas, the enzyme from Gram-positive bacteria such as *Bacillus anthracis* (Domigan et al., 2009), *Bacillus cereus* (Hoganson & Stahly, 1975), *Corynebacterium glutamicum* (Cremer et al., 1988), *Lactobacillus plantarum* (Cahyanto et al., 2006) and *Staphylococcus aureus* (Burgess et al., 2008) show little or no inhibition by lysine.

The crystal structure of DHDPS in complex with lysine from *E. coli* shows that the lysine allosteric binding site is situated in a crevice at the interface of the tight dimer, distal from the active site, but connected to the active site via a water channel (Blickling et al., 1997a). Two inhibitory lysine molecules are bound in close proximity within van der Waals contact to each other. Seven residues located within the allosteric site bind lysine, namely Ala49, His53, His56, Gly78, Asp80, Glu84, and Tyr106 (Blickling et al., 1997a).

Studies show that lysine inhibition is cooperative with the second lysine molecule binding 10^5 times more tightly than the first (Blickling et al., 1997a). The mechanism by which lysine exerts regulatory control over bacterial DHDPS is not well understood, although kinetic and structural studies suggest that it is an allosteric inhibitor, causing partial inhibition (approximately 90%) at saturating concentrations (Blickling et al., 1997a). It has recently been suggested that lysine exerts some effect on the first half reaction by attenuating proton-relay and also the function of Arg138, thought to be crucial for ASA binding (Dobson et al., 2004b). The crystal structure of the *E. coli* DHDPS-lysine complex was solved in the absence of substrate; however, thermodynamic studies have illustrated that the substrate pyruvate has a substantial effect on the nature of enzyme-inhibitor association (Blickling et al., 1997a).

2.2 Structure of DHDPS

2.2.1 Subunit and quaternary structure of DHDPS

DHDPS from *B. anthracis* (Blagova et al., 2006; Voss et al., 2010), *E. coli* (Mirwaldt et al., 1995), *Mycobacterium tuberculosis* (Kefala et al., 2008), *Thermoanaerobacter tengcongensis* (Wolterink-van Loo et al., 2008), *Thermotoga maritima* (Pearce et al., 2006), and several other species is a homotetramer in both crystal structure and solution (Fig. 3). In *E. coli*, the monomer is 292 amino acids in length and is composed of two domains (Mirwaldt et al., 1995). The N-terminal domain is a $(\beta/\alpha)_8$ TIM-barrel (residues 1-224) with the active site located within the centre of the barrel (Fig. 3). The C-terminal domain (residues 225-292) consists of three α-helices and contains several key residues that mediate tetramerisation (Dobson et al., 2005a). The association of the four monomers leaves a large water-filled cavity in the centre of the tetramer, such that each monomer has contacts with two neighbouring monomers only. The tetramer can also be described as a dimer of dimers, with strong interactions between the monomers A & B and C & D at the so-called tight dimer interface, and weaker interactions between the dimers A-B and C-D at the weak dimer interface (Dobson et al., 2005a) (Fig. 3).

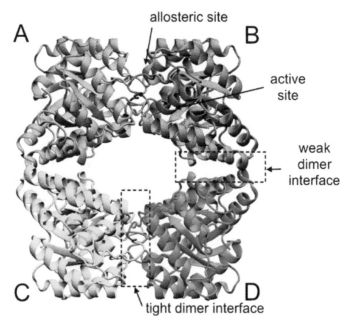

Fig. 3. *E. coli* DHDPS structure . The active sites, allosteric sites, dimerisation interface (tight dimer interface) and tetramerisation interface (weak dimer interface) are shown (PDB: 1YXC).

2.2.2 Active site

The active site is located in cavities formed by the two monomers of the dimer. A long solvent-accessible catalytic crevice with a depth of 10 Å is formed between β-strands 4 and 5 of the barrel (Mirwaldt et al., 1995). Lys161, involved in Schiff-base formation is situated in the β-barrel near the catalytic triad of three residues, namely Tyr133, Thr44 and Tyr107, which act as a proton shuttle (Blickling et al., 1997a) (Fig. 4). Thr44 is hydrogen bonded to both Tyr133 and Tyr107 and its position in the hydrogen-bonding network may play a role in Schiff base formation and cyclisation (Dobson et al., 2005a). The dihedral angles of Tyr107 fall in the disallowed region of the Ramachandran plot, suggesting an important role in the enzyme's function (Mirwaldt et al., 1995). It is believed to be involved in shuttling protons between the active site and solvent (Dobson et al., 2005a). In contrast, Tyr133 plays an important role in substrate binding, donating a proton to the Schiff base hydroxyl. It is also thought to coordinate the attacking amino group of ASA, which requires the loss of a proton subsequent to cyclisation (Fig. 2). A marked reduction in activity is observed in single substitution mutants, highlighting the importance of this catalytic triad (Dobson et al., 2004a).

Situated at the entrance to the active site, Arg138 is essential for ASA binding (Dobson et al., 2005b). In the *E. coli* DHDPS structure, a hydrogen bond is formed between Arg138 and Tyr107 (Dobson et al., 2004a) and a water mediated hydrogen bond is formed between Arg138 and Tyr133 (Dobson et al., 2005a). Arg138 is thus also important for stabilisation of the catalytic triad, both of which are highly conserved in all DHDPS enzymes (Dobson et al., 2005a).

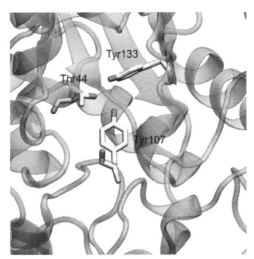

Fig. 4. *E. coli* DHDPS active site, illustrating the catalytic triad Thr44, Tyr133 and Tyr107 interdigitating from the opposing monomer (PDB: 1YXC).

2.2.3 Tight dimer interface

In *E. coli* DHDPS, 1400 Å² of surface area from one monomer in each dimer is buried at the tight dimer interface (Dobson et al., 2005a) (Fig. 3). This interface is made up of 25 residues from each monomer, with hydrogen bonds formed between Ser111 and Cys141, and hydrophobic interactions between Leu51 and Ala81, among others. In addition, Tyr107 of one monomer is coordinated with Tyr106 from the adjacent subunit, interdigitating across the monomer-monomer interface and thus forming a stabilising hydrophobic, sandwich-like stacking of aromatic rings.

2.2.4 Weak dimer interface

The tight dimer units of the *E. coli* DHDPS tetramer associate via two isologous interfaces formed between corresponding monomers (Fig. 3). This interface buries approximately 538 Å² of surface area. Nine residues from each monomer are involved in contacts at the weak dimer interface (Mirwaldt et al., 1995), situated within the α6, α7 and α9-helices. The interface is stabilised by hydrophobic contacts between Leu167, Thr168 and Leu197 (Dobson et al., 2004a). The importance of Leu197 at the interface has been demonstrated with mutations resulting in a dimeric species, unable to form a tetramer (Griffin et al., 2008, 2010). This interface is not conserved in other DHDPS structures. A greater number of contacts are observed at the weak dimer interface in DHDPS from *B. anthracis* (Blagova et al., 2006; Voss et al, 2010), *M. tuberculosis* (Kefala et al., 2008), and most strikingly, *T. maritima* (Pearce et al., 2006) with 20 residues involved in many interactions.

2.2.5 Allosteric site

As described in Section 2.1.1, lysine is an allosteric modulator of DHDPS function, partially inhibiting DHDPS activity. The lysine binding site is situated in a crevice at the interface of the tight dimer, distal from the active site, but connected via a water filled channel (Fig. 3). The crystal structure of lysine-bound *E. coli* DHDPS shows two lysine molecules bound per dimer (four per tetramer) with each molecule interacting with both monomers and the adjacent lysine molecule (Blickling et al., 1997c) (Fig. 5).

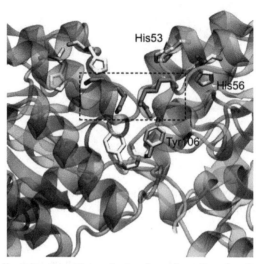

Fig. 5. *E. coli* DHDPS allosteric site with two lysine (boxed - red) molecules bound (PDB: 2ATS).

As stated earlier, seven residues are believed to be involved in binding lysine; Ser48, Ala49, His53, His56, Asn80, Glu84 and Tyr106. All these residues show slightly altered conformations in the presence of lysine, moving to accommodate the molecule (Blickling et al., 1997a). Importantly, Tyr106 moves towards the carboxyl group of lysine, which alters the aromatic stacking of Tyr106 and Tyr107. Otherwise, very few changes are observed upon lysine binding, with no significant secondary structure or quaternary structure change occurring (Dobson et al., 2005a). Most of the residues identified as important in the lysine allosteric binding site are not conserved in those DHDPS enzymes that are not inhibited by lysine (Burgess et al., 2008; Kefala et al., 2008; Voss et al., 2010; Wolterink-van Loo et al., 2008).

2.2.6 Alternative quaternary architecture

Whilst the DHDPS monomer from most bacteria has a molecular mass of approximately 31 kDa, the plant enzymes are larger. For example, DHDPS from *Nicotiana sylvestris* (Blickling et al., 1997b) has a relative molecular mass of 36 kDa, whilst DHDPS from *Pivus sativum* (Dereppe et al., 1992) has been reported to be a homotrimer of 43 kDa monomers based on gel filtration liquid chromatography studies, although this result is uncorroborated. The only plant DHDPS structure solved to date is from *N. sylvestris* (Blickling et al., 1997b). As for the bacterial enzymes, it is a homotetramer, described as a dimer of dimers. The contact areas within the tight dimer are similar within the plant and bacterial enzyme, with 13 of the 19 residues contributing to the interface conserved in both bacteria and plants. However, as Figure 6 shows, the plant dimer of dimers has an alternative architecture, namely the residues involved at the weak dimer interface are located on the opposite face of the monomer. The plant enzyme can thus be described as a "back-to-back" arrangement of dimers

(Fig. 6) compared to the "head-to-head" arrangement observed for bacterial DHDPS (Fig. 3). Compared to the bacterial interface, the weak dimer interface of *N. sylvestris* DHDPS is larger than its bacterial counterpart, burying 810 Å² surface area, which is reflected in the greater number of residues contributing to inter-subunit contacts. The additional residues of the C-terminus, as well as the novel quaternary structure of *N. sylvestris* DHDPS, reduces the central water filled cavity and results in a tetramer where all subunits are in contact with each other (Fig. 6). Despite the significant structural differences between the plant and bacterial enzymes, the position and orientation of all active site residues are conserved. Most strikingly, considering the rearrangement of dimers, lysine binds at an equivalent binding pocket at the interfaces of the two monomers of a dimer in both the *E. coli* (Blickling et al., 1997a) and *N. sylvestris* (Blickling et al., 1997b) enzymes. The lysine molecules also bind in the same orientation, with coordination of the α-amino and α-carboxyl groups almost identical.

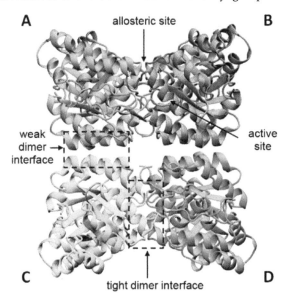

Fig. 6. *N. sylvestris* DHDPS, showing an alternate architecture ("back-to-back" arrangement of dimers) to that observed for bacterial DHDPS enzymes.

In addition, other quaternary structures of bacterial DHDPS enzymes have been reported. For example, DHDPS from methicillin-resistant *S. aureus* (MRSA) has recently been shown to be a dimer in solution (Burgess et al., 2008), with enzymatic activity similar to that of the wild-type *E. coli* tetramer. Several DHDPS enzymes have also been identified in *Agrobacterium tumefaciens*, with two forms crystallising as dimers (3B4U, 2R8W) and one as a hexamer (2HMC), although the function of this enzyme has yet to be confirmed.

2.3 Inhibition of DHDPS

A number of potential DHDPS inhibitors have been synthesised and characterised. A variety of heterocyclic analogues of DHDP and HTPA have been shown to act as moderate inhibitors of DHDPS (Hutton et al., 2007). Analogues of the cyclic lactol form of ASA, including homoserine lactone and 2-aminocyclopentanone, show non-competitive moderate inhibition with K_i = 12-24 mM (Hutton et al., 2007). Analogues of the straight chain hydrate form of ASA have also been investigated, with aspartic acid showing mixed type inhibition with K_i = 90-140 μM (Hutton et al., 2007). Product analogues have also been investigated, exhibiting moderate DHDPS inhibition. More success was achieved with inhibitors based on the acyclic enzyme-bound DHDPS intermediates, such as diethyl (E,E)-4-oxo-2,5-heptadienedioate (Turner et al., 2005) and a bis-oxime ester (Boughton et al., 2008), which irreversibly inhibit DHDPS. Interestingly, several of these compounds have displayed clear differentiation in inhibition of DHDPS enzymes from different species (Mitsakos et al., 2008), suggesting the potential for targeting compounds to specific pathogens.

3. Dihydrodipicolinate reductase

3.1 Function of DHDPR

Dihydrodipicolinate reductase (DHDPR, EC 1.3.1.26) was first isolated from *E. coli* in 1965 (Farkas & Gilvarg, 1965). Since then, the enzyme has been characterised from several species including *B. cereus* (Kimura & Goto, 1977), *Bacillus megaterium* (Kimura & Goto, 1977), *Bacillus subtilis* (Kimura, 1975), *C. glutamicum* (Cremer et al., 1988), *Methylophilus methylotrophus* (Gunji et al., 2004), *M. tuberculosis* (Cirilli et al., 2003), *S. aureus* (Dommaraju et al., 2011; Girish et al., 2011), and *T. maritima* (Pearce et al., 2008). DHDPR catalyses the second step in the lysine biosynthesis pathway (Fig. 1), the pyridine nucleotide-dependent reduction of dihydrodipicolinate (DHDP) to form *L*-2,3,4,5,-tetrahydrodipicolinate (THDP) (Dogovski et al., 2009; Hutton et al., 2007).

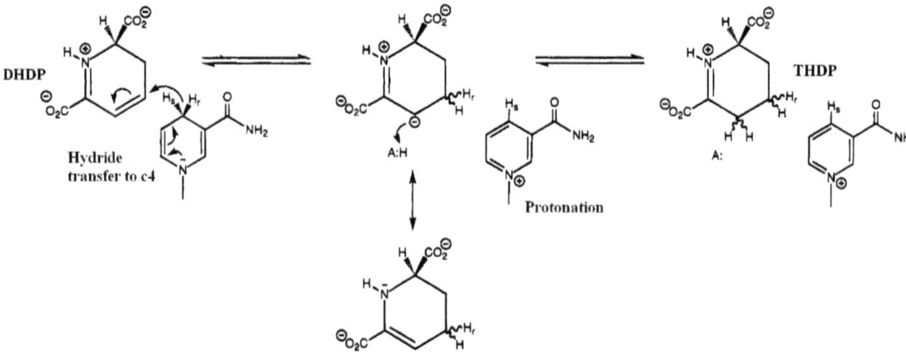

Fig. 7. Schematic representation of the catalytic mechanism of DHDPR.

In *E. coli*, DHDPR is encoded by the *dapB* gene, which is also an essential bacterial gene (Dogovski et al., 2009; Hutton et al., 2007). The open reading frame encodes a 273 amino acid polypeptide with a monomeric molecular weight of 28,758 Da. The enzyme functions by utilising either phosphorylated or non-phosphorylated pyridine nucleotides, NAD(P)H, as hydrogen donors to carry out its reaction. The kinetic mechanism of *E. coli* DHDPR is ordered and sequential (Reddy et al., 1995), involving binding of NAD(P)H followed by DHDP. The reaction is initiated by hydride transfer from the 4-pro-*R* position of NAD(P)H to the C4-position of DHDP, with the resultant enamine then undergoing tautomerisation to form THDP. Upon completion of the reaction, the release of the product THDP is followed by NAD(P)+ release (Reddy et al., 1995) (Fig. 7).

3.1.2 Nucleotide preference of bacterial DHDPR

Pyridine nucleotide-dependent dehydrogenases typically have a strong preference for either NADPH or NADH as co-factors (Cirilli et al., 2003; Pearce et al., 2008; Reddy et al., 1996). In most cases dual-cofactor enzymes preferentially utilise NADPH over NADH. In light of this observation, there has been significant interest in studying the molecular basis of nucleotide preference. All NAD-dependent dehydrogenases contain the consensus sequence GXGXXG or GXXGXXG and conserved acidic amino acids 20-30 residues downstream of this glycine rich region (Dommaraju et al., 2011). The main chain nitrogen of the second residue (X) in the consensus sequence interacts with this conserved acidic residue. E. coli DHDPR has an unusual pyridine nucleotide specificity, exhibiting only a modest selectivity for its nucleotides. Kinetic studies show that E. coli DHDPR utilises NADH only slightly more efficiently than NADPH (Reddy et al., 1996). This is consistent with the observation that the binding affinity of E. coli DHDPR to NADH (K_D = 0.26 μM) is stronger than that of NADPH (K_D = 1.8 μM) (Reddy et al., 1996). Structural studies of E. coli DHDPR show the existence of hydrogen bonds between the side-chain of the acidic residue Glu38 and that of the O3' of the adenine ribose of NADH. It is hypothesised that the basic residue Arg39, also found in the nucleotide binding pocket, can interact with the negatively charged 2' phosphate of NADPH, thus enabling the enzyme to utilise both NADH and NADPH. Kinetic analysis of DHDPR from M. tuberculosis also shows that the enzyme exhibits only a moderate preference for NADH. The crystal structures of M. tuberculosis DHDPR in two ternary complexes (DHDPR-2,6-PDC-NADH and DHDPR-2,6-PDC-NADPH) demonstrate that the number of hydrogen bonds between DHDPR and the nucleotides NADH and NADPH are very similar (Cirilli et al., 2003; Reddy et al., 1996; Scapin et al., 1997).

3.2 Structure of DHDPR

3.2.1 Subunit and quaternary structure of DHDPR

The three-dimensional structure of DHDPR has been elucidated by X-ray crystallography from five diverse bacterial species, namely, Bartonella henselae, (PDB: 3IJP), E. coli (Scapin et al., 1995, 1997), M. tuberculosis (Cirilli et al., 2003), S. aureus (Girish et al., 2011), and T. maritima (Pearce et al., 2008). DHDPR from E. coli (Fig. 8) was the first DHDPR enzyme to be extensively studied in terms of structure and function (Farkas & Gilvarg, 1965; Reddy et al., 1995; Scapin et al., 1995, 1997).

DHDPR is a tetrameric enzyme consisting of four identical monomers (Fig. 8). Each monomer is comprised of an N-terminal nucleotide binding domain and a C-terminal substrate binding domain (Fig. 9). In E. coli DHDPR, the nucleotide binding domain is formed by the first 130 and last 36 residues of the polypeptide chain, whereas the substrate binding domain is formed by residues 130-240. The nucleotide binding domain consists of four α-helices and seven β-strands, which are arranged to form a Rossmann (dinucleotide binding) fold. The substrate binding domain contains two α-helices and four β-strands, which form an open mixed β-sandwich (Scapin et al., 1995). Interactions between the four subunits of the tetramer occur exclusively between residues of the substrate binding domain. A long loop (Leu182 to Gly204) also extends from the substrate binding domain and plays an important role in maintaining the quaternary structure of the enzyme. The four monomers interact by pairing the four β-strands on the substrate binding domain to form a 16-stranded, mixed, flattened β-barrel (Fig. 8). This central barrel is anchored by the four long loops (Leu182 to Gly204) that extend from the body of the substrate binding domain of each monomer and wrap around the mixed β-sheet of the neighboring monomer. Residues 65-74 and 127-130 form flexible hinge regions between the nucleotide and substrate binding domains (Scapin et al., 1995).

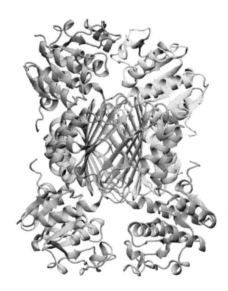

Fig. 8. Structure of *E. coli* DHDPR (PDB: 1ARZ).

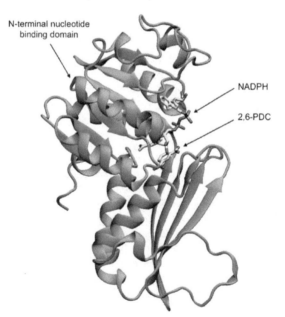

Fig. 9. Structure of the *E. coli* DHDPR monomer bound to NADH and the substrate analogue, 2,6-PDC (PDB: 1ARZ).

3.2.2 Substrate binding site

The consensus sequence, E(L/A)HHXXKXDAPSGTA is found in the substrate binding domain of all known bacterial DHDPR enzymes (Pavelka et al., 1997). This sequence is thought to contain residues involved in binding of substrate and/or catalysis. Molecular modelling studies, using the apo form (enzyme in the absence of substrate) of E. coli DHDPR as a structural template, suggest a cluster of five basic residues are the key catalytic site residues (Scapin et al., 1997), namely His159, His160, Arg161, His162 and Lys163 (all contained within the consensus sequence). These residues are located in the loop connecting β-strand B7 to α-helix A5. Structural studies of E. coli DHDPR in complex with NADH and the substrate analogue and inhibitor, 2,6-pyridinedicarboxylate (2,6-PDC), show that 2,6-PDC is bound to the substrate binding domain of DHDPR, in a spherical cavity bordered by residues from both the nucleotide binding (Gly102-Phe106 and Ala126-Ser130) and substrate binding domains (Ile155-Gly175 and Val217-His220) (Scapin et al., 1997). The bound inhibitor makes several hydrogen bonding interactions with the atoms of the conserved E(L/A)HHXXKXDAPSGTA motif. Similar interactions are observed between 2,6-PDC and DHDPR from M. tuberculosis (Cirilli et al., 2003).

3.2.3 Nucleotide binding site

The nucleotide binding domain of DHDPR adopts a Rossmann fold, which is typical of nucleotide-dependent dehydrogenases (Fig. 9). The consensus sequence (V/I)(A/G)(V/I)-XGXXGXXG located within this domain, is conserved in all NAD(P)H-dependent dehydrogenases, including DHDPR (Pavelka et al., 1997). Structural analyses of E. coli DHDPR show that this motif extends from the C-terminal end of β-strand B1 to the loop that connects B1 to α-helix A1. An acidic residue (Glu38 in E. coli DHDPR) is located approximately 20 amino acids downstream of the conserved consensus sequence. The two hydroxyl groups from the adenine ribose are known to interact with the side-chain of Glu38 and also the backbone atoms of the glycine rich motif GXXGXXG. Several hydrophobic interactions exist between the adenine ring of NADH and the residues Arg39, Gly84 and His88. The pyrophosphate group of NADH is located over the α-helix A1 and interacts with residues contained within the loop connecting β-strand B1 and α-helix A1 (Reddy et al., 1996; Scapin et al., 1997).

3.3 Inhibition of DHDPR

The substrate analogue, 2,6-PDC, is a competitive inhibitor (K_i = 26 µM) of DHDPR (Scapin et al., 1995) (Fig. 10A). Other substrate analogues such as picolinic acid (Fig. 10B), isopthalic acid (Fig. 10C), pipecolic acid (Fig. 10D) and dimethyl chelidamate (Fig. 10E), are much weaker inhibitors, each displaying an IC_{50} > 10 mM (Hutton et al., 2003). A vinylogous amide that acts as a competitive inhibitor of DHDPR (K_i = 32 µM) has been described and is one of the most potent inhibitors of DHDPR reported to date (Caplan et al., 2000). Molecular modeling in tandem with conventional drug screening strategies has identified novel inhibitors, including sulfones and sulfonamides, with K_i values ranging from 7-90 µM (Caplan et al., 2000). However, a sub-micromolar inhibitor of DHDPR has not been discovered to date.

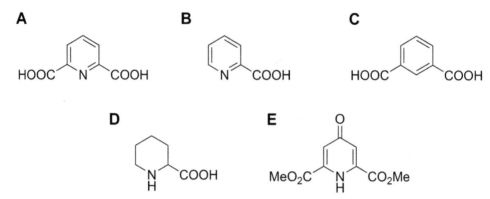

Fig. 10. Inhibitors of DHDPR.

4. Succinylase pathway

4.1 Tetrahydrodipicolinate N-succinyltransferase

Tetrahydrodipicolinate N-succinyltransferase (THPC-NST, EC 2.3.1.117) is a succinyl-coenzyme A (SCoA) dependant enzyme that catalyses the conversion of cyclic *L*-2,3,4,5,-tetrahydrodipicolinate (THDP) to acyclic N-succinyl-*L*-2-amino-6-ketopimelate (NSAKP) (Simms et al., 1984) (Fig. 1). The reaction occurs via a *L*-2-amino-6-ketopimelate (AKP) intermediate. The transfer of an acyl group functions to maintain a linear conformation of the product of the reaction (NSKAP) and exposes the 6-keto group for subsequent transamination (Beaman et al., 2002). Substrate and cofactor kinetic parameters for *E. coli* THPC-NST have been determined. Studies show that the K_M^{app} for THDP and succinyl-CoA are 20 μM and 15 μM, respectively (Berges et al., 1986b; Simms et al., 1984).

The *dapD* gene encoding THPC-NST is found in a large number of bacterial species including *E. coli* and *Mycobacterium* species (Beaman et al., 1997; Richaud et al., 1984; Schuldt et al., 2009). Expression of this gene in *E. coli* is weakly inhibited by lysine (Ou et al., 2008; Richaud et al., 1984). THPC-NST enzymes characterised to date are comprised of approximately 290 residues and show greater than 18% sequence identity (Beaman et al., 1997; Richaud et al., 1984; Schuldt et al., 2009).

The crystal structure of THPC-NST from *Mycobacterium bovis* (Fig. 11) shows that the enzyme forms a homotrimer. The monomer consists of three domains, namely, the (i) N-terminal, (ii) left handed parallel β-helix (LβH), and (iii) C-terminal domains (Beaman et al., 1997). The N-terminal domain is comprised of four α-helices and two hairpin loops. The LβH domain, comprising 50% of the subunit, contains the hexapeptide repeat motif ([LIV]-[GAED]-X₂-[STAV]-X) within each turn of the β-helix. The LβH domain is interrupted by two loops, including a flexible loop (residues 166-175) that is involved in binding substrate. The C-terminal domain consists of a β-stranded structure. All three domains contribute to inter-subunit contacts. The structure of THPC-NST from other bacterial species have since been determined and show a high degree of similarity to that of *M. bovis* THPC-NST (Nguyen et al., 2008; Schuldt et al., 2009).

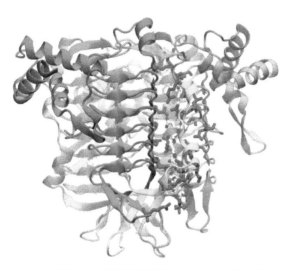

Fig. 11. Structure of trimeric *M. bovis* THPC-NST in complex with *L*-2-aminopimelate and succinamide-CoA. The N-terminal (orange), LβH (blue) and C-terminal (green) domains are indicated. The substrate *L*-2-aminopimelate (yellow) and cofactor succinamide-CoA (yellow) are bound via the THPC-NST active site residues (pink) (PDB: 1KGQ).

Crystal structures of *M. bovis* THPC-NST in complex with substrate analogs and several forms of coenzyme A have resulted in a model describing substrate binding and catalysis (Beaman et al., 1998, 2002). Self-association of the monomer subunit results in a homotrimer complex containing three active sites. The AKP and SCoA binding sites are located at the LβH domain interfaces. Binding of SCoA and possibly AKP is thought to promote a large conformational change that encloses the bound substrate and cofactor within the active site. In this state, the 2-amino group of AKP is placed in close proximity to the SCoA thioester, allowing nucleophilic attack and transfer of the succinyl group (Beaman et al., 2002).

Studies have shown that *L*-2-aminopimelic acid, an analog of AKP, is an inhibitor of THPC-NST, although it does not display antibacterial activity (Berges et al., 1986a). However, peptide derivatives of 2-aminopimelic acid show significant antibacterial activity against a range of Gram-negative bacteria (Berges et al., 1986a).

4.2 N-succinyldiaminopimelate aminotransferase

N-succinyldiaminopimelate aminotransferase (NSDAP-AT, EC 2.6.1.17) catalyses the conversion of NSKAP to N-succinyl-*L*,*L*-2,6,-diaminopimelate (NSDAP) (Fig. 1). The reaction begins by the formation of a Schiff base linkage between an active site lysine and the cofactor pyridoxal-5′-phosphate (PLP). An amino group, donated by glutamate, is transferred to PLP, to form pyridoxamine phosphate (PMP). The enzyme subsequently transfers the amino group from PMP to NSAKP to yield N-succinyl-*L*,*L*-2,6,-diaminopimelate (NSDAP) and α-ketoglutarate (Peterkofsky & Gilvarg., 1961; Ledwige & Blanchard., 1999). Studies of *E. coli* NSDAP-AT report K_M values for the substrates NSKAP and glutamate of 0.5 mM and 0.52 mM, respectively (Peterkofsky & Gilvarg., 1961).

The gene encoding NSDAP-AT (*dapC*), is found in a large number of bacterial species including *Bordetella pertussis* (Fuchs et al., 2000), *C. glutamicum* (Hartmann et al., 2003), *E. coli*, (Peterkofsky & Gilvarg., 1961) and *M. tuberculosis* (Weyand et al., 2006). In *E. coli*, the gene encoding NSDAP-AT is annotated *argD* (Ledwidge & Blanchard., 1999). This enzyme also functions as a N-acetylornithine aminotransferase, a component of the arginine biosynthesis pathway. The *dapC* gene in *B. pertussis* (Fuchs et al., 2000), *C. glutamicum*, (Hartmann et al., 2003), and *E. coli* (Bukari & Taylor., 1971) has been found to map in close proximity to the *dapD* gene on the chromosome. Sequence analyses have shown that NSDAP-AT consists of approximately 400 residues and shares greater than 26% identity across species (Fuchs et al., 2000; Hartmann et al., 2003; Peterkofsky & Gilvarg., 1961; Weyand et al., 2006). The NSDAP-AT sequence is characterised by the presence of the PLP binding sequence motif, SLS<u>K</u>XSNVXGXRAG, that includes an active site lysine residue (underlined) (Fuchs et al., 2000).

Structure studies of *M. tuberculosis* NSDAP-AT in complex with PLP shows that the enzyme forms a homodimer (Fig. 12). The structure is characteristic of the aminotransferase family of class I PLP-binding proteins (Weyand et al., 2007). The monomer subunit is comprised of (i) an α-helical N-terminal extension, (ii) a central domain comprising an 8-stranded β-sheet surrounded by 8 α-helices, and (iii) a C-terminal domain consisting of a four stranded β-sheet flanked by 4 α-helices. The active site of each subunit is located at the dimer interface with residues from both subunits contributing to the architecture of the active sites. PLP is bound to the active site Lys232, presumably via a Schiff base, and makes a number of noncovalent contacts with other residues within the active site via a hydrogen bond network.

A number of hydrazino-dipeptide analogs of NSDAP inhibit NSDAP-AT with K_i values ranging from 22-556 nM and show significant antibacterial activity against *E. coli* (Cox et al., 1998).

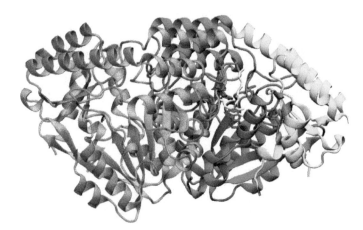

Fig. 12. Structure of dimeric *M. tuberculosis* NSDAP-AT in complex with PLP. The α-helical N-terminal extension (orange), central (blue) and C-terminal (green) domains are indicated. The cofactor PLP (yellow) is bound by the NSDAP-AT active site residues (pink) (PDB: 2O0R).

4.3 Succinyldiaminopimelate desuccinylase

Succinyldiaminopimelate desuccinylase (SDAP-DS, EC 3.5.1.18) catalyses the hydrolysis of N-succinyl-*L,L*-2,6,-diaminopimelate (NSDAP) to yield *L,L*-2,6-diaminopimelate (DAP) and succinate (Kindler & Gilvarg., 1960) (Fig. 1). Kinetic parameters for SDAP-DS from several bacterial species have been reported, with substrate K_M and k_{cat} values ranging from 0.73 - 1.3 mM and 140 - 200 s^{-1}, respectively (Bienvenue et al., 2003; Born et al., 1998; Lin et al., 1988).

The gene encoding SDAP-DS, *dap*E, is present in a large number of bacterial species including, *C. glutamicum* (Wehrmann et al., 1994), *E. coli* (Bouvier et al., 1992), *Haemophilus influenzae*, (Born et al., 1998) and *Salmonella enterica* (Broder & Miller., 2003). In general, SDAP-DS contains approximately 375 residues and shares greater than 22% sequence identity across bacterial species. Alignment of SDAP-DS amino acid sequences show conservation of histidine and glutamate metal binding residues that are characteristic of metal-dependent amidases (Born et al., 1998).

Consistent with the conservation of metal binding residues, the activity of SDAP-DS enzymes are dependent on Zn^{2+} ions (Born et al., 1998; Lin *et al.*, 1988). Futhermore, studies involving Zn K-edge extended X-ray absorption fine structure (EXAFS) analyses of *H. influenzae* SDAP-DS indicate that the enzyme contains dinuclear Zn^{2+} active sites (Cosper et al., 2003). Studies of *H. influenzae* SDAP-DS mutants by kinetics, electronic absorption spectroscopy and electron paramagnetic resonance spectroscopy showed that His67 and His349 coordinate Zn^{2+} ions, with His67 functioning in catalysis (Gillner et al., 2009). A similar study showed that residue Glu134 is also involved in catalysis, possibly functioning as an acid/base (Davis et al., 2006).

The crystal structure of zinc bound SDAP-DS has been determined (Fig. 13). Studies have shown that the enzyme forms a homodimer, with each monomer subunit containing a catalytic domain and a dimerisation domain (Nocek et al., 2010). The core of the catalytic domain is composed of an eight-stranded twisted β-sheet that is sandwiched between seven α-helices. The dimerisation domain adopts a two layer α+β sandwich fold and is comprised of a four stranded antiparallel β-sheet and two α-helices.

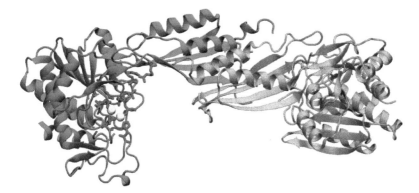

Fig. 13. Structure of dimeric *H. influenzae* SDAP-DS in complex with two zinc ions. the dimerisation (orange) and catalytic (blue) domains are indicated. Zinc ions (yellow) are bound by SDAP-DS active site residues (pink) (PDB: 3IC1).

The catalytic domain incorporates a negatively charged active site cleft, containing two zinc ions. One zinc ion is coordinated by the imidazole group and sidechain oxygens of His67 and Glu163, respectively, whilst another zinc ion is coordinated in a similar manner by His349 and Glu135. The zinc ions are bridged together by interaction with Asp100 and a water/hydroxide.

The availability of a structural model has resulted in a proposed mechanism for hydrolysis of NSDAP by SDAP-DS (Born et al., 1998; Nocek et al., 2010). It is hypothesised that NSDAP adopts an extended conformation when bound to the active site of the enzyme. The NSDAP amide carbonyl coordinates to an active site Zn^{2+} ion and becomes avaliable for nucleophilic attack. This binding event displaces a bridging water molecule, resulting in its hydrolysis by Glu134 and the generation of a zinc bound nucleophilic hydroxide. The hydroxide then attacks the target carbonyl carbon to form a η-1-µ-transition-state complex, which then resolves to release DAP and succinate.

The DAP isomers L,L-DAP and D,L-DAP are competitive inhibitors of *H. influenzae* SDAP-DS, exhibiting K_i values of 8 and 12 mM, respectively (Born et al., 1998). Studies employing Zn K-edge EXAFS suggest that the *H. influenzae* SDAP-DS inhibitor, 5-mercaptopentanoic acid, may exert its effect through binding to active site Zn^{2+} ions (Cosper et al., 2003).

5. Acetylase pathway

5.1 Tetrahydrodipicolinate N-acetyltransferase

Tetrahydrodipicolinate N-acetyltransferase (THDP-NAT, EC 2.3.1.89) is an acetyl-coenzyme A (ACoA) dependant enzyme that catalyses the conversion of cyclic THDP to acyclic N-acetyl-(S)-2-amino-6-ketopimelate (NAAKP) (Chatterjee & White., 1982) (Fig. 1). The transferred acyl group maintains the linear conformation of the product and exposes the 6-keto group for subsequent transamination (Beaman et al., 2002). Crude cell extracts from *B. megaterium* were found to contain active THDP-NAT (Chatterjee & White., 1982). THDP-NAT enzymes are thought to be largely confined to *Bacillus sp.* (Weinberger & Gilvarg., 1970).

5.2 Aminotransferase A

Aminotransferase A (ATA, EC 2.6.1) is a PLP-dependant enzyme that catalyses the conversion of NAAKP and glutamate to N-acetyl-(2S)-2,6,-diaminopimelate (NADAP) and α-ketoglutarate (Ledwidge & Blanchard., 1999; Peterkofsky & Gilvarg., 1961) (Fig. 1). It has been speculated that the ATA reaction mechanism resembles that of NSDAP-AT (Section 4.2) (Ledwidge & Blanchard., 1999; Peterkofsky & Gilvarg., 1961). Crude cell extracts from *B. megaterium* were found to contain active ATA (Chatterjee & White., 1982), with ATA activity identified by monitoring enzyme activity in the reverse direction utilising an acid ninhydrin assay (Chatterjee & White., 1982; Sundharadas & Gilvarg., 1967).

5.3 N-acetyldiaminopimelate deacetylase

N-acetyldiaminopimelate deacetylase (NAD-DAC, EC 3.5.1.47) catalyses the hydrolysis of NADAP to form DAP and acetate (Fig. 1). NAD-DAC was first identified from studies involving the isolation of a *B. megaterium* DAP auxotroph (Saleh & White., 1979; Sundharadas & Gilvarg., 1967). The mutant strain possesses a non-functional form of NAD-

DAC and consequently accumulates NADAP. Early studies of this enzyme centred on Gram-positive species, with NAD-DAC activity identified by utilising an acid ninhydrin assay to detect NADAP formation in crude cell extracts (Chatterjee & White., 1982; Weinberger & Gilvarg., 1970). The distribution of NAD-DAC has since been investigated in large number of Gram-negative and Gram-positive bacteria. Interestingly, the enzyme appears to be restricted to *Bacillus sp.* (Weinberger & Gilvarg., 1970).

6. Aminotransferase pathway

6.1 Function of diaminopimelate aminotransferase

Diaminopimelate aminotransferase (*LL*-DAP-AT, EC 2.6.1.83) is a PLP-dependant enzyme that catalyses the conversion of *L*-2,3,4,5,-tetrahydrodipicolinate (THDP) to *L*,*L*-2,6-diaminopimelate (*LL*-DAP) (Fig. 1). This transamination reaction utilises glutamate as an amino donor to yield α-ketoglutarate. (Hudson et al., 2006, 2008; Liu et al., 2010; McCoy et al., 2006)

The enzyme was first isolated from plant and cyanobacterial species and thus demonstrated a new branch of the lysine biosynthesis pathway existed (Hudson et al., 2006). Although plants are known to synthesise lysine *de novo*, components of the pathway required for conversion of THDP to *meso*-DAP had not been identified previously despite years of investigation. Studies of crude cell extracts had shown that plants do not catalyse reactions specific to the succinylase, acetylase or dehydrogenase branches of the pathway. This was subsequently confirmed with the observation that annotated plant genomes, including that from *Arabidopsis thaliana*, lack some or all genes associated with the three classical branches (Chatterjee et al., 1994; Hudson et al., 2005). The identification and characterisation of *LL*-DAP-AT from *A. thaliana* demonstrated for the first time the means by which plant species catalyse the conversion of THDP to *meso*-DAP via the aminotransferase sub-pathway (Hudson et al., 2006).

More recently *LL*-DAP-AT has been identified in algal, archaeal and bacterial species including, *Chlamydia trachomatis* (McCoy et al., 2006), *Chlamydomonas reihardtii* (Hudson et al., 2011), *Methanocaldococcus jannaschii* (Liu et al., 2010), and *Protochlamydia amoebophila* (McCoy et al., 2006). Comparative genomic analyses shows that *LL*-DAP-AT is restricted to the eubacterial lineages, *Bacteroidetes*, *Chlamydiae*, *Chloroflexi*, *Cyanobacteria*, *Desulfuromonadales*, *Firmicutes*, and *Spirochaeta*; and the archaea, *Archaeoglobaceae* and *Methanobacteriaceae* (Hudson et al., 2008). The phylogeny of *LL*-DAP-AT from these species has established the existence of two classes of *LL*-DAP-AT orthologues, namely, DapL1 and DapL2, which differ significantly in primary amino acid sequence. DapL1 and DapL2 are found predominantly in eubacteria and archaea, respectively (Hudson et al., 2008).

LL-DAP-AT enzymes are classified as members of the PLP-dependant protein superfamily of class I/II aminotransferases (Hudson et al., 2008; Jensen et al., 1996; Sung et al., 1991). Orthologues are in general 410 amino acids in length and can share as little as 29% sequence identity. Kinetic parameters for the *LL*-DAP-AT reaction have been determined for enzymes from a number of species, including *A. thaliana*, *C. trachomatis*, *Desulfitobacterium hafniense*, *Leptospira interrogans*, *Methanobacterium thermoautotrophicus*, *Morella thermoacetica*, and *P. amoebophila*. (Hudson et al., 2006, 2008; McCoy et al., 2006). In the human pathogen *C.*

trachomatis, the K_M values for the substrates THDP and glutamate have been reported as 19 µM and 2.1 µM, respectively (Hudson et al., 2008).

6.2 Structure of *LL*-DAP-AT

At present, the PDB reports twelve *LL*-DAP-AT X-ray crystal structures from three species, namely, *A. thaliana, C. trachomatis and C. reihardtii* (Watanabe et al., 2007, 2008, 2011; Dobson et al., 2011). The tertiary and quaternary structure of all three proteins are very similar with *LL*-DAP-AT existing as a homodimer (Fig. 14).

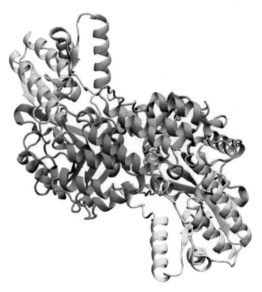

Fig. 14. Structure of dimeric *C. trachomatis LL*-DAP-AT. Monomers, indicated in blue and green, associate to form a functional dimer (PDB: 3ASA).

The subunit structure of *C. trachomatis LL*-DAP-AT is described as containing two domains, a large domain (LD) (residues 48-294) and a small domain (SD) (residues 1-47 and 295-394) (Watanabe et al., 2011; Watanabe & James, 2011). The LD is composed of α-β-α sandwich, whilst the SD assumes an α-β complex (Fig. 14). The LD is involved in binding PLP and also dimer formation, whereas the SD forms an N-terminal arm and also the C-terminal region. The active site is situated in a groove between the two domains of the monomer (Fig. 14). Importantly, the dimer structure is proposed to be essential for function as both subunits participate in substrate binding. Study of the structures of apo and ligand-bound forms of *C. trachomatis LL*-DAP-AT have revealed that the enzyme adopts an open and closed conformation (Watanabe et al., 2011). In the absence of ligand, the enzyme assumes an open state, whereby the active site is exposed to solvent. Upon PLP binding, the enzyme adopts a closed conformation. Within the active site, PLP is covalently linked to Lys236 via a Schiff base and is stabilised through an aromatic stacking interaction with Tyr128. PLP also forms a network of hydrogen bonding interactions with residues within the enzyme active site (Watanabe et al., 2011) (Fig 15).

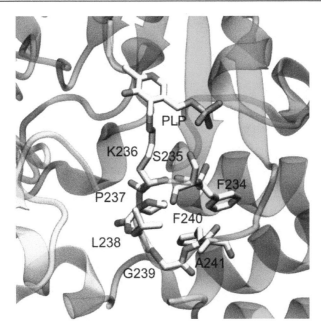

Fig. 15. Catalytic site of *LL*-DAP-AT from *C. trachomatis* (PDB: 3ASA). Ligand binding induces a closed conformation. PLP is covalently linked to Lys236 *via* a Schiff base.

6.3 Inhibition of *LL*-DAP-AT

A number of potential *LL*-DAP-AT inhibitors have been synthesised and characterised. In a screen involving 29,201 molecules, 15 compounds displayed IC_{50} values ranging from 20 μM to 60 μM, with the best hit being an aryl hydrazide showing an IC_{50} of 5 μM (Fan et al., 2010). However, the best hit appears to be an uncompetitive inhibitor and probably reacts irreversibly with PLP. Analogues of this compound have been synthesised and studies show that they fail to effectively inhibit *LL*-DAP-AT. In addition, there are two rhodanine-based molecules reported that show IC_{50} values of 41 μM and 46 μM (Fan et al., 2010).

7. Diaminopimelate epimerase

7.1 Function of DAPE

Diaminopimelate epimerase (DAPE, EC 5.1.1.7) catalyses the penultimate step in the lysine biosynthetic pathway whereby *L,L*-2,6-diaminopimelate (*LL*-DAP) is converted to *meso*-DAP (Fig. 1, Fig. 16). In *E. coli*, the enzyme is encoded by the *dapF* gene and is constitutively expressed (Neidhardt & Curtiss, 1996). DAPE was first characterised in 1957 using enzyme derived from crude extracts of *E. coli* (Work, 1957). The enzyme specifically recognises the *LL*-DAP isomer (Anita et al., 1957), whereas the *DD*-DAP isomer is not a substrate or inhibitor of the enzyme. Early studies noted that DAPE was inhibited by low concentrations of thiol-binding reagents and could be reactivated by reducing agents, suggesting the presence of an essential sulfhydryl group (Work, 1957). This finding was subsequently confirmed upon purification of DAPE to homogeneity (Wiseman, & Nichols, 1984).

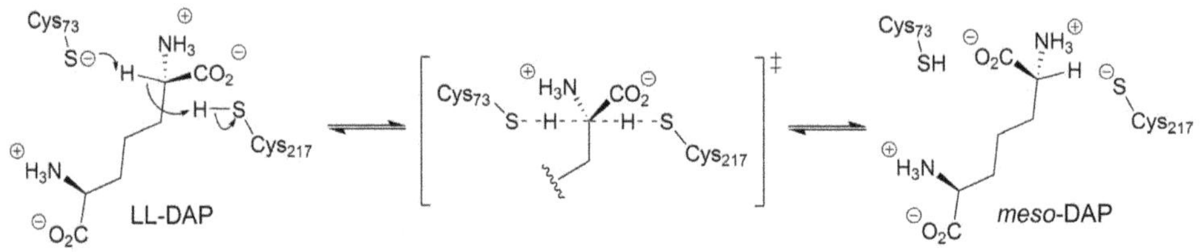

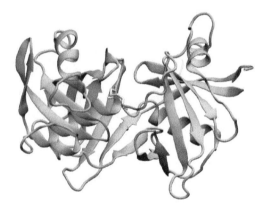

Fig. 16. DAPE catalysed reaction.

DAPE catalyses the conversion of LL-DAP to meso-DAP by employing a "two-base" mechanism (Wiseman, & Nichols, 1984). The reaction involves two active site Cys residues, where the first Cys residue (73 in H. influenzae) acts as base abstracting proton from LL-DAP, while the second Cys residue (217 in H. influenzae) re-protonates the molecule to generate meso-DAP. The enzyme is also capable of catalysing the reverse reaction, with the two Cys residues reversing their roles (Wiseman, & Nichols, 1984).

7.2 Structure of DAPE

The structures of DAPE from four species have been described. These include DAPE from B. anthracis (PDB:2OTN), H. influenza (Cirilli et al., 1998; Lloyd et al., 2004), and M. tuberculosis (Usha et al., 2009); and also the plant species A. thaliana (Pillai et al., 2009). The enzyme is a symmetrical monomer comprised of two domains containing eight β-strands and two α-helices (Cirilli et al., 1998) (Fig. 17).

Fig. 17. Structure of DAPE from H. influenzae. Domains are coloured pink and blue, active site cysteines (disulfide linked) are shown in yellow (PDB: 1BWZ).

This fold, first observed in H. influenzae DAPE, is now referred to as the DAP epimerase-like fold. The structure of DAPE from H. influenzae shows that each domain of the enzyme contributes one active site Cys (residues 73 and 217). The distal, non-reacting end of the substrate interacts via a number of hydrogen bonds to residues Asn157, Asp190, Arg209, Asn64, and Glu208 (Fig. 18). The nature of this interaction ensures that only the LL-DAP stereoisomer is recognised. Interestingly, DAPE adopts two distinct conformational states. In the absence of substrate, the enzyme exists in an open conformation, and upon binding substrate adopts a closed conformation (Pillai et al., 2007).

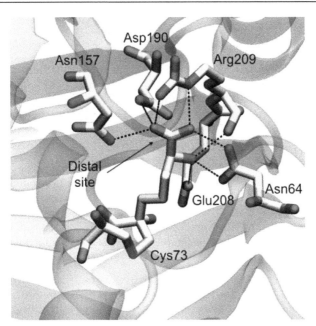

Fig. 18. Catalytic site of DAPE from *H. influenzae*. Hydrogen bond interactions (black dotted lines) at the distal site of the substrate analogue *LL*-AziDAP (arrow indicating position of the analogue) (PDB: 2GKE).

7.3 Inhibition of DAPE

Substrate analogues of DAP have been used as the basis for the generation of inhibitors of DAPE. These inhibitors take advantage of the anionic character at the α-carbon during the reaction or mimic the planar transition state. The most potent inhibitors are shown in Fig. 19 (Williams et al., 1996).

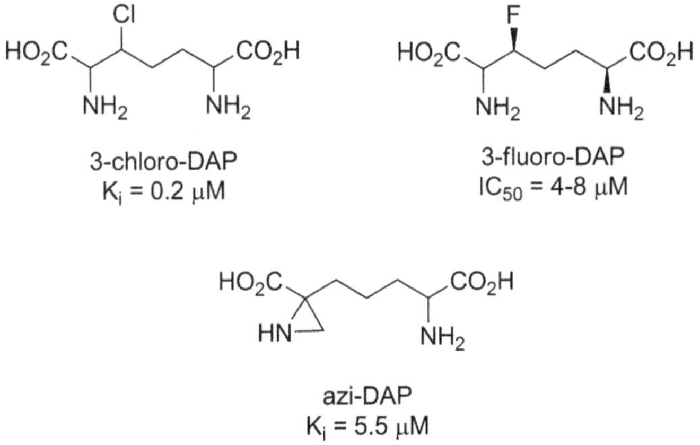

3-chloro-DAP
$K_i = 0.2 \ \mu M$

3-fluoro-DAP
$IC_{50} = 4-8 \ \mu M$

azi-DAP
$K_i = 5.5 \ \mu M$

Fig. 19. Inhibitors of DAPE.

8. Dehydrogenase pathway

8.1 *Diaminopimelate dehydrogenase*

Diaminopimelate dehydrogenase (DAPDH EC 1.4.1.16) is a NADPH dependant enzyme that catalyses the reductive amination of *L*-2-amino-6-ketopimelate (AKP), the acyclic form of *L*-2,3,4,5,-tetrahydrodipicolinate (THDP), to produce *meso*-DAP (Misono et al., 1976; Misono & Soda., 1980) (Fig. 1). It is assumed that the reaction occurs via an imine intermediate as a result of amination of *L*-2-amino-6-ketopimelate. Reduction of the imine by hydride transfer from NADPH generates *meso*-DAP (Scapin et al., 1998).

Only a small group of Gram-positive and Gram-negative bacteria posses DAPDH activity. These include *Bacillus sphaericus*, *Brevibacterium sp.*, *C. glutamicum* and *Proteus vulgaris* (Misono et al., 1979). Characterised DAPDH enzymes are comprised of approximately 320 residues and share greater than 27% sequence identity (Ishino et al., 1987; Hudson et al., 2011b). Kinetic studies of DAPDH from *C. glutamicum* has yielded K_M values for NADPH, *L*-2-amino-6-ketopimelate and ammonia of 0.13 mM, 0.28 mM and 36 mM, respectively (Misono *et al.*, 1986).

Some bacterial species possessing DAPDH activity use multiple pathways to synthesise lysine. For example, *C. glutamicium* (Schrumpf et al., 1991) can synthesise lysine by either the dehydrogenase or succinylase pathway, whilst *Bacillus macerans* (Hudson et al., 2011b) can employ enzymes of the dehydrogenase or acetylase pathways.

DAPDH from *C. glutamicum* forms a homodimer (Scapin et al., 1996) (Fig. 20). The DAPDH monomer subunit is comprised of (i) a dinucleotide binding domain, that is similar to but not identical to a classical Rossman fold, (ii) a dimerisation domain, and (iii) a C-terminal domain (Fig. 20). Monomer subunits interact via two α-helices and a three-stranded antiparallel β-sheet to form the dimer.

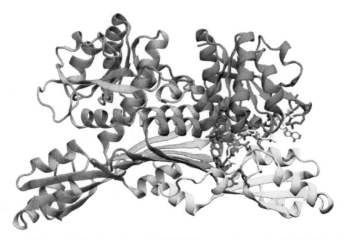

Fig. 20. Structure of dimeric *C. glutamicum* DAPDH in complex with NADPH and *L*-2-amino-6-methylene-pimelate. The dimerisation (orange), dinucleotide binding (blue), and C-terminal (green) domains are indicated. The cofactor NADPH (yellow) and inhibitor *L*-2-amino-6-methylene-pimelate (yellow) are bound by active site residues (pink) (PDB:1F06).

The crystal structure of the *C. glutamicum* DAPDH in complex with ligand shows that the oxidised cofactor, NADP+, is bound within each of the dinucleotide binding domains (Scapin et al., 1996). The domains exhibit open and closed conformations thought to represent the binding and active states of DAPDH, respectively (Scapin et al., 1996). In the closed conformation the NADP$^+$ pyrophosphate forms seven additional noncovalent contacts. Subsequent studies demonstrate the product, *meso*-DAP, binds within an elongated cavity formed at the interface of the dimerisation and dinucleotide binding domains (Scapin et al., 1998).

Crystal structures of *C. glutamicum* complexed with the inhibitors (2*S*,5*S*)-2-amino-3-(3-carboxy-2-isoxazolin-5-yl)-propanoic acid (K_i = 4.2 µM) and *L*-2-amino-6-methylene-pimelate (K_i = 5 µM) show that they form similar interactions with DAPDH as the product *meso*-DAP (Scapin et al., 1998). An additional hydrogen bond between the α-amino group of the *L*-2-amino-6-methylene-pimelate and the indole ring of DAPDH Trp144 is thought to account for the strong competitive inhibition observed (Scapin et al., 1998).

9. Diaminopimelate decarboxylase

9.1 Function of DAPDC

Diaminopimelate decarboxylase (DAPDC, EC 4. 1. 1. 20) is a PLP-dependant enzyme that is responsible for catalysing the final reaction of the lysine biosynthesis pathway (Fig. 1). In this non-reversible reaction, DAPDC converts the substrate *meso*-DAP to lysine and carbon dioxide (Fig. 21). Unlike other PLP-dependant decarboxylases that decarboxylate an *L*-stereocentre, DAPDC specifically cleaves the *D*-stereocentre carboxyl group. Thus, the enzyme possesses a means to differentiate between two stereocentres (Gokulan et al., 2003; Ray et al., 2002). DAPDC is classified as a type III class PLP enzyme, from the alanine racemase family.

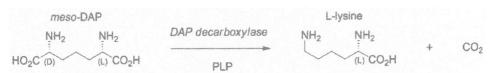

Fig. 21. DAPDC catalysed reaction.

Compared to other enzymes within the lysine biosynthesis pathway, DAPDC has not been studied extensively. Consequently, the catalytic mechanism is poorly defined. However, current understanding of the structure and function of this enzyme is based on work performed on DAPDC from *Helicobacter pylori* (Hu et al., 2008), *M. tubercolosis* (Weyand et al., 2009), and *Methanococcus jannaschii* (Ray et al., 2002).

9.2 Structure of DAPDC

The crystal structures of DAPDC from seven species have been determined. There appears to be no consensus in quaternary structure of the enzyme as monomeric, dimeric, and tetrameric forms of DAPDC have been described. This is unusual, and possibly not a true reflection of what occurs in nature. Studies have shown that the active site of DAPDC is located at the dimer interface (Hu et al., 2008; Ray et al., 2002; Weyand et al., 2009). This implies that the dimer is the minimal catalytic unit. Therefore, monomeric forms of DAPDC are likely to be non-functional; however, this does not rule out the existence of active tetrameric forms of DAPDC.

In species such as *M. jannaschii* (Ray et al., 2002) and *M. tubercolosis,* (Gokulan et al., 2003; Weyand et al., 2009) DAPDC is composed of a homodimer, whereby subunits associate to form a head-to-tail quaternary architecture (Fig. 22).

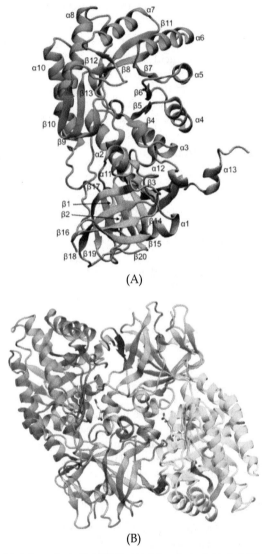

(A)

(B)

Fig. 22. Structure of *M. tuberculosis* DAPDC. (A) *M. tuberculosis* DAPDC monomer - The N-terminal (purple) and C-terminal (grey) domains are indicated. (B) *M. tuberculosis* DAPDC dimer – The active site is situated at the homodimer interface. PLP (yellow) and lysine (yellow) are located within the active site cavity (PDB: 1HKV).

The DAPDC monomer is composed of two domains, consisting of an N-terminal 8-fold α/β-barrel domain and a C-terminal β-sheet domain (Gokulan et al., 2003; Ray et al., 2002) (Fig. 22A). In *M. tuberculosis* DAPDC, the N-terminal α/β-barrel domain (residues 48-308) is comprised of β-strands $\beta4-\beta13$ and helices $\alpha2-\alpha10$ (Fig. 22A). The C-terminal domain (residues 2-47 and 309-446) is comprised of β-strands $\beta1-\beta3$, $\beta14-\beta21$ and helices $\alpha1$, $\alpha11-\alpha13$ (Gokulan et al., 2003) (Fig. 22A). The active site is located at the interface between the α/β-barrel domain of one subunit and β-sheet domain of both subunits (Gokulan et al., 2003) (Fig. 22B).

The X-ray structure of *H. pylori* DAPDC has allowed identification of key residues involved in substrate and cofactor recognition. The enzyme was crystallised in the presence of PLP and lysine. The *H. pylori* structure is very similar to that of *M. tuberculosis* DAPDC, forming a homodimer in a head-to-tail conformation. In this enzyme, PLP forms Schiff base linkages with Lys46 and lysine to produce a lysine-PLP external aldimine. This aldimine is believed to mimic the catalytic intermediate formed between *meso*-DAP and PLP (Hu et al., 2008).

9.3 Inhibition of DAPDC

Diaminopimelic acid analogues (Fig. 23) have been synthesised to study the inhibition of DAPDC from *B. sphaericus* (Kelland et al., 1986). Mixtures of isomers of N-hydroxydiaminopimelate and N-aminodiaminopimelate are potent competitive inhibitors of DAPDC, with K_i values of 0.91 mM and 0.1 mM, respectively. Lanthionine sulfoxides (Fig. 23) are good competitive inhibitors, providing about 50% inhibition at 1 mM. Weaker competitive inhibitors include the *meso* and *LL*-isomers of lanthionine sulfone and lanthionine, whereas the *DD*-isomers (Fig. 23) were less effective.

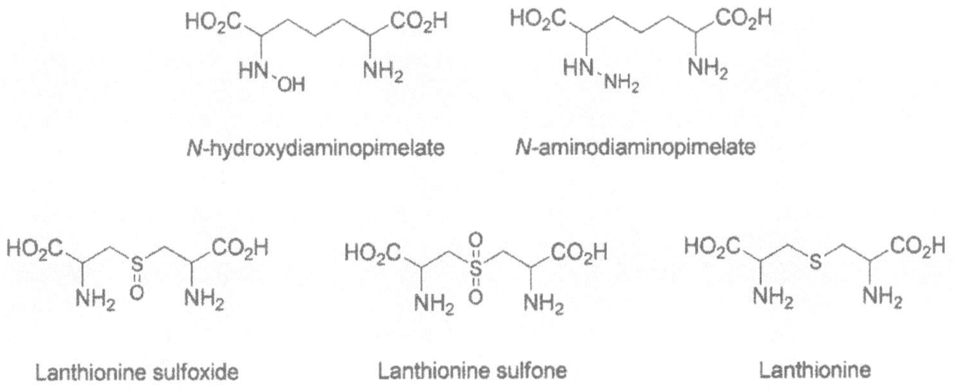

N-hydroxydiaminopimelate   *N*-aminodiaminopimelate

Lanthionine sulfoxide   Lanthionine sulfone   Lanthionine

Fig. 23. Inhibitors of DAPDC.

10. Conclusions

Significant advances in our understanding of the enzymes of the lysine biosynthetic pathway have occurred in recent years, particularly through detailed kinetic and structural studies of wild-type and mutant enzymes. While advances in inhibitor design have not been as dramatic, our increased structural knowledge augurs well for the design of potent enzyme inhibitors in the near future, and subtle differences between the structures of the enzymes from different pathogenic species offers great potential of designing pathogen-specific antibiotics. The improvements in our understanding of the lysine biosynthetic pathway in recent years will no doubt advance our efforts toward the ultimate goal of developing novel antibiotics that target this essential bacterial pathway.

11. References

Anita, M., Hoare, D. S., & Work, E. (1957). The stereoisomers of diaminopimelic acid. *Biochem. J.*, Vol.65, No.3, (March 1957), pp. 448-459, ISSN 0264-6021

Beaman, T. W., Binder, D. A., Blanchard, J. S. & Roderick, S. L. (1997). Three-dimensional structure of tetrahydrodipicolinate N-succinyltransferase. *Biochemistry*, Vol.36, No.3, (January 1997), pp. 489-494, ISSN 0006-2960

Beaman, T. W., Blanchard, J. S., & Roderick, S. L. (1998). The conformational change and active site structure of tetrahydrodipicolinate N-succinyltransferase. *Biochemistry*, Vol.37, No.29, (July 1998), pp. 10363-10369, ISSN 0006-2960

Beaman, T. W., Vogel, K. W., Drueckhammer, D. G., Blanchard, J. S., & Roderick, S. L. (2002). Acyl group specificity at the active site of tetrahydridipicolinate N-succinyltransferase. *Protein Sci.*, Vol.11, No.4, (April 2002), pp. 974-9, ISSN 0961-8368

Berges, D. A., DeWolf, W. E. Jr., Dunn, G. L., Grappel, S. F., Newman, D. J., Taggart, J. J., & Gilvarg, C. (1986a). Peptides of 2-aminopimelic acid: antibacterial agents that inhibit diaminopimelic acid biosynthesis. *J. Med. Chem.*, Vol.29, No.1, (January 1986), pp. 89-95, ISSN 0022-2623

Berges, D. A., DeWolf, W. E. Jr., Dunn, G. L., Newman, D. J., Schmidt, S. J., Taggart, J. J., & Gilvarg, C. (1986b). Studies on the active site of succinyl-CoA:tetrahydrodipicolinate N-succinyltransferase. Characterization using analogs of tetrahydrodipicolinate. *J. Biol. Chem.*, Vol.261, No.14, (May 1986), pp. 6160-6167, ISSN 0021-9258

Bienvenue, D. L., Gilner, D. M., Davis, R. S., Bennett, B., & Holz, R. C. (2003). Substrate specificity, metal binding properties, and spectroscopic characterization of the DapE-encoded N-succinyl-L,L-diaminopimelic acid desuccinylase from *Haemophilus influenzae*. *Biochemistry*, Vol.42, No.36, (September 2003), pp. 10756-63, ISSN 0006-2960

Blagova, E., Levdikov, V., Milioti, N., Fogg, M. J., Kalliomaa, A. K., Brannigan, J. A., Wilson, K. S., & Wilkinson, A. J. (2006). Crystal structure of dihydrodipicolinate synthase (BA3935) from *Bacillus anthracis* at 1.94 Å resolution. *Proteins: Structure, Function, & Bioinformatics*, Vol.62, No.1, (January 2006), pp. 297-301, ISSN 0887-3585

Blickling, S., Renner, C., Laber, B., Pohlenz, H., Holak, T. A., & Huber, R. (1997a). Reaction mechanism of *Escherichia coli* dihydrodipicolinate synthase investigated by X-ray crystallography and NMR spectroscopy. *Biochemistry*, Vol.36, No.1, (January 1997), pp. 24-33, ISSN 0006-2960

Blickling, S., Beisel, H. G., Bozic, D., Knäblein, J., Laber, B., & Huber, R. (1997b). Structure of dihydrodipicolinate synthase of *Nicotiana sylvestris* reveals novel quaternary structure. *J. Mol. Biol.*, Vol.274, No.4, (December 1997), pp. 608-621, ISSN 0022-2836

Blickling, S., & Knablein, J. (1997c). Feedback inhibition of dihydrodipicolinate synthase enzymes by L-lysine. *Biol. Chem.*, Vol.378, No.3-4, (March-April 1997), pp. 207-210, ISSN 1431-6730

Born, T. L., Zheng, R., & Blanchard, J. S. (1998). Hydrolysis of N-succinyl-L,L-diaminopimelic acid by the *Haemophilus influenzae* dapE-encoded desuccinylase: metal activation, solvent isotope effects, and kinetic mechanism. *Biochemistry*, Vol.37, No.29, (July 1998), pp. 10478-87, ISSN 0006-2960

Boughton, B. A., Dobson, R. C., Gerrard, J. A., & Hutton, C. A. (2008). Conformationally constrained diketopimelic acid analogues as inhibitors of dihydrodipicolinate synthase. *Bioorg. Med. Chem. Lett.*, Vol.18, No.2, (January 2008), pp. 460-463, ISSN 0960-894X

Bouvier, J., Richaud, C., Higgins, W., Bögler, O., & Stragier, P. (1992). Cloning, characterization, and expression of the dapE gene of *Escherichia coli*. *J. Bacteriol.*, Vol.174, No.16, (August 1992), pp. 5265-5271, ISSN 0021-9193

Broder, D. H., & Miller, C. G. (2003). DapE can function as an aspartyl peptidase in the presence of Mn2+. *J. Bacteriol.*, Vol.185, No.16, (August 2003), pp. 4748-54, ISSN 0021-9193

Bukhari, A. I., & Taylor, A. L. (1971). Genetic analysis of diaminopimelic acid- and lysine-requiring mutants of *Escherichia coli*. *J. Bacteriol.*, Vol.105, No.3, (March 1971), pp. 844-854, ISSN 0021-9193

Burgess, B. R., Dobson, R. C. J., Bailey, M. F., Atkinson, S. C., Griffin, M. D. W., Jameson, G. B., Parker, M. W., Gerrard, J. A., & Perugini, M. A. (2008). Structure and evolution of a novel dimeric enzyme from a clinically-important bacterial pathogen. *J. Biol. Chem.*, Vol.283, No.41, (October 2008), pp. 27598-27603, ISSN 0021-9258

Cahyanto, M. N., Kawasaki, H., Fujiyama, K., & Seki, T. (2006). Regulation of aspartokinase, aspartate semialdehyde dehydrogenase, dihydrodipicolinate synthase and dihydrodipicolinate reductase in *Lactobacillus plantarum*. *Microbiology*, Vol.152, No.Pt 1, (January 2006), pp. 105-112, ISSN 1350-0872

Caplan, J. F., Zheng, R., Blanchard, J. S., & Vederas, J. C. (2000). Vinylogous amide analogues of diaminopimelic acid (DAP) as inhibitors of enzymes involved in bacterial lysine biosynthesis. *Org. Lett.*, Vol.2, No.24, (November 2000), pp. 3857-60, ISSN 1523-7060

Cremer, J., Treptow, C., Eggeling, L., & Sahm, H. (1988). Regulation of enzymes of lysine biosynthesis in *Corynebacterium glutamicum*. *J. Gen. Microbiol.*, Vol.134, No.12, (December 1988), pp. 3221-3229, ISSN 0022-1287

Chatterjee, S. P., & White, P. J. (1982). Activities and regulation of the enzymes of lysine biosynthesis in a lysine-excreting strain of *Bacillus megaterium*. *J. Gen. Microbiol.*, Vol.128, (October 1982), pp. 1073-1081, ISSN 0022-1287

Chatterjee S. P., Singh B. K., & Gilvarg C. (1994). Biosynthesis of lysine in plants: the putative role of meso-diaminopimelate dehydrogenase. *Plant Mol. Biol.*, Vol.26, No.1, (October 1994), pp. 285-290, ISSN 0167-4412

Cirilli, M., Zheng, R., Scapin, G., & Blanchard, J. S. (1998). Structural symmetry: the three-dimensional structure of *Haemophilus Influenzae* diaminopimelate epimerase. *Biochemistry*, Vol.37, No.47, (November 1998), pp. 16452-16458, ISSN 0006-2960

Cirilli, M., Scapin, G., Sutherland, A., Vederas, J. C., & Blanchard, J. S. (2000). The three-dimensional structure of the ternary complex of *Corynebacterium glutamicum* diaminopimelate dehydrogenase-NADPH-L-2-amino-6-methylene-pimelate. *Protein Sci.*, Vol.9, No.10, (October 2000), pp. 2034-2037, ISSN 0961-8368

Cirilli, M., Zheng, R., Scapin, G., & Blanchard, J. S. (2003). The three-dimensional structures of the *Mycobacterium tuberculosis* dihydrodipicolinate reductase-NADH-2,6-PDC and -NADPH-2,6-PDC complexes. Structural and mutagenic analysis of relaxed nucleotide specificity. *Biochemistry*, Vol.42, No.36, (August 2003), pp. 10644-50, ISSN 0006-2960

Cosper, N. J., Bienvenue, D. L., Shokes, J. E., Gilner, D. M., Tsukamoto, T., Scott, R. A., & Holz, R. C. (2003). The dapE-encoded N-succinyl-l,l-diaminopimelic acid desuccinylase from *Haemophilus influenzae* is a dinuclear metallohydrolase. *J. Am. Chem. Soc.*, Vol.125, No.48, (December 2003), pp. 14654-14655, ISSN 0002-7863

Cox, R. J., Schouten, J. A., Stentiford, R. A., & Wareing, K. J. (1998). Peptide inhibitors of N-succinyl diaminopimelic acid aminotransferase (DAP-AT): a novel class of antimicrobial compounds. *Bioorg. Med. Chem. Lett.*, Vol.8, No.8, (April 1998), pp. 945-950, ISSN 1464-3405

Davis, R., Bienvenue, D., Swierczek, S. I., Gilner, D. M., Rajagopal, L., Bennett, B., & Holz, R. C. (2006). Kinetic and spectroscopic characterization of the E134A- and E134D-altered dapE-encoded N-succinyl-L,L-diaminopimelic acid desuccinylase from *Haemophilus influenzae*. *J. Biol. Inorg. Chem.*, Vol.11, No.2, (March 2006), pp. 206-216, ISSN 0949-8257

Dereppe, C., Bold, G., Ghisalba, O., Ebert, E., & Schar, H. (1992). Purification and Characterization of Dihydrodipicolinate Synthase from Pea. *Plant Physiol.*, Vol.98, No.3, (March 1992), pp. 813-821, ISSN 0032-0889

Devenish, S. R. A., Huisman, F. H., Parker, E. J., Hadfield, A. D., & Gerrard, J. A. (2009). Cloning and characterisation of dihydrodipicolinate synthase from the pathogen *Neisseria meningitidis*. *Biochimica et Biophysica Acta*, Vol.1794, No.8, (August 2009), pp. 1168-1174, ISSN 0006-3002

Dobson, R. C., Valegård, K., & Gerrard, J. A. (2004a). The crystal structure of three site-directed mutants of *Escherichia coli* dihydrodipicolinate synthase: further evidence for a catalytic triad. *J. Mol. Biol.*, Vol.338, No.2, (April 2004), pp. 329-339, ISSN 0022-2836

Dobson, R. C. J., Griffin, M. D. W., Roberts, S. J., & Gerrard, J. A. (2004b). Dihydrodipicolinate synthase (DHDPS) from *Escherichia coli* displays partial mixed inhibition with respect to its first substrate, pyruvate. *Biochimie*, Vol.86, No.4-5, (April-May 2004), pp. 311-315, ISSN 0300-9084

Dobson, R. C., Griffin, M. D., Jameson, G. B., Gerrard, J. A. (2005a). The crystal structures of native and (S)-lysine-bound dihydrodipicolinate synthase from *Escherichia coli* with improved resolution show new features of biological significance. *Acta Crystallogr. D Biol. Crystallogr.*, Vol.61, No.Pt8, (August 2005), pp. 1116-1124, ISSN 0907-4449

Dobson, R. C., Devenish, S. R. A., Turner, L. A., Clifford, V. R., Pearce, F. G., Jameson, G. B., & Gerrard, J. A. (2005b). Role of arginine 138 in the catalysis and regulation of

Escherichia coli dihydrodipicolinate synthase. *Biochemistry*, Vol.44, No.39, (October 2005), pp. 13007-13013, ISSN 0006-2960

Dobson, R. C. J., Griffin, M. D. W., Devenish, S. R. A., Pearce, F. G., Hutton, C. A., Gerrard, J. A., Jameson, G. B., & Perugini, M. A. (2008) Conserved main-chain peptide distortions: a proposed role for Ile203 in catalysis by dihydrodipicolinate synthase. *Protein Sci.*, Vol.17, No.12, (December 2008), pp. 2080-2090, ISSN 0961-8368

Dobson, R. C., Girón, I., & Hudson, A. O. (2011), L,L-diaminopimelate aminotransferase from *Chlamydomonas reinhardtii*: a target for algaecide development. *PLoS one.*, Vol.6 No.5, (May 2011), In press, ISSN 1932-6203

Dogovski, C., Atkinson, S. C., Dommaraju, S. R., Hor, L., Dobson, R. C. J., Hutton C. A., Gerrard, J. A., & Perugini, M. A. (2009). Lysine biosynthesis in bacteria: an unchartered pathway for novel antibiotic design. In: *Encyclopedia Of Life Support Systems*, Volume 11 (Biotechnology Part I), pp116-136, edited by H.W. Doelle and S. Rokem, Eolss Publishers, Oxford ,UK < http://www.eolss.net>

Domigan, L. J., Scally, S. W., Fogg. M. J., Hutton, C. A., Perugini, M. A., Dobson, R. C. J., Muscroft-Taylor, A., Gerrard, J. A. & Devenish, S. R. A. (2009) Characterisation of dihydrodipicolinate synthase from *Bacillus anthracis*. *BBA Proteins* Vol.1794, No.10, (October 2009) pp. 1510-1516, ISSN 1570-9639

Dommaraju, S. R., Dogovski, C., Czabotar, P. E., Hor, L., Smith, B. J., & Perugini, M. A. (2011). Catalytic Mechanism and Cofactor Preference of Dihydrodipicolinate Reductase from Methicillin-Resistant *Staphylococcus aureus*. *Arch. Biochem. Biophys.*, Vol.512, No.2, (August 2011), pp. 167-74, ISSN 0003-9861

Fan, C., Clay, M. D., Deyholos, M. K., & Vederas, J. C. (2010). Exploration of inhibitors for diaminopimelate aminotransferase. *Bioorg. Med. Chem.*, Vol.18, No.6, (March 2010), pp. 2141-2151, ISSN 0968-0896

Farkas, W., & Gilvarg, C. (1965). The reduction step in diaminopimelic acid biosynthesis. *J. Biol. Chem.*, Vol.240, No.12, (December 1965), pp. 4717-22, ISSN 0021-9258

Frisch, D. A., Gengenbach, B. G., Tommey, A. M., Seliner, J. M., Somers, D. A., & Myers, D. E. (1991). Isolation and Characterization of Dihydrodipicolinate Synthase from Maize. *Plant Physiol.*, Vol.96, No.2, (June 1991), pp. 444-452, ISSN 0032-0889

Fuchs, T. M., Schneider, B., Krumbach, K., Eggeling, L., & Gross, R. (2000). Characterization of a *Bordetella pertussis* diaminopimelate (DAP) biosynthesis locus identifies dapC, a novel gene coding for an N-succinyl-L,L-DAP aminotransferase. *J. Bacteriol.*, Vol.182, No.13, (July 2000), pp. 3626-3631, ISSN 0021-9193

Gillner, D. M., Bienvenue, D. L., Nocek, B. P., Joachimiak, A., Zachary, V., Bennett, B., & Holz, R. C. (2009). The dapE-encoded N-succinyl-L,L-diaminopimelic acid desuccinylase from *Haemophilus influenzae* contains two active-site histidine residues. *J. Biol. Inorg. Chem.*, Vol.14, No.1, (January 2009), pp. 1-10, ISSN 0949-8257

Girish, T. S., Navratna, V., & Gopal, B. (2011). Structure and nucleotide specificity of *Staphylococcus aureus* dihydrodipicolinate reductase (DapB). *FEBS Lett.*, Vol.585, No.16, (August 2011), pp. 2561-7, ISSN 0014-5793

Gokulan, K., Rupp, B., Pavelka, M. S., Jr., Jacobs, W. R., Jr., & Sacchettini, J. C. (2003). Crystal structure of *Mycobacterium tuberculosis* diaminopimelate decarboxylase, an essential enzyme in bacterial lysine biosynthesis. *J. Biol. Chem.*, Vol.278, No.20, (March 2003), pp. 18588-18596, ISSN 0021-9258

Griffin, M. D. W., Dobson, R. C. J., Pearce, F. G., Antonio, L., Whitten, A. E., Liew, C. K., Mackay, J. P., Trewhella, J., Jameson, G. B., Perugini, M. A., & Gerrard, J. A. (2008). Evolution of quaternary structure in a homotetrameric protein. *J. Mol. Biol.*, Vol.380, No.4, (July 2008), pp. 691-703, ISSN 0022-2836

Griffin, M. D. W., Dobson, R. C. J., Gerrard, J. A., & Perugini, M. A. (2010) Exploring the dimerdimer interface of the dihydrodipicolinate synthase tetramer: how resilient is the interface? *Arch. Biochem. Biophys.*, Vol.494, No.1, (February 2010), pp. 58-63, ISSN 0003-9861

Gunji, Y., Tsujimoto, N., Shimaoka, M., Ogawa-Miyata, Y., Sugimoto, S., & Yasueda, H. (2004). Characterization of the L-lysine biosynthetic pathway in the obligate methylotroph *Methylophilus methylotrophus*. *Biosci. Biotechnol. Biochem.*, Vol.68, No.7, (July 2004), pp. 1449-60, ISSN 0916-8451

Hartmann, M., Tauch, A., Eggeling, L., Bathe, B., Möckel, B., Pühler, A., & Kalinowski, J. (2003). Identification and characterization of the last two unknown genes, dapC and dapF, in the succinylase branch of the L-lysine biosynthesis of *Corynebacterium glutamicum*. *J. Biotechnol.*, Vol.104, No.1-3, (September 2003), pp. 199-211, ISSN 0168-1656

Hoganson, D.A., & Stahly, D.P. (1975). Regulation of dihydrodipicolinate synthase during growth and sporulation of *Bacillus cereus*. *J. Bacteriol.*, Vol.124, No.3, (December 1975), pp. 1344-1350, ISSN 0021-9193

Hu, T., Wu, D., Chen, J., Ding, J., & Jiang, H, Shen, X. (2008). The catalytic intermediate stabilized by a "down" active site loop for diaminopimelate decarboxylase from *Helicobacter pylori*. Enzymatic characterization with crystal structure analysis. *J. Biol. Chem.*, Vol.283, No.30 (May 2008), pp. 21284-21293, ISSN 0021-9258

Hudson, A. O., Bless, C., Macedo, P., Chatterjee, S. P., Singh, B. K., Gilvarg, C., & Leustek, T. (2005). Biosynthesis of lysine in plants: evidence for a variant of the known bacterial pathways. *Biochim. Biophys. Acta.*, Vol.1721, No.1-3, (January 2005), pp. 27-36, ISSN 0006-3002

Hudson, A. O., Singh, B. K., Leustek, T., & Gilvarg, C. (2006). An LL-Diaminopimelate Aminotransferase Defines a Novel Variant of the Lysine Biosynthesis Pathway in Plants. *Plant Physiol.*, Vol.140, No.1, (January 2006), pp. 292-301, ISSN 0032-0889

Hudson, A. O., Gilvarg, C., & Leustek, T. (2008). Biochemical and Phylogenetic Characterization of a novel diaminopimelate biosynthesis pathway in prokaryotes identifies a diverged form of LL-diaminopimelate aminotransferase. *J. Bact.*, Vol.190. No.9, (May 2008), pp. 3256–3263, ISSN 0021-9193

Hudson, A. O., Girón, I., & Dobson, R. C. (2011a). Crystallization and preliminary X-ray diffraction analysis of L,L-diaminopimelate aminotransferase (DapL) from *Chlamydomonas reinhardtii*. *Acta Cryst. F*, Vol.67, No.Pt1, (January 2011), pp. 140-3, ISSN 1744-3091

Hudson, A. O., Klartag, A., Gilvarg, C., Dobson, R. C., Marques, F. G., & Leustek, T. (2011b). Dual diaminopimelate biosynthesis pathways in *Bacteroides fragilis* and *Clostridium thermocellum*. *Biochim. Biophys. Acta.*, Vol.1814, No.9, (September 2011), pp. 1162-1168, ISSN 0006-3002

Hutton, C. A., Southwood, T. J., & Turner, J. J. (2003). Inhibitors of lysine biosynthesis as antibacterial agents. *Mini Rev. Med. Chem.*, Vol.3, No.2, (March 2003), pp. 115-27, ISSN 1389-5575

Hutton, C. A., Perugini, M. A., & Gerrard, J. A. (2007). Inhibition of lysine biosynthesis: an emerging antibiotic strategy. *Molecular BioSystems*, Vol.3, No.7, (July 2007), pp. 458-465, ISSN 1742-2051

Ishino, S., Mizukami, T., Yamaguchi, K., Katsumata, R., & Araki, K. (1987). Nucleotide sequence of the meso-diaminopimelate D-dehydrogenase gene from *Corynebacterium glutamicum*. *Nucleic Acids Res.*, Vol.15, No.9, (May 1987), pp. 3917, ISSN 0305-1048

Jensen, R. A., & Gu, W. (1996). Evolutionary recruitment of biochemically specialized subdivisions of Family I within the protein superfamily of aminotransferases. *J. Bact.*, Vol.178, No.8, (April 1996), pp. 2161-2171., ISSN 0021-9193

Karsten, W. E. (1997). Dihydrodipicolinate synthase from *Escherichia coli*: pH dependent changes in the kinetic mechanism and kinetic mechanism of allosteric inhibition by lysine. *Biochemistry*, Vol.36, No.7 (February 1997), pp. 1730–1739, ISSN 0006-2960

Kefala, G., Evans, G. L., Griffin, M. D. W., Devenish, S. R. A., Pearce, F. G., Perugini, M. A., Gerrard, J. A., Weiss, M. S., & Dobson, R. C. J. (2008). Crystal structure and kinetic study of dihydrodipicolinate synthase from *Mycobacterium tuberculosis*. *Biochem. J.*, Vol.411, No.2, (April 2008), pp. 351-360, ISSN 0264-6021

Kelland, J. G., Arnold, L. D., Palcic, M. M., Pickard, M. A., & Vederas, J. C. (1986). Analogs of diaminopimelic acid as inhibitors of meso-diaminopimelate decarboxylase from *Bacillus sphaerius* and wheat germ. *J. Biol. Chem.*, Vol.261, No.28, (October 1986), pp. 13216-13223, ISSN 0021-9258

Kimura, K. (1975). A new flavin enzyme catalyzing the reduction of dihydrodipicolinate in sporulating *Bacillus subtilis*: I. Purification and properties. *J. Biochem.*, Vol.77, No.2, (July 1975), pp. 405-13, ISSN 0021-924X

Kimura, K., & Goto, T. (1977). Dihydrodipicolinate reductases from *Bacillus cereus* and *Bacillus megaterium*. *J. Biochem.*, Vol.81, No.5, (May 1977), pp. 1367-73, ISSN 0021-924X

Kindler, S. H., & Gilvarg, C. (1960). N-Succinyl-L-2,6-diaminopimelic acid deacylase. *J. Biol. Chem.*, Vol.235, (December 1960), pp. 3532-3535, ISSN 0021-9258

Kumpaisal, R., Hashimoto, T., & Yamada, Y. (1987). Purification and Characterization of Dihydrodipicolinate Synthase from Wheat Suspension Cultures. *Plant Physiol.*, Vol.85, No.1, (September 1987), pp. 145-151, ISSN 0032-0889

Laber, B., Gomis-Rüth, F., & Romão, M. J., & Huber, R. (1992). *Escherichia coli* dihydrodipicolinate synthase. Identification of the active site and crystallization. *Biochem. J.*, Vol.288, No.Pt2, (December 1992), pp. 691-695, ISSN 0264-6021

Ledwidge, R., & Blanchard, J. S. (1999). The dual biosynthetic capability of N-acetylornithine aminotransferase in arginine and lysine biosynthesis. *Biochemistry*, Vol.38, No.10, (March 1999), pp. 3019-3024, ISSN 0006-2960

Lin, Y. K., Myhrman, R., Schrag, M. L., & Gelb, M. H. (1988). Bacterial N-succinyl-L-diaminopimelic acid desuccinylase. Purification, partial characterization, and substrate specificity. *J. Biol. Chem.*, Vol.263, No.4, (February 1988), pp. 1622-7, ISSN 0021-9258

Liu, Y., White, R. H., & Whitman, W. B. (2010). Methanococci use the diaminopimelate aminotransferase (DapL) pathway for lysine biosynthesis. *J. Bact.*, Vol.192, No.13, (July 2010), pp. 3304–3310, ISSN 0021-9193

Lloyd, A. J., Huyton, T., Turkenburg, J., & Roper, D. I. (2004). Refinement of *Haemophilus influenzae* diaminopimelic acid epimerase (DapF) at 1.75 A resolution suggests a mechanism for stereocontrol during catalysis. *Acta Cryst. Sect. D*, Vol.60, No.2, (November 1998), pp. 397-400, ISSN 1399-0047

Matthews, B. F., & Widholm, J. M. (1979). Expression of aspartokinase, dihydrodipicolinic acid synthase and homoserine dehydrogenase during growth of carrot cell suspension cultures on lysine- and threonine-supplemented media. *Z Naturforsch [C]*, Vol.34, No.12, (December 1979), pp. 1177-1185, ISSN 0939-5075

McCoy, A. J., Adams, N. E., Hudson, A. O., Gilvarg, C., Leustek, T., & Maurelli, A. T. (2006). L,L-diaminopimelate aminotransferase, a trans-kingdom enzyme shared by Chlamydia and plants for synthesis of diaminopimelate/lysine. *PNAS*, Vol.103, No.47, (November 2006), pp. 17909-17914, ISSN 0027-8424

Mirwaldt, C., Korndorfer, I., & Huber, R. (1995). The crystal structure of dihydrodipicolinate synthase from *Escherichia coli* at 2.5 Å resolution. *J. Mol. Biol.*, Vol.246, No.1, (February 1995), pp. 227-239, ISSN 0022-2836

Misono, H., Togawa, H., Yamamoto, T., & Soda, K. (1976). Occurrence of meso-alpha, epsilon-diaminopimelate dehydrogenase in *Bacillus sphaericus*. *Biochem. Biophys. Res. Commun.*, Vol.72, No.1, (September 1976), pp. 89-93, ISSN 0006-291X

Misono, H., Togawa, H., Yamamoto, T., & Soda, K. (1979). Meso-alpha,epsilon-diaminopimelate D-dehydrogenase: distribution and the reaction product. *J. Bacteriol.*, Vol.137, No.1, (January 1979), pp. 22-27, ISSN 0021-9193

Misono, H., & Soda, K. (1980). Properties of meso-alpha,epsilon-diaminopimelate D-dehydrogenase from *Bacillus sphaericus*. *J. Biol. Chem.*, Vol.255, No.22, (November 1980), pp. 10599-10605, ISSN 0021-9258

Misono, H., Ogasawara, M., & Nagasaki, S. (1986). Characterization of meso-diaminopimelate dehydrogenase from *Corynebacterium glutamicum* and its distribution in bacteria. *Agric. Biol. Chem.*, Vol.50, No.11, (March 1986), pp. 2729-2734, ISSN 0002-1369

Mitsakos, V., Dobson, R. C. J., Pearce, F. G., Devenish, S. R., Evans, G. L., Burgess, B. R., Perugini, M. A., Gerrard, J. A., & Hutton, C. A. (2008). Inhibiting dihydrodipicolinate synthase across species: towards specificity for pathogens? *Bioorg. Med. Chem. Lett.*, Vol.18, No.2, (January 2008), pp. 842-844, ISSN 0960-894X

Neidhardt, F. C., & Curtiss, R. (1996). *Escherichia coli and Salmonella Cellular and Molecular Biology* (2nd), ASM Press, ISBN 9781555810849, Washington D.C.

Nguyen, L., Kozlov, G., & Gehring, K. (2008). Structure of *Escherichia coli* tetrahydrodipicolinate N-succinyltransferase reveals the role of a conserved C-terminal helix in cooperative substrate binding. *FEBS Lett.*, Vol.582, No.5, (March 2008), pp. 623-626, ISSN 0014-5793

Nocek, B. P., Gillner, D. M., Fan, Y., Holz, R. C., & Joachimiak, A. (2010). Structural basis for catalysis by the mono- and dimetalated forms of the dapE-encoded N-succinyl-L,L-diaminopimelic acid desuccinylase. *J. Mol. Biol.*, Vol.397, No.3, (April 2010), pp. 617-626, ISSN 0022-2836

Ou, J., Yamada, T., Nagahisa, K., Hirasawa, T., Furusawa, C., Yomo, T., & Shimizu, H. (2008). Dynamic change in promoter activation during lysine biosynthesis in *Escherichia coli* cells. *Mol. Biosyst.*, Vol.4, No.2, (February 2008), pp. 128-34, ISSN 1742-2051

Pavelka Jr., M. S., Weisbrod, T. R., & Jacobs Jr., W. R. (1997). Cloning of the dapB gene, encoding dihydrodipicolinate reductase, from *Mycobacterium tuberculosis. J. Bacteriol.*, Vol.179, No.8, (April 1997), pp. 2777-82, ISSN 0021-9193

Pearce, F. G., Perugini, M. A., Mckerchar, H. J., & Gerrard, J. A. (2006). Dihydrodipicolinate synthase from *Thermotoga maritima. Biochem J.*, Vol.400, No.2, (December 2006), pp. 359-366, ISSN 0264-6021

Pearce, F. G., Sprissler, C., & Gerrard, J. A. (2008). Characterization of dihydrodipicolinate reductase from *Thermotoga maritima* reveals evolution of substrate binding kinetics. *J. Biochem.*, Vol.143, No.5, (May 2008), pp. 617-23, ISSN 0021-924X

Peterkofsky, B., & Gilvarg, C. (1961). N-Succinyl-L-diaminopimelic-glutamic transaminase. *J. Biol. Chem.*, Vol.236, (May 1961), pp. 1432-1438, ISSN 0021-9258

Phenix, C. P., & Palmer, D. R. (2008). Isothermal titration microcalorimetry reveals the cooperative and noncompetitive nature of inhibition of *Sinorhizobium meliloti* L5-30 dihydrodipicolinate synthase by (S)-lysine. *Biochemistry*, Vol.47, No.30, (July 2008), pp. 7779-7781, ISSN 0006-2960

Pillai, B., Cherney, M., Diaper, C. M., Sutherland, A., Blanchard, J. S., Vederas, J. C., & James, M. N. G. (2007). Dynamics of catalysis revealed from the crystal structures of mutants of diaminopimelate epimerase. *Biochem. Bioph. Res. Co.*, Vol.363, No.3, (November 2007), pp. 547-553, ISSN 0006-291X

Pillai, B., Moorthie, V. A., Van Belkum, M. J., Marcus, S. L., Cherney, M. M., Diaper, C. M., Vederas, J. C., & James, M. N. G. (2009). Crystal Structure of Diaminopimelate Epimerase from *Arabidopsis thaliana*, an Amino Acid Racemase Critical for L-Lysine Biosynthesis. *J. Mol. Biol.*, Vol.385, No.2, (January 2009), pp. 580-594, ISSN 0022-2836

Ray, S. S., Bonanno, J. B., Rajashankar, K. R., Pinho, M. G., He, G., & De Lencastre, H. (2002). Cocrystal structures of diaminopimelate decarboxylase: mechanism, evolution, and inhibition of an antibiotic resistance accessory factor. *Structure*, Vol.10, No.11, (November 2002), pp. 1499-1508, ISSN 0969-2126

Reddy, S. G., Sacchettini, J. C., & Blanchard, J. S. (1995). Expression, purification, and characterization of *Escherichia coli* dihydrodipicolinate reductase. *Biochemistry*, Vol.34, No.11, (March 1995), pp. 3492-501, ISSN 0006-2960

Reddy, S. G., Scapin, G., & Blanchard, J. S. (1996). Interaction of pyridine nucleotide substrates with *Escherichia coli* dihydrodipicolinate reductase: thermodynamic and structural analysis of binary complexes. *Biochemistry*, Vol.35, No.41, (October 1996), pp. 13294-302, ISSN 0006-2960

Richaud, C., Richaud, F., Martin, C., Haziza, C., & Patte, J. C. (1984). Regulation of expression and nucleotide sequence of the *Escherichia coli* dapD gene. *J. Biol. Chem.*, Vol.259, No.23, (December 1984), pp. 14824-14828, ISSN 0021-9258

Saleh, F., & White, P. J. (1979). Metabolism of DD-2,6-diaminopimelic acid by a diaminopimelate-requiring mutant of *Bacillus megaterium. J. Gen. Microbiol.*, Vol.115, (February 1979), pp. 95-100, ISSN 0022-1287

Scapin, G., Blanchard, J. S., & Sacchettini, J. C. (1995). Three-Dimensional Structure of *Escherichia coli* Dihydrodipicolinate Reductase. *Biochemistry*, Vol.34, No.11, (March 1995), pp. 3502–3512, ISSN 0006-2960

Scapin, G., Reddy, S. G., & Blanchard, J. S. (1996). Three-dimensional structure of meso-diaminopimelic acid dehydrogenase from *Corynebacterium glutamicum. Biochemistry*, Vol.35, No.42, (October 1996), pp. 13540-13551, ISSN 0006-2960

Scapin, G., Reddy, S. G., Zheng, R., & Blanchard, J. S. (1997). Three-Dimensional Structure of *Escherichia coli* Dihydrodipicolinate Reductase in Complex with NADH and the Inhibitor 2,6-Pyridinedicarboxylate. *Biochemistry*, Vol.36, No.49, (December 1997), pp. 15081–15088, ISSN 0006-2960

Scapin, G., Cirilli, M., Reddy, S. G., Gao, Y., Vederas, J. C., & Blanchard, J. S. (1998). Substrate and inhibitor binding sites in *Corynebacterium glutamicum* diaminopimelate dehydrogenase. *Biochemistry*, Vol.37, No.10, (March 1998), pp. 3278-3285, ISSN 0006-2960

Schrumpf, B., Schwarzer, A., Kalinowski, J., Pühler, A., Eggeling, L., & Sahm, H. (1991). A functionally split pathway for lysine synthesis in *Corynebacterium glutamicium. J. Bacteriol.*, Vol.173, No.14, (July 1991), pp. 4510-4516, ISSN 0021-9193

Schuldt, L., Weyand, S., Kefala, G., & Weiss, M. S. (2009). The three-dimensional Structure of a mycobacterial DapD provides insights into DapD diversity and reveals unexpected particulars about the enzymatic mechanism. *J. Mol. Biol.*, Vol.389, No.5, (April 2009), pp. 863-879, ISSN 0022-2836

Simms, S. A., Voige, W. H., & Gilvarg, C. (1984). Purification and characterization of succinyl-CoA: tetrahydrodipicolinate N-succinyltransferase from *Escherichia coli. J. Biol. Chem.*, Vol.259, No.5, (March 1984), pp. 2734-2741, ISSN 0021-9258

Sundharadas, G., & Gilvarg, C. (1967). Biosynthesis of alpha,epsilon-diaminopimelic acid in *Bacillus megaterium. J. Biol. Chem.*, Vol.242, No.17, (September 1967), pp. 3983-3984, ISSN 0021-9258

Sung, M. H., Tanizawa, K., Tanaka, H., Kuramitsu, S., Kagamiyama, H., Hirotsu, K., Okamoto, A., Higuchi, T., & Soda, K. (1991). Thermostable aspartate aminotransferase from a thermophilic Bacillus species. Gene cloning, sequence determination, and preliminary x-ray characterization. *J. Biol. Chem.*, Vol.266, No.4, (February 1991), pp. 2567-2572, ISSN 0021-9258

Turner, J. J., Healy, J. P., Dobson, R. C. J., Gerrard, J. A., & Hutton, C. A. (2005). Two new irreversible inhibitors of dihydrodipicolinate synthase: diethyl (E,E)-4-oxo-2,5-heptadienedioate and diethyl (E)-4-oxo-2-heptenedioate. *Bioorg. Med. Chem. Lett.*, Vol.15, No.4, (February 2005), pp. 995-998, ISSN 0960-894X

Usha, V., Dover, L. G., Roper, D. I., Futterer, K., & Besra, G. S. (2009). Structure of the diaminopimelate epimerase DapF from *Mycobacterium tuberculosis. Acta Cryst. Sect. D*, Vol.65, No 4, (April 2009), pp. 383-387, ISSN 0907-4449

Voss, J. E., Scally, S. W., Taylor, N. L., Atkinson, S. C., Griffin, M. D. W., Hutton, C. A., Parker, M. W., Alderton, M. R., Gerrard, J. A., Dobson, R. C. J., Dogovski, C., & Perugini, M. A. (2010). Substrate-mediated stabilization of a tetrameric drug target reveals achilles heel in anthrax. *J. Biol. Chem.*, Vol.285, No.8, (February 2010), pp. 5188-5195, ISSN 0021-9258

Wallsgrove, R. M., & Mazelis, M. (1980). The enzymology of lysine biosynthesis in higher plants: complete localization of the regulatory enzyme dihydrodipicolinate synthase in the chloroplasts of spinach leaves. *FEBS Lett.*, Vol.116, No.2, (July 2008), pp. 189-192, ISSN 0014-5793

Watanabe, N., Cherney, M. M., van Belkum, M. J., Marcus, S. L., Flegel, M. D., Clay, M. D., Deyholos, M. K., Vederas, J. C., & James, M. N. (2007). Crystal structure of LL-diaminopimelate aminotransferase from *Arabidopsis thaliana*: a recently discovered enzyme in the biosynthesis of L-lysine by plants and Chlamydia. *J. Mol. Biol.*, Vol.371, No.3, (August 2007), pp. 685-702, ISSN 0022-2836

Watanabe, N., Clay, M. D., van Belkum M. J., Cherney, M. M., Vederas, J. C., & James, M. N. (2008). Mechanism of substrate recognition and PLP-induced conformational changes in LL-diaminopimelate aminotransferase from *Arabidopsis thaliana*. *J. Mol. Biol.*, Vol.384, No.5, (December 2008), pp. 1314-1329, ISSN 0022-2836

Watanabe, N., Clay, M. D., van Belkum, M. J., Fan, C., Vederas, J. C., & James, M. N. (2011). The Structure of LL-Diaminopimelate Aminotransferase from *Chlamydia trachomatis*: Implications for Its Broad Substrate Specificity. *J. Mol. Biol.*, Vol.411, No.3, (August 2011), pp. 649-660, ISSN 0022-2836

Watanabe, N., & James, M. N. (2011). Structural insights for the substrate recognition mechanism of LL-diaminopimelate aminotransferase. *Biochim. Biophys. Acta.*, (March 2011), In press, ISSN 0006-3002

Wehrmann, A., Eggeling, L., & Sahm, H. (1994). Analysis of different DNA fragments of Corynebacterium glutamicum complementing dapE of *Escherichia coli*. *Microbiology*, Vol.140, No.12, (December 1994), pp. 3349-3356, ISSN 1350-0872

Weinberger, S., & Gilvarg, C. (1970). Bacterial distribution of the use of succinyl and acetyl blocking groups in diaminopimelic acid biosynthesis. *J. Bacteriol.*, Vol.101, No.1, (January 1970), pp. 323-324, ISSN 0021-9193

Weyand, S., Kefala, G. & Weiss, M. S. (2006). Cloning, expression, purification, crystallization and preliminary X-ray diffraction analysis of DapC (Rv0858c) from *Mycobacterium tuberculosis*. *Acta Crystallogr. Sect. F Struc.t Biol. Cryst. Commun.*, Vol.62, No.8, (August 2006), pp. 794-797, ISSN 1744-3091

Weyand, S., Kefala, G., & Weiss, M. S. (2007). The three-dimensional structure of N-succinyldiaminopimelate aminotransferase from *Mycobacterium tuberculosis*. *J. Mol. Biol.*, Vol.367, No.3, (March 2007), pp. 825-838, ISSN 0022-2836

Weyand, S., Kefala, G., Svergun, D. I., & Weiss, M. S. (2009). The three-dimensional structure of diaminopimelate decarboxylase from *Mycobacterium tuberculosis* reveals a tetrameric enzyme organisation. *J. Struct. Funct. Genomics*, Vol.10, No.3, (September 2009), pp. 209-217, ISSN 1345-711X

Williams, R. M., Fegley, G. J., Gallegos, R., Schaefer, F., & Pruess, D. L. (1996). Asymmetric Syntheses of (2S,3S,6S)-, (2S,3S,6R)-, and (2R,3R,6S)-2,3-Methano-2,6-diaminopimelic Acids. Studies Directed to the Design of Novel Substrate-based Inhibitors of L,L-Diaminopimelate Epimerase. *Tetrahedron*, Vol.52, No.4, (January 1996), pp. 1149-1164, ISSN 0040-4020

Wiseman, J. S., & Nichols, J. S. (1984). Purification and properties of diaminopimelic acid epimerase from *Escherichia coli*. *J. Biol. Chem.*, Vol.259, No.14, (July 1984), pp. 8907-8914, ISSN 0021-9258

Wolterink-van Loo, S., Levisson, M., Cabrières, M.C., Franssen, M.C.R., & van der Oost, J. (2008). Characterization of a thermostable dihydrodipicolinate synthase from *Thermoanaerobacter tengcongensis*. *Extremophiles*, Vol.12, No.3, (May 2008), pp. 461-469, ISSN 1431-0651

Work, E. (1962). Diaminopimelic racemase. In: *Methods in Enzymology*, Colowick, S. P. & Kaplan, N. O., pp. 858-864, Academic Press, ISBN 0076-6879, New York

Yugari, Y., & Gilvarg C. (1965). The condensation step in diaminopimelate synthesis. *J. Biol. Chem.*, Vol.240, No.12, (December 1965), pp. 4710-4716, ISSN 0021-9258

3

Glucose Metabolism and Cancer

Lei Zheng[1,2], Jiangtao Li[2] and Yan Luo[3]
*[1]The Sidney Kimmel Comprehensive Cancer Center, The Skip Viragh Center for
Pancreatic Cancer, The Sol Goldman Pancreatic Cancer Center
Department of Oncology, and Department of Surgery, The Johns Hopkins University
School of Medicine, Baltimore, Maryland
[2]Department of Surgery, The Second Affiliated Hospital
Zhejiang University College of Medicine, Hangzhou
[3]Section of Biochemistry and Genetics, School of Basic Medical Sciences
and Cancer Institute, the Second Affiliated Hospital
Zhejiang University College of Medicine, Hangzhou
[1]USA
[2,3]China*

1. Introduction

An outstanding biochemical characteristic of neoplastic tissues is that despite ample oxygen supply, glycolysis is the dominant pathway for adenosine 5'-triphosphate (ATP) production, a phenomenon termed "the Warburg effect" (Warburg et al., 1927; Warburg, 1956). This aerobic glycolysis seems unexpected as one would imagine that cancer cells should have, given sufficient oxygen, adapted to utilize the complete oxidative phosphorylation to maximize ATP production. In addition, for producing same level of ATP, glycolysis would consume >15-fold more glucose, resulting in an addiction for glucose during active tumor growth. There must be a biological logic for cancer cells to prefer utilization of glycolysis for ATP production: they obtain and/or sustain growth advantage at the cost of an "addiction" to glycolysis. Mechanistically, there have been many hypotheses proposed to explain this phenomenon and two are more prominent: one is that by consuming excessive glucose, cancer cells may change the tumor's microenvironment to gain growth advantage over normal cells; the second is that, to sustain a dominant glycolysis process, cancer cells alter the expression or functions of glucose metabolic enzymes, which may have resulted in additional changes that promote cancer development (Vander Heiden et al., 2009). A growing body of evidence has supported both hypotheses and will be reviewed here.

2. Altered glycolytic enzymes in cancer cells

In cancer cells, the activities or expression levels of many enzymes participating in glucose metabolism are altered, those involved in glycolysis in particular. The glycolysis commonly refers to the reactions that covert glucose into pyruvate or lactate (Figure 1). Usually, one or more isoforms of glycolytic enzymes have altered expression patterns or activities in cancer cells, which was previously reviewed (Herling et al., 2011; Porporato et al., 2011).

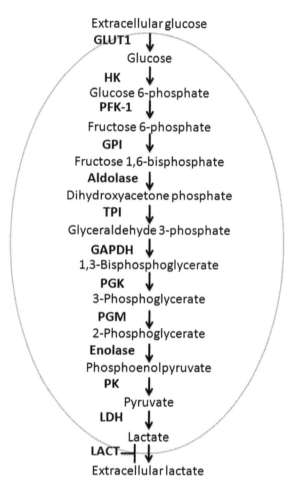

Fig. 1. Glycolysis in cancer cells

- GLUT1, an isoform of **glucose transporters** (GLUTs) is overexpressed in many types of human malignancies, whereas the insulin-sensitive GLUT4 is downregulated in cancer cells. The imbalanced expression of GLUT1 versus GLUT4 in cancer cells may have contributed to the insulin-independent glucose uptake in cancer cells. In addition, GLUT3 was also reported to be overexpressed in cancer cells (Smith, 1999; Medina and Owen, 2002; Noguchi et al., 1998).

- HK-2, an isoform of **hexokinase** that is one of the three rate limiting glycolytic enzymes, is overexpressed in cancer cells and was shown to contribute to the Warburg effect. The high glycolytic rate characteristic of hypoxic solid tumors is attributed to the overexpression of HK-2 (Mathupala et al., 2009; Wolf et al.).

- The expression level of the **glucose 6-phosphate isomerase** (GPI) has also been reported to be elevated in its mRNA level in different human cancer cell lines (Funasaka et al., 2005).

- The alteration of **phosphofructokinase-1** (PFK-1) in cancer cells is not at expression level, but at the level of the enzyme activity. PFK in cancer cells is less sensitive to the inhibition by its allosteric regulators such as citrate and ATP (Meldolesi et al., 1976). Additionally, the expression of all four genes of the **Phosphofructokinase-2** (PFK-2/FBPase/PFKFB) family is inducible by hypoxia, among which the gene encoding *PFKFB3* is highly expressed in several types of human neoplasm (Minchenko et al., 2005; Atsumi et al., 2002; Kessler et al., 2008).
- The expression of **aldolase** isoenzymes is downregulated in some cancer types such as hepatocellular carcinoma (Song et al., 2004) and upregulated in some other tumor types such as pancreatic ductal adenocarcinoma (Cui et al., 2009).
- **Triosephosphate isomerase** (TPI) was detected in the plasma of cancer patients (Robert et al., 1961) and autoantibodies against TPI was also detected in sera from breast cancer patients (Tamesa et al., 2009).
- **Phosphoglycerate kinase-1** (PGK-1) is overexpressed in majority of the pancreatic ductal adenocarcinomas and can also be detected in sera of patients with these tumors (Hwang et al., 2006).
- **Phosphoglycerate mutase** (PGM/PGAM) M type subunit (PGM-M) is overexpressed in many cancers including lung, colon, liver and breast (Durany et al., 2000; Durany et al., 1997).
- The expression of **Enolase 1** (ENOA) is upregulated at the transcriptional and / or translational level in multiple types of tumor including brain, breast, cervix, colon, eye, gastric, head and neck, kidney, leukemia, liver, lung, muscle, ovary, pancreas, prostate, skin and testis(Capello et al.). It is also differentially regulated at the post-translational level in cancer cells as compared with normal cells. ENOA in tumor cells is subjected to more acetylation, methylation and phoshorylation than in normal tissues (Capello et al.). Specific acetylated residues of ENOA were found in cervix, pancreatic, and colon cancers. Five aspartate and five glutamate residues were found to be specifically methylated in pancreatic cancer. Several serine and threonine residues were found to be specifically phosphorylated in leukemia, cervix, and lung cancers. Although ENOA is phosphorylated at Serine 419 in both normal and malignant pancreatic tissues, this phosphorylated form of ENOA is overexpressed in pancreatic cancer(Zhou et al., 2010).
- **Pyruvate kinase** has two isoforms, M and L, which have tissue-specific expression. Normal proliferating cells including embryonic cells and adult stem cells selectively express the M2 isoform (PKM2) (Reinacher and Eigenbrodt, 1981; Yamada and Noguchi, 1999). During tissue differentiation in development, embryonic PKM2 is replaced by tissue-specific isoforms. However, the tissue-specific expression pattern of PK is disrupted during tumorigenesis (Hacker et al., 1998). PKM2 is re-expressed and becomes the only predominant isoform in cancer cells (Mazurek et al., 2005).
- The **lactate dehydrogenase** (LDH) family of tetrameric enzymes catalyzes the pyruvate reduction into lactate. LDHs are formed by four subunits of two different isozymes of either LDH-H or LDH-M. LDH-H is encoded by the *LDH-B* gene and is ubiquitously expressed, and LDH-M is encoded by *LDH-A*. LDH-M has a higher Km for pyruvate and a higher $Vmax$ for pyruvate reduction than LDH-H (Markert et al., 1975). Consequently, LDHs predominantly comprising of the LDH-M subunits drive the reduction of pyruvate to lactate; LDHs predominantly comprising of the LDH-H subunits drive the oxidation of lactate to pyruvate. Many types of tumor cells manifest a high expression of the *LDH-A* gene. Elevated expression of LDH5, which comprises of

four LDH-M subunits, is an unfavorable prognostic factor for many human malignancies (Koukourakis et al., 2003; Koukourakis et al., 2005; Koukourakis et al., 2009). In addition, the glycolysis process in cancer cells is reportedly to be associated with hypermethylation of the *LDH-B* gene promoter, which is linked to gene silencing (Leiblich et al., 2006; Thangaraju et al., 2009).

• Plasma membrane **lactate transport** (LACT) is facilitated by the family of proton-linked moncarboxylate transporters (MCTs) or by SMCT1, a sodium coupled lactate transporter. The MCT4 isoform is upregulated in many cancer types. The expression of MCT2, an isoform mainly implicated in lactate import, is decreased in tumor cell lines. SMCT1, also implicated in lactate import, is downregulated in a number of cancer types including colon, thyroid, and stomach(Herling et al., 2011; Porporato et al., 2011).

3. Glucose metabolic pathways switched by oncogenes and tumor suppressors

It has been long proposed that the altered expression or enzyme activities of glycolytic enzymes are regulated by oncogenes and tumor suppressor genes.

• The expression of GLUT1 has been demonstrated to be controlled by the hypoxia-inducible transcription factor HIF-1, c-Myc, and Akt (Chen et al., 2001; Osthus et al., 2000; Rathmell et al., 2003).

• The gene encoding HK-2, but not HK-1, is known to be a transcriptional target of HIF-1(Rempel et al., 1996). It was later shown that HIF-1 cooperates with c-Myc to transactivate HK-2 under hypoxia (Kim et al., 2007). The phosphorylated form of HK-2 interacts with the voltage-dependent anion channel (VDAC) at the outer mitochondrial membrane (Bustamante and Pedersen, 1977; Nakashima et al., 1986; Gottlob et al., 2001). The interaction of HK-2 with VDAC interferes with the binding of the pro-apoptotic protein Bax to VDAC thus preventing the formation of the channel through which cytochrome c can escape from mitochondria to trigger apoptosis (Pastorino et al., 2002). Therefore, overexpression of HK-2 in cancer cells leads to a switch from HK-1 to HK-2 and offers a metabolic advantage by protecting cancer cells against apoptosis(Porporato et al., 2011).

• Overexpression of GPI can be induced by HIF-1 and VEGF(Funasaka et al., 2005).

• The expression of PFK-2 genes in tumor cells is shown to be regulated by Ras and src (Yalcin et al., 2009) and that of PFKFB3 is demonstrated to be induced by HIF-1, c-myc, ras, src, and loss of function of p53 (Minchenko et al., 2002). Among the four PFK-2 genes, PFKFB3 is the most significantly induced in response to hypoxia. Hypoxia-induced PFK-2 activity of human PFKFB3 is further enhanced through phosphorylation of the serine 462 residue (Marsin et al., 2002). This phosphorylation process may involve AMP-activated protein kinase (AMPK) and Akt(Shaw and Cantley, 2006; Yun et al., 2005).

• Although no evidence has suggested that the expression level of **Glyceraldehyde-3-phosphate dehydrogenase** (GAPDH) is altered in cancer cells, its expression has been demonstrated to be highly dependent on the proliferative state of the cells and can by regulated by HIF-1, p53, and c-jun(Colell et al., 2009; Colell et al., 2007).

• The gene encoding ENOA is a target of, and its expression is upregulated by, c-Myc (Sedoris et al. 2010). Notably, a growing body of evidence has suggested that ENOA is a tumor-associated antigen(Capello et al., 2011). In patients of many different cancer types, including pancreatic, leukemia, melanoma, head and neck, breast and lung, anti-ENOA autoantibodies have been detected(Capello et al., 2011). In pancreatic cancer patients, the anti-ENOA autoantibodies are directed against phosphorylated Serine 419(Tomaino et al., 2011). One study has shown that, in pancreatic cancer, ENOA elicits a CD4+ and CD8+ T cell response both in vitro and in vivo (Cappello et al., 2009). In pancreatic cancer patients, production of anti-ENOA IgG is correlated with the ability of T cells to be activated in response to ENOA(Cappello et al., 2009). In patients with oral squamous cell carcinoma, an HLA-DR8-restricted peptide (amino acid residues 321–336) of human ENOA recognized by CD4+ T cell has been identified (Kondo et al., 2002).

- As described above, PKM2 is a predominant isoform of PKM in cancer cells. PKM2 is less active than other PKs but is however the only PK subject to regulation by the allosteric activator fructose-1,6-bi-phosphate (FBP) and possessing the capability to bind phosphotyrosine proteins (Christofk et al., 2008). Binding of phosphotyrosine peptides to PKM2 leads to dissociation of FBP hence lowering the PKM2 enzyme activity, which may provide a link between cell growth signals and the Warburg effect given that many growth signals and oncogenic pathways involve tyrosine kinases. The unique feature of the PKM2 isoform provides a mechanism by which oncogenes regulate glycolysis and the Warburg effect, and offers an advantage in metabolic plasticity by equipping cancer cells with an exquisitely regulated switch between promoting ATP production and cell proliferation; and lowering the PKM2 activity upon growth signaling was proposed to allow efficient biomass building (anabolic) pathways branched from glycolysis (Christofk et al., 2008). It was suggested that PKM2 is regulated by HIF-1 at the transcriptional level (Discher et al., 1998). Recently, it was demonstrated that, after nuclear translocation, PKM2 cooperates with HIF-1 to transactivate genes, the products of which are involved in promoting the glycolysis and tumor angiogenesis (Luo et al., 2011). Despite these advancements, the regulatory mechanism of PKM2 in cancer cells is not fully understood.
- The expression of PGM/PGAM was found to be downregulated by p53 (Kondoh et al., 2005). Thus, loss of p53 function is anticipated to induce the expression of PGM/PGAM. Phosphoenolpyruvatue, the substrate for PK in cells, transfers the phosphate to the histidine residue located in the catalytic center of the human PGAM1; however, this reaction occurs only in those PKM2-expressing cells but is independent of the PKM2 enzyme activity. Thus, histidine phosphorylation of PGAM1 may provide an alternate glycolytic step in proliferating cancer cells where low PK activity accompanies the expression of PKM2 (Vander Heiden et al., 2010b).
- LDH-A is a target gene of c-Myc and HIF-1 (Dang et al., 2009). Loss of LDH-A functions results in diminished cellular transformation, anchorage independent tumor growth under hypoxic conditions, or xenograft tumor growth (Shim et al., 1997).

Therefore, activation of oncogenes and loss of tumor suppressors are believed to underlie the metabolic switch in cancer cells. Many cancer associated gene products are involved, including c-Myc, NF-kB, Akt, and multiple types of tyrosine kinase including epidermal growth factors (EGFs) and insulin-like growth factor 1 (IGF-1) receptor (Levine and Puzio-Kuter, 2010). Several pathways have been shown to be important for the regulation of glucose metabolism including the PI3K-AKT-mTOR and c-Myc pathways and both have

HIF-1 as their downstream effector. Consistently, among the HIF-1 regulated genes, most are those encoding glycolytic enzymes (Semenza, 2003). The loss of PTEN and concurrent increase of Akt and mTOR lead to the HIF-1 activation and the Warburg effect (Arsham et al., 2002; Zundel et al., 2000).

Another oncogene, K-Ras, can alter glucose metabolism so as to provide tumor cells with a selective advantage. In cells with mutated K-Ras, GLUT1 is upregulated, leading to an augmented glucose uptake, glycolysis and lactate production. Interestingly, in these cells mitochondrial functions and oxidative phosphorylation are not compromised, which allows increased survival rate of the K-Ras mutant cells during glucose deprivation (Annibaldi and Widmann, 2010).

Loss of p53 functions also leads to the Warburg effect. As described above, p53 represses transcription of the genes encoding GLUT1 and 4, and induces transcription of the TIGAR gene, which in turn lowers the intracellular level of PFK/FBPase (Bensaad et al., 2006). In addition, p53 inhibits the PI3K-Akt-mTOR pathways. This appears to be mediated by the transcriptional targets of p53 including PTEN, IGF-binding protein 3, tuberous sclerosis protein TSC-2, and the beta subunit of AMPK (Feng et al., 2007). Conceivably, loss of p53 functions and subsequent loss of expression of these target genes lead to a high HIF level and establishment of the Warburg effect.

4. Tumor "friendly" microenvironment attributed to altered glucose metabolism

One would ask: what is the advantage for cancer cells to use energy-inefficient glycolysis under adequate oxygen supply. A potential advantage, as a result of Warburg effect, is high production of lactic acid due to enhanced glycolysis. Positive correlation between lactate serum levels and tumor burden in cancer patients has been well documented, implicating a role of acidic microenvironment in promoting tumor growth and development (McCarty and Whitaker, 2011).

First, accumulated evidence has suggested that acidic environment amplifies the capacity of invasion and metastasis of cancer cells. For instance, acid pretreatment of tumor cells enhance their ability to form metastases in tumor-transplanted mice (Rofstad et al., 2006). Consistently, it has been shown that increasing tumor pH via bicarbonate therapy significantly reduces the number and the size of metastases in a mouse model of breast cancer (Robey et al., 2009).

Second, the extracellular pH of solid tumors is significantly more acidic than that of normal tissues, thus impairing the uptake of weakly basic chemotherapeutic drugs (Raghunand et al., 1999). Several anticancer drugs such as doxorubicin, mitoxantrone and vincristine are weak bases that are protonated in slightly acid tumor microenvironments. The protonated forms of the drugs cannot easily diffuse across the plasma membrane and therefore their cellular uptake is suppressed. It has been demonstrated that the addition of sodium bicarbonate in the drinking water enhanced the anti-tumor effect of doxorubicin on xenotransplanted tumors presumably by enhancing the intracellular drug delivery through raising the pH of the extracellular milieu in mice (Raghunand et al., 1999). The reverse situation was also demonstrated in another study showing that glucose administration to mice led to a lower efficacy of doxorubicin on tumors presumably due to a decrease in the extracellular pH (Gerweck et al., 2006).

Third, acidic microenvironment inhibits anti-tumor immune response. For instance, lactic acid suppressed the proliferation and cytokine production of human cytotoxic T lymphocytes (CTLs) up to 95% and led to a 50% decrease in cytotoxic activity (Fischer et al., 2007). Activated lymphocytes themselves use glycolysis, which relies on the efficient secretion of lactic acid. Export of lactic acid from lymphocytes depends on a gradient between intracellular and extracellular lactic acid concentration. High extracelullar acidity would diminish this gradient and block the secretion of lactic acid from lymphocytes. The accumulation of intracellular lactic acid eventually disturbs the glycolysis process hence affecting the activity of lymphocytes. Acidification similarly inhibits the activity of other immune cells such as dendritic cells.

5. Coupled biological and metabolic processes and the logic of a mammalian metabolic cycle

Glycolytic enzymes have multiple cellular functions. For instance, GAPDH has been implicated in numerous non-glycolytic functions (Colell et al., 2007; McKnight, 2003). In 2003, we published a paper that describes the isolation and characterization of OCA-S, which is a transcription cofactor complex that directly stimulates the transcription of the histone H2B gene in an S-phase-specific manner (Zheng et al., 2003). Surprisingly, a key component of the OCA-S complex represents a nuclear form of GAPDH, which regulates H2B transcription in a redox dependent manner. LDH was later shown to be an essential OCA-S component as well and can exercise the enzyme activity to reverse in vitro inhibition of H2B transcription by converting NADH to NAD$^+$ in the presence of substrate pyruvate (Dai et al., 2008). Conceivably, the participation of these glycolytic enzymes in such a cell cycle event would subject cell cycle regulation to altered glucose metabolism in cancer cells, providing yet another mechanistic explanation of cancer growth and development. The "moonlighting" participation of the glycolytic enzymes in a cellular process would in theory impose a dynamic modulation of the redox status in the cellular compartment where this process is executed and subsequently affect the functions of other redox-sensitive proteins in the same intracellular compartment. Thus, in addition to histone expression, our study has suggested that other cellular processes including cell cycle regulation, DNA replication and damage repair are potentially all coupled through the redox signals (Yu et al., 2009). The coupling of these cellular processes is apparently crucial for the maintenance of chromatin integrity during cell cycle, and thus altered glucose metabolism in cancer cells potentially would disrupt the coupling of these processes and make cancer genomes more error-prone.

It is well delineated that in yeast, quite a few biological and metabolic processes are known to be compartmentalized in time, termed the yeast metabolic cycle (YMC) that is in sync with the cell cycle progression (Tu et al., 2005). The YMC has oxidative, reductive/building and reductive/charging phases, and the S-phase of yeast cell is synchronized with the most reductive stage of YMC. We subsequently found that the oxidative and reductive phases in mammalian cells are also synchronized with the cell cycle, and our study demonstrated that the free NAD$^+$/NADH ratio fluctuated in a defined manner during cell cycle(Yu et al., 2009). At G1 phase, the intracellular NAD$^+$/NADH ratio is high, suggesting that G1 cells maintain an oxidative status. Upon entering S phase, the ratio becomes lower, corresponding to a reductive status. When the cells exit S phase and enter G2 phase, the NAD$^+$/NADH ratio becomes higher again. This oxidative status appears to be maintained until cells enter the next S phase. This phenomenon has been dubbed mammalian metabolic cycle (MMC) for its similarity with YMC. The fluctuating $NAD^+/NADH$ ratios in a mammalian cell cycle must reflect overall oscillatory cellular metabolism; whether and how glucose metabolism is synchronized with the cell cycle remains to be explored. Nonetheless, this synchronization must have been very precisely regulated. Conceivably, if glucose metabolism is altered, the cell cycle must be coordinately modulated, and vice versa. Therefore, cancer cells may have acquired the growth and proliferative advantage over normal cells through alteration in glucose metabolism.

6. Targeting glycolysis for cancer treatment

The aberrant metabolic pathways underlying the Warburg effect are being considered as novel targets for cancer therapy. Several strategies have been employed to target glucose metabolic pathways for cancer treatment.

First, inhibitors of glycolytic enzymes or glycolytic pathways are being searched to identify therapeutic agents that can inhibit cancer growth and development.

A number of small molecules have been reported to target glycolysis although none to date has been shown to have specific molecular targets. For example, 3-bromopyruvate, a highly active alkylating agent, was reported to target HK-2 and/or GAPDH (Dang et al., 2009). 2-deoxyglucose (2-DG) can be phosphorylated by HK-2, which in turn inhibits HK-2 (Ralser et al., 2008). It is also shown to be a GLUT inhibitor.

Many efforts have been made to identify specific inhibitors. Lonidamine has been described as a specific inhibitor of mitochondria-bound HK (Floridi et al., 1981) and has been tested in multiple clinical trials including a phase II study in combination with diazepam for the treatment of glioblastoma patients (Porporato et al., 2011). Unfortunately, none of these clinical trials have successfully shown its therapeutic benefit in terms of time-to-progression and overall survival (Oudard et al., 2003); one study showed its severe hepatic adverse effects.

Drugs that target more specific metabolic control points of glycolysis in cancer cells, such as PKM2 or LDH-A, warrant investigation as potential cancer therapies. gossypol/AT-101, a natural product and a non-specific LDH inhibitor that has more preferential inhibitory activity on malarial and spermoctye LDHs, has already been tested in human clinical trials for its anti-cancer effect(Porporato et al., 2011). A selective competitive inhibitor of LDH5, 3-dihydroxy-6-methyl-7-(phenylmethyl)-4-propylnaphthalene-1-carboxylicacid (FX11), has been identified through screening a library of compounds derived from gossypol (Yu et al., 2001). FX11 has been shown to suppress in vivo xenograft tumor growth of human B lymphoid tumor and pancreatic cancer cells (Le et al. 2010), providing a strong rationale for clinical development of therapeutic agents targeting LDH-A. Recently, N-Hydroxy-2-carboxy-substituted indole compounds have been identified as LDH5-specific inhibitors (Granchi et al., 2011).

Several clinical trials with the PKM2 inhibitor TLN-232/CAP-232, a seven amino-acid peptide, have been initiated. Encouraging preliminary results demonstrated that it is safe, well tolerated, and may offer disease control (Porporato et al., 2011). New small molecule inhibitors of PKM2 were also screened. Among them, the most potent one resulted in decreased glycolysis and increased cell death in respond to loss of growth factor signaling, supporting the feasibility and viability of targeting glucose metabolism as a novel strategy to treating human cancers (Vander Heiden et al., 2010a).

Conceivably, ENOA, as a tumor-associated antigen with an ability of eliciting both B cell and T cell immune response (Capello et al., 2011), is an ideal target of cancer vaccine and immunotherapy. Comparing to small molecule inhibitors, immunotherapy offers superior target specificity and may be used as an alternative approach to target other glycolysis enzymes.

Second, inhibition of glycolytic enzyme or glycolysis pathways serves as a strategy to enhance the sensitivity of tumor cells to conventional cytotoxic chemotherapy agents.

Inhibition of LDH-A has been shown to re-sensitize Taxol-resistant cancer cells to Taxol (Zhou et al.). 2-DG is another example. The safety of using it as an anti-cancer agent has been questioned notably because of brain toxicity (Tennant et al., 2010). However, it has a proven efficacy in sensitizing human osteosarcoma and non-small cell lung cancers to adriamycin and paclitaxel (Maschek et al., 2004). Recently, a Phase I clinical study for prostate cancer has defined a maximum tolerance dose of 45mg/kg for Phase II trials (Stein et al., 2010). It will be interesting to test whether 2-DG can enhance the efficacy of chemotherapy agents even if it cannot offer anti-cancer activity by itself at this dose level.

As aforementioned, chemosensitivity is enhanced by counteracting the acidification of tumor's microenvironment. Inhibitors of glycolytic enzymes may impose an alkalizing effect in tumor's microenvironment particularly at the tumor tissue level and thus may have a more specific and powerful role in enhancing the sensitivity of tumor cells to basic chemotherapy drugs. Major targets for counteracting the acidification of tumor's microenvironment include carbonic anhydrasesis (CA)-9 and -12, sodium–proton exchanger 1 (NHE1), sodium bicarbonate cotransporter (NBC), vacuolar ATPase (V-ATPase), sodium–potassium (NaK) ATPase, and MCT4 (Porporato et al., 2011). Indisulam is a leading compound for CA9 inhibition. It is a sulfonamide derivative and shown to inhibit CA9 at nanomolar concentrations (Abbate et al., 2004; Supuran, 2008; Owa et al., 2002). It has been tested in multiple clinical trials for the treatment of melanoma, lung, pancreatic and metastatic breast cancers and has not been found in the completed clinical trials to have antitumor efficacy as a single agent (Talbot et al., 2007). Girentuximab, a specific antibody targeting CA9, is now being tested in Phase III clinical trials for the treatment of clear-cell renal cell carcinoma (Reichert, 2011). Several inhibitors of membrane-bound V-ATPase have been reported to have antitumor activity in preclinical studies (Perez-Sayans et al., 2009). Other enzymes involved in the cellular export of protons are also studied as targets of anti-cancer therapeutic development. However, future studies should emphasize on combining anti-acidification therapies with cytotoxic chemotherapy or immunotherapy to achieve the effective anti-cancer treatment.

Third, combination of inhibitors of glucose metabolic enzymes with inhibitors of oncogenic pathways may result in synergistic anti-tumor effects.

Inhibitors of oncogenic pathways have been extensively tested for cancer therapy, with only moderate success in a few types of human cancers. Among them, inhibitors of the Ras pathway are essentially not effective. As K-ras mutated cancer cells have an enhanced survival when glucose is deprived, a combinatorial treatment with both glycolysis inhibitors and Ras pathway inhibitors may target Ras-mutated cancer cells more effectively and more specifically. Supporting this hypothesis, the hexokinase inhibitor 3-bromopyruvate was demonstrated to be highly toxic specifically to cancer cells with K-ras mutation, but not to cancer cells with wild-type K-ras (Yun et al., 2009). BAY87-2243, which is a small molecule inhibitor of HIF-1 activity and of HIF-1α stability, and EZN-2968, which is an antisense oligonucleotide targeting HIF-1α, had been tested in clinical trials (Greenberger et al., 2008). Metformin is an AMPK-activating drug and is currently used for type-2 diabetes treatment. Epidemiological studies have shown reduced incidence of cancer in diabetic patients treated with metformin (Evans et al., 2005; Jalving et al., 2010; Libby et al., 2009). It is highly intriguing to test whether this clinically safe and known glucose metabolism modulating drug can enhance anti-cancer activity of cytotoxic chemotherapy and/or further lower the cancer recurrence following adjuvant chemotherapy. Similarly, other combination treatments with inhibitors of both glucose metabolisms and oncogenic pathways such as Akt, mTOR, etc. also warrant investigation.

7. References

Abbate, F., Casini, A., Owa, T., Scozzafava, A., and Supuran, C.T. (2004). Carbonic anhydrase inhibitors: E7070, a sulfonamide anticancer agent, potently inhibits cytosolic isozymes I and II, and transmembrane, tumor-associated isozyme IX. *Bioorg Med Chem Lett* 14:217-223.

Annibaldi, A., and Widmann, C. (2011). Glucose metabolism in cancer cells. *Current opinion in clinical nutrition and metabolic care* 13:466-470.

Arsham, A.M., Plas, D.R., Thompson, C.B., and Simon, M.C. (2002). Phosphatidylinositol 3-kinase/Akt signaling is neither required for hypoxic stabilization of HIF-1 alpha nor sufficient for HIF-1-dependent target gene transcription. *The Journal of biological chemistry* 277:15162-15170.

Atsumi, T., Chesney, J., Metz, C., Leng, L., Donnelly, S., Makita, Z., Mitchell, R., and Bucala, R. (2002). High expression of inducible 6-phosphofructo-2-kinase/fructose-2,6-bisphosphatase (iPFK-2; PFKFB3) in human cancers. *Cancer research* 62:5881-5887.

Bensaad, K., Tsuruta, A., Selak, M.A., Vidal, M.N., Nakano, K., Bartrons, R., Gottlieb, E., and Vousden, K.H. (2006). TIGAR, a p53-inducible regulator of glycolysis and apoptosis. *Cell* 126:107-120.

Bustamante, E., and Pedersen, P.L. (1977). High aerobic glycolysis of rat hepatoma cells in culture: role of mitochondrial hexokinase. *Proceedings of the National Academy of Sciences of the United States of America* 74:3735-3739.

Capello, M., Ferri-Borgogno, S., Cappello, P., and Novelli, F. (2011). alpha-Enolase: a promising therapeutic and diagnostic tumor target. *The FEBS journal* 278:1064-1074.

Cappello, P., Tomaino, B., Chiarle, R., Ceruti, P., Novarino, A., Castagnoli, C., Migliorini, P., Perconti, G., Giallongo, A., Milella, M., *et al.* (2009). An integrated humoral and cellular response is elicited in pancreatic cancer by alpha-enolase, a novel pancreatic ductal adenocarcinoma-associated antigen. *Int J Cancer* 125:639-648.

Chen, C., Pore, N., Behrooz, A., Ismail-Beigi, F., and Maity, A. (2001). Regulation of glut1 mRNA by hypoxia-inducible factor-1. Interaction between H-ras and hypoxia. *The Journal of biological chemistry* 276: 9519-9525.

Christofk, H.R., Vander Heiden, M.G., Harris, M.H., Ramanathan, A., Gerszten, R.E., Wei, R., Fleming, M.D., Schreiber, S.L., and Cantley, L.C. (2008). The M2 splice isoform of pyruvate kinase is important for cancer metabolism and tumour growth. *Nature* 452:230-233.

Colell, A., Green, D.R., and Ricci, J.E. (2009). Novel roles for GAPDH in cell death and carcinogenesis. *Cell death and differentiation* 16:1573-1581.

Colell, A., Ricci, J.E., Tait, S., Milasta, S., Maurer, U., Bouchier-Hayes, L., Fitzgerald, P., Guio-Carrion, A., Waterhouse, N.J., Li, C.W., *et al.* (2007). GAPDH and autophagy preserve survival after apoptotic cytochrome c release in the absence of caspase activation. *Cell* 129:983-997.

Cui, Y., Tian, M., Zong, M., Teng, M., Chen, Y., Lu, J., Jiang, J., Liu, X., and Han, J. (2009). Proteomic analysis of pancreatic ductal adenocarcinoma compared with normal adjacent pancreatic tissue and pancreatic benign cystadenoma. *Pancreatology* 9:89-98.

Dai, R.P., Yu, F.X., Goh, S.R., Chng, H.W., Tan, Y.L., Fu, J.L., Zheng, L., and Luo, Y. (2008). Histone 2B (H2B) expression is confined to a proper NAD+/NADH redox status. *The Journal of biological chemistry* 283:26894-26901.

Dang, C.V., Le, A., and Gao, P. (2009). MYC-induced cancer cell energy metabolism and therapeutic opportunities. *Clin Cancer Res* 15:6479-6483.

Discher, D.J., Bishopric, N.H., Wu, X., Peterson, C.A., and Webster, K.A. (1998). Hypoxia regulates beta-enolase and pyruvate kinase-M promoters by modulating Sp1/Sp3 binding to a conserved GC element. *The Journal of biological chemistry* 273:26087-26093.

Durany, N., Joseph, J., Campo, E., Molina, R., and Carreras, J. (1997). Phosphoglycerate mutase, 2,3-bisphosphoglycerate phosphatase and enolase activity and isoenzymes in lung, colon and liver carcinomas. *British journal of cancer* 75:969-977.

Durany, N., Joseph, J., Jimenez, O.M., Climent, F., Fernandez, P.L., Rivera, F., and Carreras, J. (2000). Phosphoglycerate mutase, 2,3-bisphosphoglycerate phosphatase, creatine kinase and enolase activity and isoenzymes in breast carcinoma. *British journal of cancer* 82:20-27.

Evans, J.M., Donnelly, L.A., Emslie-Smith, A.M., Alessi, D.R., and Morris, A.D. (2005). Metformin and reduced risk of cancer in diabetic patients. *BMJ* 330:1304-1305.

Feng, Z., Hu, W., de Stanchina, E., Teresky, A.K., Jin, S., Lowe, S., and Levine, A.J. (2007). The regulation of AMPK beta1, TSC2, and PTEN expression by p53: stress, cell and tissue specificity, and the role of these gene products in modulating the IGF-1-AKT-mTOR pathways. *Cancer research* 67:3043-3053.

Fischer, K., Hoffmann, P., Voelkl, S., Meidenbauer, N., Ammer, J., Edinger, M., Gottfried, E., Schwarz, S., Rothe, G., Hoves, S., *et al.* (2007). Inhibitory effect of tumor cell-derived lactic acid on human T cells. *Blood* 109:3812-3819.

Floridi, A., Paggi, M.G., Marcante, M.L., Silvestrini, B., Caputo, A., and De Martino, C. (1981). Lonidamine, a selective inhibitor of aerobic glycolysis of murine tumor cells. *J Natl Cancer Inst* 66:497-499.

Funasaka, T., Yanagawa, T., Hogan, V., and Raz, A. (2005). Regulation of phosphoglucose isomerase/autocrine motility factor expression by hypoxia. *Faseb J* 19:1422-1430.

Gerweck, L.E., Vijayappa, S., and Kozin, S. (2006). Tumor pH controls the in vivo efficacy of weak acid and base chemotherapeutics. *Molecular cancer therapeutics* 5:1275-1279.

Gottlob, K., Majewski, N., Kennedy, S., Kandel, E., Robey, R.B., and Hay, N. (2001). Inhibition of early apoptotic events by Akt/PKB is dependent on the first committed step of glycolysis and mitochondrial hexokinase. *Genes & development* 15:1406-1418.

Granchi, C., Roy, S., Giacomelli, C., Macchia, M., Tuccinardi, T., Martinelli, A., Lanza, M., Betti, L., Giannaccini, G., Lucacchini, A., *et al.* (2011). Discovery of Nhydroxyindole-based inhibitors of human lactate dehydrogenase isoform A (LDH-A) as starvation agents against cancer cells. *J Med Chem 54*:1599-1612.

Greenberger, L.M., Horak, I.D., Filpula, D., Sapra, P., Westergaard, M., Frydenlund, H.F., Albaek, C., Schroder, H., and Orum, H. (2008). A RNA antagonist of hypoxia-inducible factor-1alpha, EZN-2968, inhibits tumor cell growth. *Molecular cancer therapeutics 7*:3598-3608.

Hacker, H.J., Steinberg, P., and Bannasch, P. (1998). Pyruvate kinase isoenzyme shift from L-type to M2-type is a late event in hepatocarcinogenesis induced in rats by a choline-deficient/DL-ethionine-supplemented diet. *Carcinogenesis 19*:99-107.

Herling, A., Konig, M., Bulik, S., and Holzhutter, H.G. (2011). Enzymatic features of the glucose metabolism in tumor cells. *The FEBS journal 278*:2436-2459.

Hwang, T.L., Liang, Y., Chien, K.Y., and Yu, J.S. (2006). Overexpression and elevated serum levels of phosphoglycerate kinase 1 in pancreatic ductal adenocarcinoma. *Proteomics 6*:2259-2272.

Jalving, M., Gietema, J.A., Lefrandt, J.D., de Jong, S., Reyners, A.K., Gans, R.O., and de Vries, E.G. (2010). Metformin: taking away the candy for cancer? *Eur J Cancer 46*:2369-2380.

Kessler, R., Bleichert, F., Warnke, J.P., and Eschrich, K. (2008). 6-Phosphofructo-2-kinase/fructose-2,6-bisphosphatase (PFKFB3) is up-regulated in high-grade astrocytomas. *J Neurooncol 86*:257-264.

Kim, J.W., Gao, P., Liu, Y.C., Semenza, G.L., and Dang, C.V. (2007). Hypoxia-inducible factor 1 and dysregulated c-Myc cooperatively induce vascular endothelial growth factor and metabolic switches hexokinase 2 and pyruvate dehydrogenase kinase 1. *Molecular and cellular biology 27*:7381-7393.

Kondo, H., Sahara, H., Miyazaki, A., Nabeta, Y., Hirohashi, Y., Kanaseki, T., Yamaguchi, A., Yamada, N., Hirayama, K., Suzuki, M., *et al.* (2002). Natural antigenic peptides from squamous cell carcinoma recognized by autologous HLA-DR8-restricted CD4+ T cells. *Jpn J Cancer Res 93*:917-924.

Kondoh, H., Lleonart, M.E., Gil, J., Wang, J., Degan, P., Peters, G., Martinez, D., Carnero, A., and Beach, D. (2005). Glycolytic enzymes can modulate cellular life span. *Cancer research 65*:177-185.

Koukourakis, M.I., Giatromanolaki, A., Simopoulos, C., Polychronidis, A., and Sivridis, E. (2005). Lactate dehydrogenase 5 (LDH5) relates to up-regulated hypoxia inducible factor pathway and metastasis in colorectal cancer. *Clin Exp Metastasis 22*:25-30.

Koukourakis, M.I., Giatromanolaki, A., Sivridis, E., Bougioukas, G., Didilis, V., Gatter, K.C., and Harris, A.L. (2003). Lactate dehydrogenase-5 (LDH-5) overexpression in non-small-cell lung cancer tissues is linked to tumour hypoxia, angiogenic factor production and poor prognosis. *British journal of cancer 89*: 877-885.

Koukourakis, M.I., Giatromanolaki, A., Winter, S., Leek, R., Sivridis, E., and Harris, A.L. (2009). Lactate dehydrogenase 5 expression in squamous cell head and neck cancer relates to prognosis following radical or postoperative radiotherapy. *Oncology 77*:285-292.

Le, A., Cooper, C.R., Gouw, A.M., Dinavahi, R., Maitra, A., Deck, L.M., Royer, R.E., Vander Jagt, D.L., Semenza, G.L., and Dang, C.V. (2010). Inhibition of lactate dehydrogenase A induces oxidative stress and inhibits tumor progression. *Proceedings of the National Academy of Sciences of the United States of America 107*:2037-2042.

Leiblich, A., Cross, S.S., Catto, J.W., Phillips, J.T., Leung, H.Y., Hamdy, F.C., and Rehman, I. (2006). Lactate dehydrogenase-B is silenced by promoter hypermethylation in human prostate cancer. *Oncogene 25*:2953-2960.

Levine, A.J., and Puzio-Kuter, A.M (2010). The control of the metabolic switch in cancers by oncogenes and tumor suppressor genes. *Science 330*:1340-1344.

Libby, G., Donnelly, L.A., Donnan, P.T., Alessi, D.R., Morris, A.D., and Evans, J.M. (2009). New users of metformin are at low risk of incident cancer: a cohort study among people with type 2 diabetes. *Diabetes Care 32*:1620-1625.

Luo, W., Hu, H., Chang, R., Zhong, J., Knabel, M., O'Meally, R., Cole, R.N., Pandey, A., and Semenza, G.L. (2011). Pyruvate kinase M2 is a PHD3-stimulated coactivator for hypoxia-inducible factor 1. *Cell 145*:732-744.

Markert, C.L., Shaklee, J.B., and Whitt, G.S. (1975). Evolution of a gene. Multiple genes for LDH isozymes provide a model of the evolution of gene structure, function and regulation. *Science 189*:102-114.

Marsin, A.S., Bouzin, C., Bertrand, L., and Hue, L. (2002). The stimulation of glycolysis by hypoxia in activated monocytes is mediated by AMP-activated protein kinase and inducible 6-phosphofructo-2-kinase. *The Journal of biological chemistry 277*:30778-30783.

Maschek, G., Savaraj, N., Priebe, W., Braunschweiger, P., Hamilton, K., Tidmarsh, G.F., De Young, L.R., and Lampidis, T.J. (2004). 2-deoxy-D-glucose increases the efficacy of adriamycin and paclitaxel in human osteosarcoma and non-small cell lung cancers in vivo. *Cancer research 64*:31-34.

Mathupala, S.P., Ko, Y.H., and Pedersen, P.L. (2009). Hexokinase-2 bound to mitochondria: cancer's stygian link to the "Warburg Effect" and a pivotal target for effective therapy. *Seminars in cancer biology 19*:17-24.

Mazurek, S., Boschek, C.B., Hugo, F., and Eigenbrodt, E. (2005). Pyruvate kinase type M2 and its role in tumor growth and spreading. Seminars in cancer biology *15*, 300-308.

McCarty, M.F., and Whitaker, J. Manipulating tumor acidification as a cancer treatment strategy. *Altern Med Rev 15*:264-272.

McKnight, S. (2003). Gene switching by metabolic enzymes--how did you get on the invitation list? *Cell 114*:150-152.

Medina, R.A., and Owen, G.I. (2002). Glucose transporters: expression, regulation and cancer. *Biological research 35*: 9-26.

Meldolesi, M.F., Macchia, V., and Laccetti, P. (1976). Differences in phosphofructokinase regulation in normal and tumor rat thyroid cells. *The Journal of biological chemistry 251*: 6244-6251.

Minchenko, A., Leshchinsky, I., Opentanova, I., Sang, N., Srinivas, V., Armstead, V., and Caro, J. (2002). Hypoxia-inducible factor-1-mediated expression of the 6-phosphofructo-2-kinase/fructose-2,6-bisphosphatase-3 (PFKFB3) gene. Its possible role in the Warburg effect. *The Journal of biological chemistry 277*: 6183-6187.

Minchenko, O.H., Ochiai, A., Opentanova, I.L., Ogura, T., Minchenko, D.O., Caro, J., Komisarenko, S.V., and Esumi, H. (2005). Overexpression of 6-phosphofructo-2-kinase/fructose-2,6-bisphosphatase-4 in the human breast and colon malignant tumors. *Biochimie 87*: 1005-1010.

Nakashima, R.A., Mangan, P.S., Colombini, M., and Pedersen, P.L. (1986). Hexokinase receptor complex in hepatoma mitochondria: evidence from N,N'-

dicyclohexylcarbodiimide-labeling studies for the involvement of the pore-forming protein VDAC. *Biochemistry 25*: 1015-1021.

Noguchi, Y., Yoshikawa, T., Marat, D., Doi, C., Makino, T., Fukuzawa, K., Tsuburaya, A., Satoh, S., Ito, T., and Mitsuse, S. (1998). Insulin resistance in cancer patients is associated with enhanced tumor necrosis factor-alpha expression in skeletal muscle. *Biochemical and biophysical research communications 253*:887-892.

Osthus, R.C., Shim, H., Kim, S., Li, Q., Reddy, R., Mukherjee, M., Xu, Y., Wonsey, D., Lee, L.A., and Dang, C.V. (2000). Deregulation of glucose transporter 1 and glycolytic gene expression by c-Myc. *The Journal of biological chemistry 275*:21797-21800.

Oudard, S., Carpentier, A., Banu, E., Fauchon, F., Celerier, D., Poupon, M.F., Dutrillaux, B., Andrieu, J.M., and Delattre, J.Y. (2003). Phase II study of lonidamine and diazepam in the treatment of recurrent glioblastoma multiforme. *J Neurooncol 63*:81-86.

Owa, T., Yokoi, A., Yamazaki, K., Yoshimatsu, K., Yamori, T., and Nagasu, T. (2002). Array-based structure and gene expression relationship study of antitumor sulfonamides including N-[2-[(4-hydroxyphenyl)amino]-3-pyridinyl]-4-methoxy-benzenesulfonamide and N-(3-chloro-7-indolyl)-1,4-benzenedisulfo-namide. *J Med Chem 45*:4913-4922.

Pastorino, J.G., Shulga, N., and Hoek, J.B. (2002). Mitochondrial binding of hexokinase II inhibits Bax-induced cytochrome c release and apoptosis. *The Journal of biological chemistry 277*:7610-7618.

Perez-Sayans, M., Somoza-Martin, J.M., Barros-Angueira, F., Rey, J.M., and Garcia-Garcia, A. (2009). V-ATPase inhibitors and implication in cancer treatment. *Cancer Treat Rev 35*:707-713.

Porporato, P.E., Dhup, S., Dadhich, R.K., Copetti, T., and Sonveaux, P. (2011). Anticancer targets in the glycolytic metabolism of tumors: a comprehensive review. *Front Pharmacol 2*:49.

Raghunand, N., He, X., van Sluis, R., Mahoney, B., Baggett, B., Taylor, C.W., Paine-Murrieta, G., Roe, D., Bhujwalla, Z.M., and Gillies, R.J. (1999). Enhancement of chemotherapy by manipulation of tumour pH. *British journal of cancer 80*:1005-1011.

Ralser, M., Wamelink, M.M., Struys, E.A., Joppich, C., Krobitsch, S., Jakobs, C., and Lehrach, H. (2008). A catabolic block does not sufficiently explain how 2-deoxy-D-glucose inhibits cell growth. *Proceedings of the National Academy of Sciences of the United States of America 105*:17807-17811.

Rathmell, J.C., Fox, C.J., Plas, D.R., Hammerman, P.S., Cinalli, R.M., and Thompson, C.B. (2003). Akt-directed glucose metabolism can prevent Bax conformation change and promote growth factor-independent survival. *Molecular and cellular biology 23*:7315-7328.

Reichert, J.M. (2011). Antibody-based therapeutics to watch in 2011. MAbs *3*, 76-99.

Reinacher, M., and Eigenbrodt, E. (1981). Immunohistological demonstration of the same type of pyruvate kinase isoenzyme (M2-Pk) in tumors of chicken and rat. *Virchows Arch B Cell Pathol Incl Mol Pathol 37*:79-88.

Rempel, A., Mathupala, S.P., Griffin, C.A., Hawkins, A.L., and Pedersen, P.L. (1996). Glucose catabolism in cancer cells: amplification of the gene encoding type II hexokinase. *Cancer research 56*:2468-2471.

Robert, J., Van Rymenant, M., and Lagae, F. (1961). Enzymes in cancer. III. Triosephosphate isomerase activity of human blood serum in normal individuals and in individuals with various pathological conditions. *Cancer 14*:1166-1174.

Robey, I.F., Baggett, B.K., Kirkpatrick, N.D., Roe, D.J., Dosescu, J., Sloane, B.F., Hashim, A.I., Morse, D.L., Raghunand, N., Gatenby, R.A., *et al.* (2009). Bicarbonate increases tumor pH and inhibits spontaneous metastases. *Cancer research 69*:2260-2268.

Rofstad, E.K., Mathiesen, B., Kindem, K., and Galappathi, K. (2006). Acidic extracellular pH promotes experimental metastasis of human melanoma cells in athymic nude mice. *Cancer research 66*:6699-6707.

Sedoris, K.C., Thomas, S.D., and Miller, D.M. (2010). Hypoxia induces differential translation of enolase/MBP-1. *BMC cancer 10*:157.

Semenza, G.L. (2003). Targeting HIF-1 for cancer therapy. Nature reviews *3*, 721-732.

Shaw, R.J., and Cantley, L.C. (2006). Ras, PI(3)K and mTOR signalling controls tumour cell growth. *Nature 441*:424-430.

Shim, H., Dolde, C., Lewis, B.C., Wu, C.S., Dang, G., Jungmann, R.A., Dalla-Favera, R., and Dang, C.V. (1997). c-Myc transactivation of LDH-A: implications for tumor metabolism and growth. *Proceedings of the National Academy of Sciences of the United States of America 94*:6658-6663.

Smith, T.A. (1999). Facilitative glucose transporter expression in human cancer tissue. *British journal of biomedical science 56*:285-292.

Song, H., Xia, S.L., Liao, C., Li, Y.L., Wang, Y.F., Li, T.P., and Zhao, M.J. (2004). Genes encoding Pir51, Beclin 1, RbAp48 and aldolase b are up or down-regulated in human primary hepatocellular carcinoma. *World J Gastroenterol 10*:509-513.

Stein, M., Lin, H., Jeyamohan, C., Dvorzhinski, D., Gounder, M., Bray, K., Eddy, S., Goodin, S., White, E., and Dipaola, R.S. (2010). Targeting tumor metabolism with 2-deoxyglucose in patients with castrate-resistant prostate cancer and advanced malignancies. *Prostate 70*:1388-1394.

Supuran, C.T. (2008). Development of small molecule carbonic anhydrase IX inhibitors. *BJU Int 101 Suppl* 4:39-40.

Talbot, D.C., von Pawel, J., Cattell, E., Yule, S.M., Johnston, C., Zandvliet, A.S., Huitema, A.D., Norbury, C.J., Ellis, P., Bosquee, L., *et al.* (2007). A randomized phase II pharmacokinetic and pharmacodynamic study of indisulam as second-line therapy in patients with advanced non-small cell lung cancer. *Clin Cancer Res 13*:1816-1822.

Tamesa, M.S., Kuramitsu, Y., Fujimoto, M., Maeda, N., Nagashima, Y., Tanaka, T., Yamamoto, S., Oka, M., and Nakamura, K. (2009). Detection of autoantibodies against cyclophilin A and triosephosphate isomerase in sera from breast cancer patients by proteomic analysis. *Electrophoresis 30*:2168-2181.

Tennant, D.A., Duran, R.V., and Gottlieb, E. (2010). Targeting metabolic transformation for cancer therapy. *Nature reviews 10*:267-277.

Thangaraju, M., Carswell, K.N., Prasad, P.D., and Ganapathy, V. (2009). Colon cancer cells maintain low levels of pyruvate to avoid cell death caused by inhibition of HDAC1/HDAC3. *Biochem J 417*:379-389.

Tomaino, B., Cappello, P., Capello, M., Fredolini, C., Sperduti, I., Migliorini, P., Salacone, P., Novarino, A., Giacobino, A., Ciuffreda, L., *et al.* (2011). Circulating autoantibodies to phosphorylated alpha-enolase are a hallmark of pancreatic cancer. *J Proteome Res 10*:105-112.

Tu, B.P., Kudlicki, A., Rowicka, M., and McKnight, S.L. (2005). Logic of the yeast metabolic cycle: temporal compartmentalization of cellular processes. *Science 310*:1152-1158.

Vander Heiden, M.G., Cantley, L.C., and Thompson, C.B. (2009). Understanding the Warburg effect: the metabolic requirements of cell proliferation. *Science 324*:1029-1033.

Vander Heiden, M.G., Christofk, H.R., Schuman, E., Subtelny, A.O., Sharfi, H., Harlow, E.E., Xian, J., and Cantley, L.C. (2010a). Identification of small molecule inhibitors of pyruvate kinase M2. *Biochem Pharmacol 79*:1118-1124.

Vander Heiden, M.G., Locasale, J.W., Swanson, K.D., Sharfi, H., Heffron, G.J., Amador-Noguez, D., Christofk, H.R., Wagner, G., Rabinowitz, J.D., Asara, J.M., *et al.* (2010b). Evidence for an alternative glycolytic pathway in rapidly proliferating cells. *Science 329*:1492-1499.

Warburg, O. (1956). On the origin of cancer cells. *Science 123*:309-314.

Warburg, O., Wind, F., and Negelein, E. (1927). The Metabolism of Tumors in the Body. *J Gen Physiol 8*:519-530.

Wolf, A., Agnihotri, S., Micallef, J., Mukherjee, J., Sabha, N., Cairns, R., Hawkins, C., and Guha, A. Hexokinase 2 is a key mediator of aerobic glycolysis and promotes tumor growth in human glioblastoma multiforme. *The Journal of experimental medicine 208*:313-326.

Yalcin, A., Telang, S., Clem, B., and Chesney, J. (2009). Regulation of glucose metabolism by 6-phosphofructo-2-kinase/fructose-2,6-bisphosphatases in cancer. *Experimental and molecular pathology 86*:174-179.

Yamada, K., and Noguchi, T. (1999). Regulation of pyruvate kinase M gene expression. Biochemical and biophysical research communications *256*:257-262.

Yu, F.X., Dai, R.P., Goh, S.R., Zheng, L., and Luo, Y. (2009). Logic of a mammalian metabolic cycle: an oscillated NAD+/NADH redox signaling regulates coordinated histone expression and S-phase progression. *Cell cycle 8*: 773-779.

Yu, Y., Deck, J.A., Hunsaker, L.A., Deck, L.M., Royer, R.E., Goldberg, E., and Vander Jagt, D.L. (2001). Selective active site inhibitors of human lactate dehydrogenases A4, B4, and C4. *Biochem Pharmacol 62*:81-89.

Yun, H., Lee, M., Kim, S.S., and Ha, J. (2005). Glucose deprivation increases mRNA stability of vascular endothelial growth factor through activation of AMP-activated protein kinase in DU145 prostate carcinoma. *The Journal of biological chemistry 280*:9963-9972.

Yun, J., Rago, C., Cheong, I., Pagliarini, R., Angenendt, P., Rajagopalan, H., Schmidt, K., Willson, J.K., Markowitz, S., Zhou, S., *et al.* (2009). Glucose deprivation contributes to the development of KRAS pathway mutations in tumor cells. *Science 325*:1555-1559.

Zheng, L., Roeder, R.G., and Luo, Y. (2003). S phase activation of the histone H2B promoter by OCA-S, a coactivator complex that contains GAPDH as a key component. *Cell 114*:255-266.

Zhou, M., Zhao, Y., Ding, Y., Liu, H., Liu, Z., Fodstad, O., Riker, A.I., Kamarajugadda, S., Lu, J., Owen, L.B., *et al.* Warburg effect in chemosensitivity: targeting lactate dehydrogenase-A re-sensitizes taxol-resistant cancer cells to taxol. *Molecular cancer 9*: 33.

Zhou, W., Capello, M., Fredolini, C., Piemonti, L., Liotta, L.A., Novelli, F., and Petricoin, E.F. (2010). Mass spectrometry analysis of the post-translational modifications of alpha-enolase from pancreatic ductal adenocarcinoma cells. *J Proteome Res 9*:2929-2936.

Zundel, W., Schindler, C., Haas-Kogan, D., Koong, A., Kaper, F., Chen, E., Gottschalk, A.R., Ryan, H.E., Johnson, R.S., Jefferson, A.B., *et al.* (2000). Loss of PTEN facilitates HIF-1-mediated gene expression. *Genes & development 14*:391-396.

4

Distinct Role for ARNT/HIF-1β in Pancreatic Beta-Cell Function, Insulin Secretion and Type 2 Diabetes

Renjitha Pillai and Jamie W. Joseph
School of Pharmacy, University of Waterloo, Waterloo
Canada

1. Introduction

Diabetes mellitus is a common metabolic syndrome that has become an epidemic in modern society and is characterized by either a near-complete lack of insulin production due to autoimmune destruction of pancreatic beta-cells as in type 1 diabetes or abnormal insulin secretion, beta-cell dysfunction and insulin resistance as in type 2 diabetes (T2D). T2D is a complex heterogeneous disease that is characterized by elevated fasting and postprandial blood glucose levels that can result in severe complications including renal failure, cardiovascular disease, blindness and slow wound healing (Lin and Sun, 2010). Pancreatic islet beta-cells play a critical role in maintaining blood glucose levels by secreting the hormone insulin following a meal. Insulin maintains blood glucose levels in the normal physiological range by promoting glucose uptake in muscles, liver and adipose tissue, and by inhibiting hepatic glucose production. Therefore, any defect in insulin secretion in response to a meal or defects in insulin action in peripheral tissues can lead to increased blood glucose levels (Tripathy and Chavez, 2010; Muoio & Newgard, 2008).

Abnormal insulin secretion is a hallmark of T2D. Despite the central role of insulin in maintaining glucose homeostasis, the fundamental biochemical mechanism regulating nutrient-stimulated insulin secretion from pancreatic beta-cells is still incompletely understood. Insulin secretion from the pancreatic beta-cells is regulated by nutrients, neurotransmitters and hormones. Among these three factors, nutrients, particularly glucose is the most dominant stimulatory signal for insulin secretion. Insulin secretion is biphasic with a first acute phase occurring within 10 minutes after a glucose load and a second more sustained phase that reaches a plateau very quickly as seen in mice or more gradually as seen in rats and humans (Gerich, 2002). Numerous models have been proposed over the last several decades to explain the mechanism governing glucose-stimulated insulin secretion (GSIS) from pancreatic beta-cells. The current model of GSIS holds that glucose enters beta-cells via the low affinity, high capacity glucose transporter 2 (GLUT2) and becomes phosphorylated by glucokinase (GK or hexokinase IV), which is the rate-limiting step in glycolysis. The glycolytic end product pyruvate then enters the tricarboxylic acid cycle (TCA), where oxidative phosphorylation occurs, leading to increased ATP production. The subsequent rise in the cytosolic ATP/ADP levels promotes closure of ATP-sensitive potassium channels (K_{ATP} channels) causing beta-cell membrane depolarization and activation of voltage-dependent Ca^{2+} channels (VDCC). The opening of VDCCs facilitates influx of extracellular Ca^{2+}, leading to a rise in the beta-cell cytosolic Ca^{2+} levels, which triggers exocytosis of the insulin-containing secretory granules (Figure 1) (Jensen et al., 2008; Prentki & Matchinsky, 1987; Ashcroft & Rorsman, 1989; Newgard & Matchinsky, 2001; Newgard & McGarry, 1995). This so-called "K_{ATP} channel-dependent" mechanism appears to be particularly important for the first, acute phase of insulin release. However, in the second and more sustained phase of insulin secretion a "K_{ATP} channel-independent" pathway also appears to play a key role in the regulation of GSIS in conjunction with the K_{ATP} channel-dependent pathway (Henquin et al., 2003, Ravier et al., 2009). Important support for "K_{ATP} channel-independent" pathway of GSIS comes from studies showing that glucose still causes a significant increase in insulin secretion in conditions where K_{ATP} channels are held open by application of diazoxide followed by membrane depolarization with high K^+, or in animals lacking functional K_{ATP} channels (Nenquin et al., 2004; Shiota et al., 2002; Szollozi et al., 2007; Ravier et al., 2009). These and more recent studies suggest that mitochondrial metabolism of glucose generates signals other than changes in the ATP/ADP ratio that are important for normal insulin secretion. Several molecules, including glutamate, malonyl-CoA/LC-CoA and NADPH, have been proposed as candidate coupling factors in GSIS (Maechler & Wollheim, 1999; Ivarsson et al., 2005; Corkey et al., 1989; Prentki et al., 1992).

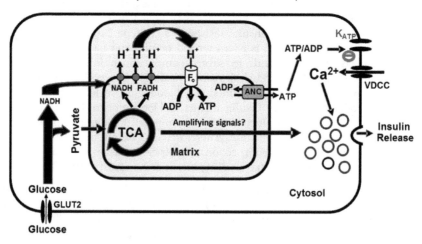

Fig. 1. Current model of glucose stimulated insulin secretion (GSIS) from pancreatic beta-cells. Glucose equilibrates across the plasma membrane through glucose transporter GLUT2, which initiates glycolysis. Pyruvate produced by glycolysis preferentially enters the mitochondria and is metabolized in the TCA cycle, producing reducing equivalents in the form of NADH and $FADH_2$. The transfer of electrons from these reducing equivalents through the mitochondrial electron transport chain is coupled with the pumping of protons from the mitochondrial matrix to the inter membrane space, leading to the generation of ATP. ATP is transferred to the cytosol through adenine nucleotide carrier (ANC), raising the ATP/ADP ratio. This results in the closure of the ATP sensitive K^+ channels (K_{ATP}), which in turn leads to membrane depolarization, opening of the voltage-sensitive Ca^{2+} channels, promoting calcium entry and increase in cytoplasmic Ca^{2+} leading to exocytosis of insulin granules. Glucose also generates amplifying signals other than ATP, which plays a significant role in the secretion of insulin from pancreatic beta-cells.

Maintenance of a functional mature beta-cell phenotype requires optimal expression of key transcription factors. Transcription factors regulate a variety of pancreatic beta-cell processes including cell differentiation, proliferation, cell signaling and apoptosis. By regulating the expression of specific sets of genes, transcription factors determine the spatio-temporal specificity of gene expression in most organisms, including mammals. Numerous studies have shown that transcription factors act synergistically to achieve normal beta-cell development and function (Cerf, 2006; Mitchell & Frayling, 2002; Lyttle et al., 2008). Development of the endocrine pancreas is initiated from multipotent precursor cells, which differentiate to form five different cell types in the pancreatic islet namely the α-cells (glucagon), β-cells (insulin), δ-cells (somatostatin), PP (pancreatic polypeptide) cells and ε-cells (ghrelin) (Steiner et al., 2010). The development of the islet architecture is regulated by an ordered system of transcriptional events activated by a hierarchy of transcription factors. Some of the major transcription factors represented in islets include several homeodomain factors like pancreatic and duodenal homeobox-1 (Pdx-1), paired box gene (Pax) Pax 4, Pax 6, Nkx 2.1 and Nkx 6.1 which are expressed in both progenitor as well as differentiated beta-cells. Pdx-1 and Nkx 2.2 are required for both early beta-cell differentiation and maintenance of a mature beta-cell phenotype (Habner et al., 2005). In addition, other transcription factors are important for maintenance of a mature beta-cell phenotype and their impairment may account for various pathophysiological abnormalities observed in type 2 diabetics. Among these, Pdx-1, neurogenin differentiation (NeuroD/BETA-2), foxhead box protein (FoxO-1), sterol regulatory element binding protein (SREBP-1c), and musculoaponeurotic fibrosarcoma oncogene homolog A (MafA) are the most studied (Johnson et al., 1994; Diraison et al., 2004; Kitamura et al., 2005).

In the context of T2D, it is a well-known fact that abnormal gene expression contributes to a myriad of beta-cell abnormalities. Support for this comes from studies of maturity-onset diabetes of the young (MODY), a monogenic form of T2D characterized by an early onset and defects in insulin secretion leading to hyperglycemia. With the exception of MODY-2, which is caused by a mutation in GK, MODY-1, 3, 4, 5 and 6 result from mutations in genes encoding transcription factors, hepatocyte nuclear factor (HNF) HNF-4α, HNF-1α, HNF-1β, Pdx-1, and NeuroD/BETA-2 respectively. These transcription factors regulate the expression of key genes involved in various aspects of beta-cell function (Stoffer & Zinkin, 1997; Habener et al., 1998; Fajans et al., 2001; Yamagata et al, 2003). Although there have been significant advancements in understanding the basic transcriptional network that exists in beta-cells, the exact mechanism of action of many of these factors still remains to be further defined. In this chapter we provide an overview of one of the recently described transcription factor in the context of impaired insulin secretion and beta-cell dysfunction, Aryl hydrocarbon receptor nuclear translocator (ARNT)/ hypoxia inducible factor 1β (HIF-1β), which is a master regulator of pancreatic beta-cell transcriptional network that regulates glucose metabolism and insulin secretion.

2. ARNT/HIF-1β

2.1 ARNT/HIF-1β structure and function

ARNT/HIF-1β belongs to a group of transcription factors, known as the basic helix loop helix - PER/ARNT/Sim (bHLH-PAS) family, which has a characteristic N-terminal bHLH motif for DNA binding, a central PAS domain which facilitates heterodimerization and a C-terminal transactivation domain for the recruitment of transcriptional coactivators such as CBP/p300 (Jain et al., 1994; Kobayashi et al., 1997). Recent evidence suggest that the PAS domain may also provide an additional binding site for coactivators and thereby recruiting them in a step necessary for transcriptional responses to hypoxia (Partch & Gardener, 2011). ARNT/HIF-1β acts as a common binding partner for most of the bHLH-PAS family of transcription factors and bind specific DNA sequences in the regulatory regions of the responsive genes. The half-site for ARNT/HIF-1β is on the 3' side of the 5'-GTG-3' recognition sequence. The sequence of the other half of the binding site depends upon the identity of the ARNT/HIF-1β dimerization partner (Swanson et al., 1995). DNA binding of ARNT/HIF-1β is mediated by its bHLH region and may also involve the PAS region. Dimerization between ARNT/HIF-1β and other bHLH-PAS proteins is mediated by their bHLH and PAS regions (Jiang et al., 1996, Lindebro et al., 1995). The human ARNT/HIF-1β gene is about 65 Kb in size, has 22 exons and is well conserved on an evolutionary scale (Scheel & Schrenk, 2000).

ARNT/HIF-1β was originally cloned as a factor required for the activity of the aryl hydrocarbon receptor (AhR). AhR induces a transcriptional response to various environmental pollutants, such as polycyclic aromatic hydrocarbons, heterocyclic amines, and polychlorinated aromatic compounds (Reyes et al., 1992). ARNT/HIF-1β was also identified as the β-subunit of a heterodimeric transcription factor, hypoxia-inducible factor 1α (HIF-1α) (Wang et al., 1995 (a)). Similar to HIF-1α, ARNT/HIF-1β gene expression and protein levels are significantly increased under hypoxic conditions suggesting that this gene plays an important role in the transcriptional response to low oxygen tension (Wang et al., 1995 (b)). Consistent with this idea, it has been shown that ARNT/HIF-1β is essential for the hypoxic induction of vascular endothelial growth factor (VEGF) and the glycolytic enzymes aldolase A (ALDO) and phosphoglycerate kinase (PGK) in a mouse hepatoma (Hepa 1c1c7) cell line (Li et al., 1996; Salceda et al., 1996). Unlike HIF-1α, which is exclusively expressed under hypoxic conditions, ARNT/HIF-1β is constitutively expressed in a number of tissues, such as the brain, heart, kidney, muscles, thymus, retina, olfactory epithelium and beta-cells of pancreas (Hirose et al., 1996).

2.2 ARNT/HIF-1β localization, binding partners, mechanism of action and lessons from knockout animals

ARNT/HIF-1β is a nuclear protein in most cell types, although it may also be located in the cytosol, particularly during embryogenesis. Studies conducted by Holmes and Pollenz (1997) in hepatic and non-hepatic cell lines derived from rat, mouse, human, and canine tissues confirm ARNT/HIF-1β as a nuclear transcription factor and showed that its physical interaction with DNA requires entry into the nucleus.

ARNT/HIF-1β serves as an obligatory binding partner for a number of other bHLH-PAS proteins, whose activity is modulated either by exogenous chemicals (AhR), hypoxia (HIF-1α, HIF-2α and HIF-3α), or which show tissue-specific expression pattern (e.g. SIM-1) (Salceda et al., 1996; Swanson et., 1995; Woods & Whitelaw, 2002). In addition to forming heterodimers, ARNT/HIF-1β appears to be capable of forming homodimers and bind to an E-box sequence 5'-CACGTG-3' (Antonsson et al., 1995). It was also shown that ARNT/HIF-1β homodimer regulates the transcription of murine cytochrome P450 (Cyp) 2a5 gene through a palindromic E-box element in the 5' regulatory region of Cyp2a5 gene in primary hepatocytes (Arpiainen et al., 2007). Two ARNT-related genes, ARNT-2 and ARNT-3 (also called BMAL-1 or MOP3) have been identified. ARNT-2 is more restricted in expression than ARNT/HIF-1β, but appears to dimerize with the same partner proteins as ARNT/HIF-1β (Hirose et al., 1996). ARNT-3 appears to have different dimerization potential than ARNT/HIF-1β (Ikeda & Nomura, 1997). The transactivation potential of ARNT/HIF-1β is not only determined through the recruitment of transcriptional cofactors, but also by signaling input from several protein kinases, such as PKC (Long et al., 1999).

The ARNT/HIF-1β/AhR heterodimer activates transcription of several genes involved in metabolism of foreign chemicals, including CYP1A1, CYP1B1, and NADP(H):oxidoreductase (NQO1) (Sogawa & Kuriayama, 1997; Beischlag et al., 2008). Transcriptional activation of these genes depends upon prior binding of AhR to xenobiotic ligands, including 2,3,7,8-tetrachlorodibenzo-p-dioxin (dioxin) and benzopyrene. The ARNT/HIF-1β/AhR heterodimer and ARNT/HIF-1β can have an impact on estrogen receptor (ER) activity. ARNT/HIF-1β interacts and functions as a potent coactivator of both ER-α and ER-β dependent transcription and it is believed that the C-terminal domain of ARNT/HIF-1β is essential for the transcriptional enhancement of ER activity (Lim et al., 2011;Rüegg et al., 2008).

ARNT/HIF-1β/HIF-1α heterodimer activity is primarily regulated by HIF-1α protein stability. Under normoxia, HIF-1α is hydroxylated by an oxygen requiring enzyme, prolyl hydroxylase (PHD), which is then targeted for ubiquitination by the E3 ubiquitin ligase, followed by binding to von Hippel-Lindau tumor suppressor (VHL) which leads to degradation of HIF-1α by the proteasome pathway. Conversely, under hypoxic conditions, a lack of oxygen inhibits hydroxylation, leading to stabilization of the HIF-1α protein and translocation of HIF-1α from the cytoplasm to the nucleus. In the nucleus, heterodimerization of HIF-1α with ARNT/HIF-1β is followed by binding to hypoxia response elements (HRE) in the promoter region of the target genes (Fedele et al., 2002) (Figure 2). Like HIF-1α, HIF-2α and 3α are stabilized by hypoxia and hypoglycemia, and activate transcription of genes involved in adapting to these adverse conditions, including the genes for erythropoietin (EPO), VEGF, and a number of enzymes of glycolysis including ALDO, phosphofructokinase (PFK) and lactate dehydrogenase (LDH) (Maltepe et al., 1997; Fraisl et al., 2009; Fedele et al., 2002). These studies suggest ARNT/HIF-1β is a central player in a number of signaling pathways and alterations in its activity can have serious impact on cellular responses to hypoxia, dioxin response and estrogen signaling in mammalian cells.

Observations from the ARNT/HIF-1β conditional knockout mice and whole body knockout mice have provided a wealth of information regarding the functional significance of this transcription factor in mammalian cells. Results obtained from the ARNT/HIF-1β null mice suggest that it plays a central role in embryonic development and physiological homeostasis as these mice are embryonic lethal due to severe defects in angiogenesis and placental development (Maltepe et al., 1997; Kozak et al., 1997). Data obtained from tissue specific ARNT/HIF-1β knockout mice demonstrates that disruption of ARNT/HIF-1β expression in liver and heart results in loss of AhR-stimulated gene transcription and that ARNT/HIF-1β is key to AhR function in these two mammalian tissues. It was also observed that ARNT/HIF-1β affects HIF-1α mediated target gene expression as several key genes including the expression of heme-oxygenase and glucose transporter-1 mRNA was abolished after treatment with $CoCl_2$, an agent that is thought to mimic hypoxia (Tomita et al., 2000).

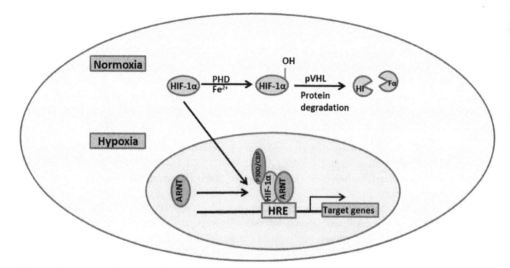

Fig. 2. Overview of gene regulation by ARNT/HIF-1β/HIF-1α complex in mammalian cells under normoxic and hypoxic conditions. In normoxic conditions, HIF-1α protein undergoes oxygen dependent hydroxylation by prolyl hydroxylases (PHD) and the hydroxylation site is recognized by pVHL, which targets the protein for ubiquitination by ubiquitin ligase, followed by degradation through ubiquitin proteasome pathway. During hypoxia, HIF-1α protein is not targeted for degradation and can translocate to the nucleus, where it heterodimerizes with ARNT/HIF-1β to form a stable transcriptional complex. The ARNT/HIF-1β/HIF-1α heterodimer then binds to the hypoxia response element (HRE) of target genes.

3. ARNT/HIF-1β and type 2 diabetes

3.1 ARNT/HIF-1β is reduced in human diabetic islets

In 2005, a study published in *Cell* (Gunton et al., 2005) suggested that ARNT/HIF-1β, a transcription factor with previously unknown functions in beta-cells, plays a significant role in mediating human beta-cell dysfunction in type 2 diabetics. Genome-wide gene expression profiling of islets obtained from human non-diabetics and type 2 diabetics revealed that the expression levels of ARNT/HIF-1β was reduced by 90% under the diabetic conditions. This was associated with reduced expression levels of several ARNT/HIF-1β target genes involved in glycolysis and insulin signaling. Several enzymes in glycolysis, including phosphoglucomutase (PGM), phosphoglucose isomerase (G6PI), PFK and ALDO were expressed at significantly lower levels as compared to those observed in normal islets. The low ARNT/HIF-1β expression levels observed under diabetic conditions was also associated with low gene expression levels of several key regulators in insulin signaling, such as the insulin receptor (IR), insulin receptor substrate 2 (IRS2), and protein kinase B (Akt2). Another interesting observation made in this study was that MODY genes, HNF-1α and HNF-4α, were poorly expressed in human islets obtained from type 2 diabetics. HNF-4α, the gene mutated in MODY1, has been shown to interact with ARNT/HIF-1β possibly providing a connection between the two transcription factors (Tsuchiya et al., 2002).

In order to rule out the possibility that the profound ARNT/HIF-1β down regulation in pancreatic beta-cells is not caused by the diabetic environment, Gunton and co-workers demonstrated that an identical gene profile was observed in a beta-cell-specific ARNT/HIF-1β knockout mouse (β-ARNT KO). β-ARNT KO mice exhibited impaired GSIS and glucose intolerance, with no significant change in beta-cell insulin content and islet mass. The finding that ARNT/HIF-1β knockout mice have normal islet mass suggests that this transcription factor does not play a role in beta-cell differentiation.

Overall, a combination of *in vivo* and *in vitro* studies in humans and rodents have provided us with convincing evidence that reduction in ARNT/HIF-1β expression in human pancreatic beta-cells has negative consequences in terms of beta-cell function and insulin secretion. However, the extent of ARNT/HIF-1β mediated regulation of gene transcription is complex since it has the potential to bind with multiple partners affecting a multitude of signaling pathways.

3.2 ARNT/HIF-1β is reduced in human diabetic hepatic cells

In both rodents and humans, the liver plays a critical role in maintaining glucose and lipid homeostasis. During fasting, hepatic glucose production is critical for providing glucose for the brain, the kidneys and red blood cells. In liver, glucose is produced by glycogenolysis during the initial stages of fasting, however, after several hours of fasting, glucose production is primarily from gluconeogenesis, a process by which the liver produces glucose from precursors such as lactate and pyruvate (Michael et al., 2000; Saltiel & Kahn, 2001). Wang et al showed that ARNT/HIF-1β was severely reduced in the livers of human type 2 diabetics (Wang et al., 2009). Gene expression profiling of liver specimens from normal, obese and obese diabetic patients revealed a 30% reduction in the expression of ARNT/HIF-1β gene in obese diabetic individuals. The study demonstrated that the reduced expression of ARNT/HIF-1β in the livers of humans with T2D was associated with high glucose levels, high insulin levels, and insulin resistance. This study also suggested that insulin, not glucose regulates the expression of ARNT/HIF-1β gene and that ARNT/HIF-1β expression is reduced in both insulin-deficient and insulin-resistant states.

Wang et al (2009) also looked at the effects of liver-specific deletion of ARNT/HIF-1β gene in mice (L-ARNT KO) and demonstrated that there was an increase in gluconeogenesis, lipogenesis and increased serum insulin levels, all characteristic of human type 2 diabetics. The increase in hepatic gluconeogenesis and lipogenesis in L-ARNT KO mice was associated with the upregulation of several important gluconeogenic and lipogenic genes including PEPCK, G6Pase, SCD1 and FAS. Expression of C/EBPα and SREBP-1C, was also induced by 2-folds in L-ARNT KO mice. C/EBPα plays a major role in kick-starting hepatic glucose production at birth, and disruption of the C/EBPα gene in mice is known to cause hypoglycemia associated with the impaired expression of the gluconeogenic enzymes PEPCK and G6Pase (Pedersen et al., 2007; Qiao et al., 2006). SREBP-1C, on the other hand is a major player in lipogenesis (Horton et al., 2002). ARNT/HIF-1β may act as an upstream regulator of these transcription factors and play a key role in maintaining whole body glucose and lipid homeostasis. However, as seen in pancreatic beta-cells, the exact pathways targeted by ARNT/HIF-1β in liver cells are not clearly understood and is complicated by the fact that ARNT/HIF-1β has multiple binding partners.

3.3 ARNT/HIF-1β regulates glucose metabolism and insulin secretion in beta-cells

The central role played by ARNT/HIF-1β/HIF-1α heterodimer in the regulation of glucose homeostasis, particularly glycolysis has been well studied (Dery et al., 2005; Semenza et al., 1994) with a focus in cancer cell metabolism (Song et al., 2009; Semenza et al., 2000; Semenza, 2003). It is widely accepted that ARNT/HIF-1β/HIF-1α heterodimer plays a role in the Warburg effect, where cancer cells undergo a high rate of anaerobic glycolysis compared to normal cells. It has been suggested that the observed increase in glycolytic enzymes in these cancer cells is associated with increased HIF-1 activity, thus aiding in tumor formation and progression. Studies conducted in ARNT/HIF-1β mutant clonal cells indicate that it is an essential component of the HIF-1α complex and that absence of ARNT/HIF-1β leads to reduced cellular responses to stimuli such as hypoxia (Woods et al., 1996).

In pancreatic beta-cells, metabolism of glucose through aerobic glycolysis and oxidative phosphorylation plays a significant role in maintaining a normal secretory capacity. Beta-cells sense glucose and secrete appropriate amounts of insulin to promote glucose uptake by muscles and adipose tissue. Insulin also inhibits hepatic glucose production. Abnormal insulin secretion is one of the earliest detectable defects at the onset of T2D and despite its relevance, the mechanisms underlying GSIS are not completely understood. The generally accepted model of GSIS holds that metabolism of glucose in the beta-cells leads to a rise in the cytosolic ATP/ADP levels, which promotes closure of the K_{ATP} channel, increased cytosolic Ca^{2+} and triggers exocytosis of insulin-containing secretory granules (Henquin et al., 2003; Jensen et al., 2008). In beta-cells glucose derived pyruvate is directed mostly towards TCA for the production of ATP, since both the pentose phosphate pathway and anaerobic glycolysis is relatively inactive (Schuit et al., 1997). This exceptionally high dependence of beta-cells on the TCA cycle suggests that hypoxia or mechanisms reducing the aerobic capacity of beta-cells would probably have profound effects on GSIS.

It has been shown that down regulation of ARNT/HIF-1β in pancreatic beta-cells leads to loss of GSIS (Gunton et al., 2005; Pillai et al 2011). Our group has demonstrated that beta-cells with reduced ARNT/HIF-1β expression levels exhibit a 31% reduction in glycolytic flux without significant changes in glucose oxidation or the ATP/ADP ratio. Metabolomics analysis revealed that clonal beta-cells (832/13) treated with siRNAs against the ARNT/HIF-1β gene have lower levels of glycolytic, TCA cycle and fatty acid intermediates (Figure 3). It was also shown that the reduced levels of glycolysis, TCA and fatty acid intermediates were associated with a corresponding decrease in the expression of key genes in all three metabolic pathways including GLUT2, GK, PC, PDH, MEc, CIC DIC, CPT1a and FAS (Figure 4). The novel finding that reducing ARNT/HIF-1β levels leads to a profound reduction in PC, DIC, and OGC expression levels and a reduction in glycoslysis and TCA metabolites, even though glucose oxidation and ATP production were unaltered, is an unexpected result. These collective changes in metabolite levels suggest that the oxidative entry of pyruvate into the TCA cycle is preserved in the absence of ARNT/HIF-1β at the expense of a loss of anaplerosis. A key role for anaplerosis in insulin secretion is supported by the finding that pyruvate flows into mitochondrial metabolic pathways, in roughly equal proportions, through the anaplerotic (PC) and oxidative (PDH) entry points. Glucose carbon entering through the PC reaction leads to an increase in TCA intermediates (called anaplerosis) (Schuit et al., 1997; Khan et al., 1996). In addition, beta-cells contain enzymes that allow "cycling" of pyruvate via its PC-catalyzed conversion to oxaloacetate (OAA), metabolism of OAA to malate, citrate, or isocitrate in the TCA cycle, and subsequent recycling of these metabolites to pyruvate via several possible

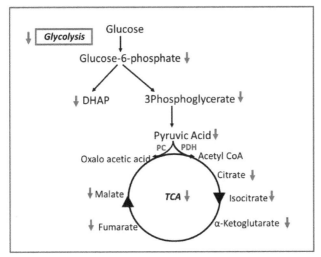

Fig. 3. Summary of the effects of siRNA mediated suppression of ARNT/HIF-1β in beta-cells. Several key metabolites in both glycolysis and TCA cycle were negatively affected by the knockdown of ARNT/HIF-1β. TCA, tricarboxylic acid cycle; DHAP, dihydroxyacetone phosphate; PC, pyruvate carboxylase; PDH, pyruvate dehydrogenase.

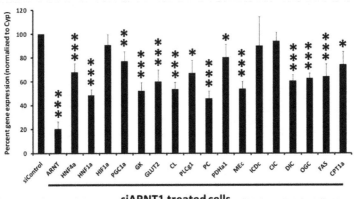

Fig. 4. Effects of siRNA mediated suppression of ARNT/HIF-1β on key genes involved in the metabolic regulation of β-cell function in 832/13 cells. Gene expression is expressed as a percentage of the target gene from siControl treated cells and corrected for by an internal control gene cyclophilin E (Cyp). There was no significant difference seen between treatment groups for Cyp (n=5). HNF4a, hepatocyte nuclear factor-4α; HNF1a, hepatocyte nuclear factor-1α; HIF1a, hypoxia inducible factor 1α; GK, glucokinase; GLUT2, glucose transporter-2; PLGg1, phospholipase γ-1; PC, pyruvate carboxylase; PDHa1, pyruvate dehydrogenase (α1 subunit); MEc, cytosolic malic enzyme; ICDc, cytosolic isocitrate dehydrogenase; CIC, citrate carrier; DIC, dicarboxylate carrier; OGC, α-ketoglutarate carrier; FAS, fatty acid synthase; CPT1a, Carnitine palmitoyl transferase 1α. * P<0.05, ** P<0.01, *** P<0.001 siControl vs siARNT1.

combinations of cytosolic and mitochondrial pathways (MacDonald et al., 1995). Numerous groups have shown that both pyruvate cycling and anaplerosis are important to maintain normal secretory capacity of beta-cells (Lu et al., 2002; Joseph et al., 2006; MacDonald et al., 2005; Ronnebaum et al., 2006). Since the amount of pyruvate is substantially lower in ARNT/HIF-1β depleted beta-cells, our data also suggests that this gene may play an important role in maintaining pyruvate cycling. However a direct link between ARNT/HIF-1β and pyruvate cycling has not yet been established. Figure 5 shows the diagrammatic representation of the transcriptional network regulated by ARNT/HIF-1β and its involvement glucose-stimulated anaplerosis and insulin release.

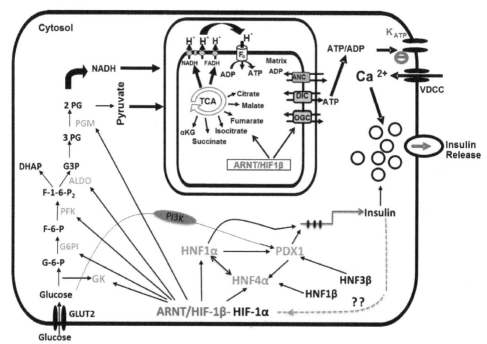

Fig. 5. Schematic of the transcriptional network regulated by ARNT/HIF-1β and its involvement in glucose-stimulated anaplerosis and insulin release from beta-cells. ARNT/HIF-1β regulates key genes in glycolysis and TCA cycle (shown in blue), including key metabolite carriers such as DIC and OGC. MODY genes regulated by ARNT/HIF-1β are shown in green. Interestingly, ARNT/HIF-1β does not seem to play a significant role in the regulation ATP production in beta-cells, however, it seems to be very important for glucose-induced anaplerosis, which provides crucial signals for GSIS.

3.4 ARNT/HIF-1β regulates beta-cell and hepatic transcriptional networks

Studies in ARNT/HIF-1β deficient beta-cells suggest that it plays a crucial role in the regulation of key genes involved in glucose metabolism and insulin secretion (Gunton et al., 2005, Pillai et al., 2011). ARNT/HIF-1β target genes in beta-cells include the MODY1 and MODY3 genes HNF4α and HNF1α, glucose metabolism genes GK, G6PI, PFK, aldolase, PC, PDH, MEc, DIC, OGC and insulin signaling genes IR, IRS2 and AKT2. In non-beta-cells it has been shown that ARNT/HIF-1β is essential for the normal function of HIF-1α, HIF2α, and AhR. These heterodimeric complexes are required for cellular responses to hypoxia (HIF proteins) and environmental toxins (AhR), respectively (Kozak et al., 1997; Kewley et al., 2004). It has been estimated that there are more than 13,000 putative ARNT/HIF-1β binding sites in promoters in the human genome (Gunton et al., 2005). Many of the target gene promoters have multiple potential binding sites. Thus it is reasonable to estimate that a substantial decrease in ARNT/HIF-1β would affect the expression of a large number of genes in humans. Although there is a lack of direct biochemical evidence, many of the genes found to be altered in association with decreased ARNT/HIF-1β gene expression have putative ARNT/HIF-1β-dimer consensus binding sites in their promoters (including HNF4α, HNF1α, Akt2, G6PI, PFK, and aldolase), suggesting a direct role for ARNT/HIF-1β containing dimers in the regulation of their expression (Figure 6).

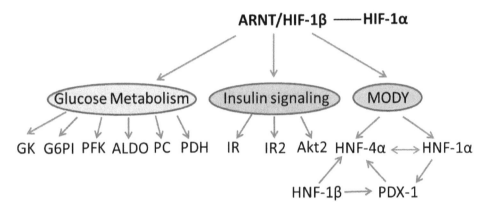

Fig. 6. Transcriptional network in pancreatic beta-cells regulated by ARNT/HIF-1β. ARNT/HIF-1β regulates several key genes involved in glucose metabolism, insulin signaling and MODY. Beta-cell specific knockout of ARNT/HIF-1β in mice leads to reduced expression of a number of important beta-cell genes including HNF-4α, HNF-1α, insulin receptor (IR), insulin receptor substrate-2 (IRS2), protein kinase b (Akt2), glucokinase (GK), glucose-6-phosphoisomerase (G6PI), phosphofructokinase (PFK), aldolase (ALDO), pyruvate carboxylase (PC) and pyruvate dehydrogenase (PDH).

In liver cells ARNT/HIF-1β has been shown to regulate the expression of several genes involved in glucose and lipid homeostasis. Support for the involvement of ARNT/HIF-1β in liver glucose homeostasis was provided by experiments showing that basal and insulin-induced expression of GLUT1, GLUT3, ALDO, PGK and VEGF were significantly reduced in ARNT/HIF-1β-defective HepG2 cells (Salceda et al., 1996). Wang *et al* (2009) demonstrated that a reduction of ARNT/HIF-1β in liver cells was associated with an increase in the expression of several important gluconeogenic and lipogenic genes including PEPCK, G6Pase, SCD1, FXR, C/EBPα, SREBP-1C, FBP-1 and FAS. The discovery that ARNT/HIF-1β may contribute to the regulation of beta-cell and hepatic genes suggests an essential role for this transcription factor in the regulation of glucose and lipid homeostasis (Figure 7).

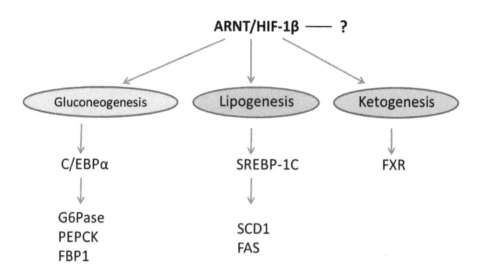

Fig. 7. ARNT/HIF-1β regulates the expression of several key genes involved in gluconeogenesis, lipogenesis and ketogenesis in the liver cells. Liver-specific knockout of ARNT/HIF-1β in mice leads to increased hepatic gluconeogensis and lipogenesis with a corresponding increase in the expression of phosphoenolpyruvate carboxykinase (PEPCK), Glucose-6-phosphatase (G6Pase), Fructose-1,6-biphosphatase (FBP1), Steroyl–CoA-desaturase (SCD1), Fatty acid synthase (FAS), CCAAT enhancer binding protein α (C/EBPα), sterol regulatory element binding protein (SREBP-1C) and Farsenoid X receptor (FXR). Adapted from Wang et al. (2009).

3.5 Regulation of ARNT/HIF-1β in beta-cells

Since ARNT/HIF-1β appears to be a major player in the pathogenesis of T2D, several attempts have been made to identify the upstream regulators of the gene in beta-cells. In 2008, Dror et al., showed that glucose and endoplasmic reticulum Ca^{2+} channels regulate the expression of ARNT/HIF-1β in beta-cells via presinilin. Presinilin is a protein that has been implicated in the cellular response to reduced metabolic activity (Koo and Koppen, 2004). Overexpression of presenilin-1 in clonal Min6 beta-cells increased ARNT/HIF-1β suggesting that ARNT/HIF-1β may be a downstream target of presenilin (Dror et al., 2008). They demonstrated that this pathway is controlled by Ca^{2+} flux through intracellular channels. ARNT/HIF-1β has also recently been shown to be regulated by the carbohydrate-responsive element-binding protein (ChREBP), which is a transcription factor shown to regulate carbohydrate metabolism in the liver and pancreatic beta-cells in response to elevated glucose concentrations (Noordeen et al., 2009). In a genome-wide approach using high-density oligonucleotide arrays, the study showed that ChREBP binds directly to ARNT/HIF-1β promoter in Min6 clonal beta-cells. Accordingly, knockdown of ChREBP using siRNA resulted in an increase ARNT/HIF-1β mRNA levels whereas overexpression of ChREBP resulted in a decrease in ARNT/HIF-1β mRNA levels in rat. They also showed that incubating INS-1 (832/13) cells with glucose led to a substantial decrease in ARNT/HIF-1β mRNA levels.

Interestingly, it has also been shown that HIF-1α, the highly regulated binding partner for ARNT/HIF-1β, may be active under normoxic conditions in mouse and human beta-cells (Cheng et al., 2010). Coimmunoprecipitation studies demonstrated that HIF-1α was bound to ARNT/HIF-1β at the promoter region providing evidence for an interaction between HIF-1α and ARNT/HIF-1β in beta-cells. Treatment of diabetic mice with deferasirox (DFS), an agent that increases HIF-1α protein levels, improved glucose tolerance, normalized the expression of ARNT/HIF-1β and its target genes in human T2D islets. The same study also showed that HIF-2α, but not AhR, is another possible binding partner for ARNT/HIF-1β in pancreatic beta-cells. These studies provide a novel mechanism to regulate ARNT/HIF-1β gene expression in beta-cells.

Three studies published from independent laboratories studying the impact of increasing HIF-1α levels in beta-cells indicate that one has to be extremely cautious when using pharmacological agents, such as DFS, to activate HIF-1α in the islets (Puri et al., 2008; Zehetner et al., 2008; Cantley et al., 2009). These studies used the Cre-loxP system to conditionally delete VHL gene in beta-cells and showed that there were adverse effects associated with an increase in HIF-1α levels on beta-cell function. In all the three studies, increased HIF-1α levels were accompanied by severely impaired GSIS and increased lactate production, indicating a switch from aerobic to anaerobic glycolysis. Thus there appears to be a dose-response curve for the affects HIF-1α protein levels on beta-cell function (Cheng *et al.*, 2010). Although complete lack of HIF-1α seems deleterious to GSIS in mice and Min6 cells, milder increases are beneficial for beta-cell function. As seen in VHL knockout mice, very high levels of HIF-1α are detrimental for normal beta-cell function. Therefore, before we begin to develop a novel treatment regime that enhances HIF-1α or ARNT/HIF-1β activity in human diabetic islets, it is imperative that we understand the expected outcomes of such changes to avoid any detrimental effects.

4. Concluding remarks

It is well known that ARNT/HIF-1β plays a role in the cellular responses to hypoxia, however recent research has demonstrated a broader role for this transcription factor in maintaining glucose and lipid homeostasis in type 2 diabetics. It is now clear that a significant decrease in ARNT/HIF-1β gene expression in both the pancreatic beta-cells and the liver cells is deleterious and can result in T2D. Conversely, targeted disruption of ARNT/HIF-1β gene expression in the adipocytes followed by treatment of mice with a high fat diet improves insulin sensitivity and decreases adiposity (Jiang *et al.*, 2011). A central role for ARNT/HIF-1β in the regulation of key genes involved in glucose sensing, GSIS and insulin signaling in rodents as well in human islets suggest it plays an important role in maintaining normal beta-cell function. Current studies support the idea that ARNT/HIF-1β could act as an upstream regulator of many of the key genes involved in glucose and lipid homeostasis. Clearly, the transcriptional network regulated by ARNT/HIF-1β and genes that are under direct or indirect control of this transcription factor is very broad and hence any change in the regulation of ARNT/HIF-1β may have an impact on many signaling pathways. The fact that ARNT/HIF-1β is a binding partner for several other Per/ARNT/Sim transcription factor family members like HIF-1α, HIF-2α, HIF-3α and AhR makes it a significant member of this family of transcription factors. Improving our understanding of the beta-cell transcription factors, establishing their mechanism of action and hierarchy and finding ways to regulate their expression could prove beneficial in developing novel tools to prevent or correct beta-cell dysfunction in T2D.

5. References

Antonsson C, Arulampalam V, Whitelaw ML, Pettersson S & Poellinger L (1995): Constitutive function of the basic helix-loop-helix/PAS factor Arnt. Regulation of target promoters via the E box motif. *Journal of Biological Chemistry*, Vol. 270, pp 13968-13972.

Arpiainen S, Lämsä V, Pelkonen O, Yim SH, Gonzalez FJ & Hakkola J (2007). Aryl hydrocarbon receptor nuclear translocator and upstream stimulatory factor regulate Cytochrome P450 2a5 transcription through a common E-box site. *Journal of Molecular Biology*, Vol. 369, No. 3, pp 640-652.

Ashcroft FM & Rorsman P (1989). Electrophysiology of the pancreatic beta-cell. *Progress in Biophysics and Molecular Biology*, Vol. 54, pp 87-143.

Beischlag TV, Luis Morales J, Hollingshead BD & Perdew GH (2008). The aryl hydrocarbon receptor complex and the control of gene expression. *Critical Reviews in Eukaryotic Gene Expression*, Vol. 18, No. 3, pp 207-250.

Cantley J, Selman C, Shukla D, Abramov AY, Forstreuter F, Esteban MA, Claret M, Lingard SJ, Clements M, Harten SK, Asare-Anane H, Batterham RL, Herrera PL, Persaud SJ, Duchen MR, Maxwell PH, Withers DJ (2009). Deletion of the von Hippel-Lindau gene in pancreatic beta cells impairs glucose homeostasis in mice. *Journal of Clinical Investigation*, Vol. 119, pp 125–135.

Cerf ME (2006).Transcription factors regulating β-cell function. *European Journal of Endocrinology*, Vol. 155, pp 671-679.

Cheng K, Ho K, Stokes R, Scott C, Lau SM, Hawthorne WJ, O'Connell PJ, Loudovaris T, Kay TW, Kulkarni RN, Okada T, Wang XL, Yim SH, Shah Y, Grey ST, Biankin AV, Kench JG, Laybutt DR, Gonzalez FJ, Kahn CR & Gunton JE (2010). Hypoxia-inducible factor-1alpha regulates beta-cell function in mouse and human islets. *Journal of clinical investigation*, Vol. 120, No.6, pp 2171-2183.

Corkey BE, Glennon MC, Chen KS, Deeney JT, Matschinsky FM & Prentki M (1989). A role for malonyl-CoA in glucose-stimulated insulin secretion from clonal pancreatic beta-cells. *Journal of Biological Chemistry*, Vol. 264, pp 21608-21612.

Déry MA, Michaud MD & Richard DE (2005). Hypoxia-inducible factor 1: regulation by hypoxic and non-hypoxic activators. *International Journal of Biochemistry & Cell Biology*, Vol. 37, No. 3, pp 535-540.

Diraison F, Parton L, Ferré P, Foufelle F, Briscoe CP, Leclerc L & Rutter GA (2004). Over-expression of sterol-regulatory-element-binding protein-1c (SREBP1c) in rat pancreatic islets induces lipogenesis and decreases glucose-stimulated insulin release: modulation by 5-aminoimidazole-4-carboxamide ribonucleoside (AICAR), *Biochemical Journal*, Vol. 378, pp 769–778.

Dror V, Kalynyak TB, Bychkivska Y, Frey MH, Tee M, Jeffrey KD, Nguyen V, Luciani DS & Johnson JD (2008). Glucose and ER-calcium channels regulate HIF-1 via presenilin in pancreatic β–cells. *Journal of Biological Chemistry*, Vol. 283, No. 15, pp 9909–9916.

Fajans SS, Bell GI & Polonsky KS (2001). Molecular mechanisms and clinical pathophysiology of maturity-onset diabetes of the young. *New England Journal of Medicine*, Vol. 345, pp 971–980.

Fedele AO, Whitelaw ML & Peet DJ (2002). Regulation of gene expression by the hypoxia-inducible factors. *Molecular Interventions*, Vol. 2, pp 229–243.

Fraisl P, Mazzone M, Schmidt T & Carmeliet P (2009). Regulation of angiogenesis by oxygen and metabolism. *Developmental Cell*, Vol. 16, No. 2, pp 167-179.

Gerich JE (2002) Is reduced first-phase insulin release the earliest detectable abnormality in individuals destined to develop type 2 diabetes? *Diabetes*, Vol. 51, No.1, pp S117-S121.

Gunton JE, Kulkarni RN, Yim S, Okada T, Hawthorne WJ, Tseng YH, Roberson RS, Ricordi C, O'Connell PJ, Gonzalez FJ & Kahn CR (2005). Loss of ARNT/HIF-1beta mediates altered gene expression and pancreatic-islet dysfunction in human type 2 diabetes. *Cell*, Vol.122, pp 337-349.

Habener JF, Stoffers DA, Stanojevic & Clarke VW (1998). A newly discovered role of transcription factors involved in pancreas development and the pathogenesis of diabetes mellitus. *Proceedings of the associations of American physicians*, Vol. 110, pp 12–21.

Habener JF, Kemp DM & Thomas MK (2005). Minireview: transcriptional regulation in pancreatic development. *Endocrinology* Vol. 146, pp 1025 – 1034.

Henquin JC, Ravier MA, Nenquin M, Jonas JC & Gilon P (2003). Hierarchy of the beta-cell signals controlling insulin secretion. *European Journal of Clinical Investigation*, Vol. 33, pp 742-750.

Hirose K, Morita M, Ema M, Mimura J, Hamada H, Fujii H, Saijo Y, Gotoh O, Sogawa K & Fujii-Kuriyama Y (1996). cDNA cloning and tissue-specific expression of a novel basic helix-loop-helix/PAS factor (Arnt2) with close sequence similarity to the aryl hydrocarbon receptor nuclear translocator (Arnt). *Molecular Cell Biology* Vol. 16, pp 1706-1713.

Holmes JL & Pollenz RS (1997). Determination of aryl hydrocarbon receptor nuclear translocator protein concentration and subcellular localization in hepatic and nonhepatic cell culture lines: development of quantitative Western blotting protocols for calculation of aryl hydrocarbon receptor and aryl hydrocarbon receptor nuclear translocator protein in total cell lysates. *Molecular Pharmacology*, Vol. 52, No. 2, pp 202-11.

Horton, JD, Goldstein, JL & Brown M.S. (2002). SREBPs: activators of the complete program of cholesterol and fatty acid synthesis in the liver. *Journal of Clinical Investigation*, Vol. 109, pp 1125–1131.

Hussain K (2010). Mutations in pancreatic ß-cell Glucokinase as a cause of hyperinsulinaemic hypoglycaemia and neonatal diabetes mellitus. *Reviews in Endocrine & Metabolic Disorders*, Vol. 11, No. 3, pp 179-183.

Ikeda M & Nomura M (1997). cDNA cloning and tissue-specific expression of a novel basic helix-loop-helix/PAS protein (BMAL1) and identification of alternatively spliced variants with alternative translation initiation site usage. Biochemical & Biophysical Research Communications, Vol. 233, No. 1, pp 258-264.

Ivarsson R, Quintens R, Dejonghe S Tsukamoto K, in 't Veld P, Renström E & Schuit FC (2005). Redox control of exocytosis: regulatory role of NADPH, thioredoxin, and glutaredoxin. Diabetes Vol. 54, pp 2132-2142.

Jain S, Dolwick KM, Schmidt JV & Bradfield CA (1994). Potent transactivation domains of the Ah receptor and the Ah receptor nuclear translocator map to their carboxyl termini. Journal of Biological Chemistry, Vol. 269, pp 31518-31524.

Jensen MV, Joseph JW, Ronnebaum SM, Burgess SC, Sherry AD & Newgard CB (2008). Metabolic cycling in control of glucose-stimulated insulin secretion. American journal of physiology endocrinology & metabolism, Vol. 295, pp E1287-E1297.

Jiang BH, Rue E, Wang GL, Roe R & Semenza GL (1996). Dimerization, DNA binding, and transactivation properties of hypoxia-inducible factor 1. Journal of Biological Chemistry, Vol. 271, pp 17771-17778.

Jiang C, Qu A, Matsubara T, Chanturiya T, Jou W, Gavrilova O, Shah YM, & Gonzalez FJ (2011). Disruption of Hypoxia-Inducible Factor 1 in Adipocytes Improves Insulin Sensitivity and Decreases Adiposity in High-Fat Diet-Fed Mice. Diabetes, (Epub ahead of print).

Johnson J, Carlsson L, Edlund T & Edlund H (1994). Insulin-promoter-factor 1 is required for pancreas development in mice. Nature, Vol. 371, pp 606–609.

Joseph JW, Jensen MV, Ilkayeva O, Palmieri F, Alárcon C, Rhodes CJ & Newgard CB (2006). The mitochondrial citrate/isocitrate carrier plays a regulatory role in glucose-stimulated insulin secretion. Journal of Biological Chemistry, Vol. 281, pp 35624-35632.

Kazuya Y, Yoshida K, Murao K, Imachi H, Cao WM, Yu X, Li J, Ahmed RA, Kitanaka N, Wong NC, Unterman TG, Magnuson MA & Ishida T (2007). Pancreatic glucokinase is activated by insulin-like growth factor-1, Endocrinology, Vol.148, pp 2904–2913.

Kewley RJ, Whitelaw ML & Chapman-Smith A (2004). The mammalian basic helix-loop-helix/PAS family of transcriptional regulators. International Journal of Biochemistry & Cell Biology, Vol. 36, No. 2, pp 189-204.

Khan A, Ling Z & Landau BR (1996). Quantifying the carboxylation of pyruvate in pancreatic islets. Journal of Biological Chemistry, Vol. 271, 2539-2542.

Kitamura YI, Kitamura T, Kruse JP,. Raum JC, Stein R, Gu W & Accili D (2005). FoxO1 protects against pancreatic beta cell failure through NeuroD and MafA induction. Cell Metabolism, Vol. 2, pp 153–163.

Kobayashi A, Numayama-Tsuruta K, Sogawa K & Fujii-Kuriyama Y (1997). CBP/p300 functions as a possible transcriptional coactivator of Ah receptor nuclear translocator (Arnt). Journal of Biological Chemistry, Vol. 122, pp 703-710.

Koo EH, Kopan R (2004). Potential role of presenilin-regulated signaling pathways in sporadic neurodegeneration. Nature Medicine, Vol. 10 pp S26-33.

Kozak KR, Abbott B & Hankinson O (1997). ARNT-deficient mice and placental differentiation. Developmental Biology,. Vol. 191, No. 2, pp 297-305.

Li H, Dong L & Whitlock JP Jr (1994).Transcriptional activation function of the mouse Ah receptor nuclear translocator. Journal of Biological Chemistry, Vol. 269, pp 28098-28105.

Li H, Ko HP & Whitlock JP (1996). Induction of phosphoglycerate kinase 1 gene expression by hypoxia. Roles of Arnt and HIF-1alpha. Journal of Biological Chemistry Vol. 271, pp 21262-21267.

Lim W, Park Y, Cho J, Park C, Park J, Park Y-K, Park H & Lee Y (2011).Estrogen receptor beta inhibits transcriptional activity of hypoxia inducible factor-1 through the downregulation of arylhydrocarbon receptor nuclear translocator. *Breast Cancer Research* , Vol. 13, R32.

Lin Y & Sun Z. Current views on type 2 diabetes (2010). *Journal of Endocrinology, Vol.* 204, No. 1, pp 1–11.

Lindebro MC, Poellinger L & Whitelaw ML (1995). Protein-protein interaction via PAS domains: role of the PAS domain in positive and negative regulation of the bHLH/PAS dioxin receptor-Arnt transcription factor complex. *EMBO Journal,* Vol. 14, No. 14, pp 3528-3539.

Long WP, Chen X & Perdew GH (1999). Protein kinase C modulates aryl hydrocarbon receptor nuclear translocator protein-mediated transactivation potential in a dimer context. *Journal of Biological Chemistry,* Vol. 274, pp 12391-12400.

Lu D, Mulder H, Zhao P, Burgess SC, Jensen MV, Kamzolova S, Newgard CB, & Sherry AD (2002). 13C NMR isotopomer analysis reveals a connection between pyruvate cycling and glucose-stimulated insulin secretion (GSIS). *Proceedings of National Academy of Sciences,* Vol. 99, pp 2708-2713.

Lyttle BM, Li J, Krishnamurthy M, Fellows F, Wheeler MB, Goodyer CG & Wang R (2008). Transcription factor expression in the developing human fetal endocrine pancreas. *Diabetologia,* Vol. 51, No. 7, pp 1169-1180.

MacDonald MJ (1995) Influence of glucose on pyruvate carboxylase expression in pancreatic islets. *Archives of Biochemistry &.Biophysics.* Vol. 319, pp 128-132.

MacDonald MJ, Fahien LA, Brown LJ, Hasan NM, Buss JD & Kendrick MA (2005) Perspective: emerging evidence for signaling roles of mitochondrial anaplerotic products in insulin secretion. *American journal of physiology endocrinology & metabolism,* Vol. 288, pp E1-15.

Maechler P & Wollheim CB (1999) Mitochondrial glutamate acts as a messenger in glucose-induced insulin exocytosis. *Nature,* Vol. 402, pp 685-689.

Maltepe E, Schmidt JV, Baunoch D, Bradfield CA & Simon, MC (1997). Abnormal angiogenesis and responses to glucose and oxygen deprivation in mice lacking the protein ARNT. *Nature,* Vol. 386, pp 403-407.

Michael MD, Kulkarni RN, Postic C, Previs SF, Shulman GI, Magnuson MA & Kahn CR (2000). Loss of insulin signaling in hepatocytes leads to severe insulin resistance and progressive hepatic dysfunction. *Molecular Cell,* Vol. 6, pp 87–97.

Mitchell SM & Frayling TM (2002). The role of transcription factors in maturity-onset diabetes of the young. *Molecular Genetics & Metabolism,* Vol. 77 No. 1-2, pp 35-43.

Muoio DM & Newgard CB (2008). Mechanisms of disease: molecular and metabolic mechanisms of insulin resistance and beta-cell failure in type 2 diabetes. Nature *Review Molecular Cell Biology,* Vol. 9, No. 3, pp 193-205.

Nenquin M, Szollosi A, guilar-Bryan L, Bryan J & Henquin JC (2004) Both triggering and amplifying pathways contribute to fuel-induced insulin secretion in the absence of sulfonylurea receptor-1 in pancreatic beta-cells. *Journal of Biological Chemistry,* Vol. 279, pp 32316-32324.

Newgard CB & Matchinsky FM (2001). Substrate control of insulin release. In: *Handbook of physiology,* section 7, The endocrine system volume II: The endocrine pancreas and regulation of metabolism, Jefferson AC, Cherrington A (eds), pp 125-151 Oxford University Press.

Newgard CB & McGarry JD (1995). Metabolic coupling factors in pancreatic b-cell signal transduction. *Annual Review of Biochemistry*, Vol. 64, pp 689-719.

Noordeen N, Khera TK, Sun G, Longbottom ER, Pullen TJ, Xavier GD, Rutter GA & Leclerc I (2009). ChREBP is a Negative Regulator of ARNT/HIF-1β Gene Expression in Pancreatic Islet β-Cells. *Diabetes*, Vol. 59, pp 153–160.

Partch CL & Gardner KH (2011). Coactivators necessary for transcriptional output of the hypoxia inducible factor, HIF, are directly recruited by ARNT PAS-B. *Proceedings of the National Academy of Sciences*, Vol. 108, No. 19, pp 7739-7744.

Pillai R, Huypens P, Huang M, Schaefer S, Sheinin T, Wettig SW & Joseph JW (2011). Aryl Hydrocarbon Receptor Nuclear Translocator/Hypoxia-inducible Factor-1β Plays a Critical Role in Maintaining Glucose-stimulated Anaplerosis and Insulin Release from Pancreatic β-Cells. *Journal of Biological Chemistry*, Vol. 286, No. 2, pp 1014-1024.

Pedersen TA, Bereshchenko O, Garcia-Silva, Ermakova, O, Kurz, E., Mandrup, S, Porse BT & Nerlov C (2007). Distinct C/EBPalpha motifs regulate lipogenic and gluconeogenic gene expression in vivo. *EMBO Journal* , Vol. 26, pp 1081–1093.

Prentki M & Matschinsky FM (1987). Ca2+, cAMP, and phospholipid-derived messengers in coupling mechanisms of insulin secretion. *Physiology Review*, Vol. 67, pp 1185-1248.

Prentki M, Vischer S, Glennon MC, Regazzi R, Deeney JT & Corkey BE (1992) Malonyl-CoA and long chain acyl-CoA esters as metabolic coupling factors in nutrient-induced insulin secretion. *Journal of Biological Chemistry*, Vol. 267, pp 5802-5810.

Puri S, Cano DA, Hebrok M (2009). A role for von Hip- pel-Lindau protein in pancreatic beta-cell function. *Diabetes*, Vol. 58, No. 2, pp 433–441.

Qiao L, MacLean PS, You H, Schaack J & Shao J. (2006). Knocking down liver CCAAT/enhancer-binding protein alpha by adenovirus-transduced silent interfering ribonucleic acid improves hepatic gluconeogenesis and lipid homeostasis in db/db mice. *Endocrinology*, Vol. 147, pp 3060–3069.

Ravier MA, Nenquin M, Miki T, Seino S, Henquin JC (2009). Glucose controls cytosolic Ca2+ and insulin secretion in mouse islets lacking adenosine triphosphate-sensitive K+ channels owing to a knockout of the pore-forming subunit Kir6.2. *Endocrinology* , Vol. 150, pp 33-45.

Reyes H, Reisz-Porszasz S & Hankinson O (1992). Identification of the Ah receptor nuclear translocator protein (Arnt) as a component of the DNA binding form of the Ah receptor. *Science*, Vol. 256, pp 1193-1195.

Ronnebaum SM, Ilkayeva O, Burgess SC, Joseph JW, Lu D, Stevens RD, Becker TC, Sherry AD, Newgard CB & Jensen MV (2006). A pyruvate cycling pathway involving cytosolic NADP-dependent isocitrate dehydrogenase regulates glucose-stimulated insulin secretion. *Journal of Biological Chemistry*, Vol. 281, pp 30593-30602.

Rüegg J, Swedenborg E, Wahlström D, Escande A, Balaguer P, Pettersson K, & Pongratz I (2008). The transcription factor aryl hydrocarbon receptor nuclear translocator functions as an estrogen receptor beta-selective coactivator, and its recruitment to alternative pathways mediates antiestrogenic effects of dioxin. *Molecular Endocrinology*, Vol. 22, pp 304-316.

Salceda S, Beck I & Caro J (1996). Absolute requirement of aryl hydrocarbon receptor nuclear translocator protein for gene activation by hypoxia. *Archives of Biochemistry & Biophysics*, Vol. 334, pp 389-394.

Saltiel AR & Kahn CR (2001). Insulin signalling and the regulation of glucose and lipid metabolism. *Nature*, Vol. 414, pp 799–806.

Schuit F, De Vos A, Farfari S, Moens K, Pipeleers D, Brun T & Prentki M (1997). Metabolic fate of glucose in purified islet cells. Glucose-regulated anaplerosis in beta cells. *Journal of Biological Chemistry*, Vol. 272, No. 30, pp 18572-18579.

Semenza GL, Roth PH, Fang HM & Wang GL (1994). Transcriptional regulation of genes encoding glycolytic enzymes by hypoxia-inducible factor 1. *Journal of Biological Chemistry*, Vol. 269, No. 38, pp 23757-23763.

Semenza GL, Agani F, Feldser D, Iyer N, Kotch L, Laughner E & Yu A (2000). Hypoxia, HIF-1, and the pathophysiology of common human diseases. *Advances in Experimental Medicine and Biology*, Vol. 475, pp 123–130.

Semenza, GL (2003). Targeting HIF-1 for cancer therapy. *Nature Review Cancer*, Vol. 3, pp 721-732.

Shiota C, Larsson O, Shelton KD, Shiota M, Efanov AM, Hoy M, Lindner J, Kooptiwut S, Juntti-Berggren L, Gromada J, Berggren PO & Magnuson MA (2002). Sulfonylurea receptor type 1 knock-out mice have intact feeding-stimulated insulin secretion despite marked impairment in their response to glucose. *Journal of Biological Chemistry*, Vol. 277: 37176-37183.

Sogawa K, Fujii-Kuriyama Y (1997). Ah receptor, a novel ligand-activated transcription factor. *J Biochem.* Vol. 122 No. 6, pp 1075-9.

Song IS, Wang AG, Yoon SY, Kim JM, Kim JH, Lee DS & Kim NS (2009). Regulation of glucose metabolism-related genes and VEGF by HIF-1α and HIF-1β, but not HIF-2α, in gastric cancer. *Experimental & Molecular Medicine*, Vol. 41, No. 1, pp 51-58.

Steiner DJ, Kim A, Miller M & Hara M (2010). Pancreatic islet plasticity Interspecies comparison of islet architecture and composition. *Islets* Vol. 2, No. 3, pp 135-145.

Stoffers DA & Zinkin NT (1997), Pancreatic agenesis attributable to a single nucleotide deletion in the human IPF1 gene coding sequence, Nature Genetics, Vol. 15, pp 106– 110.

Swanson HI, Chan WK & Bradfield CA (1995). DNA binding specificities and pairing rules of the Ah receptor, ARNT, and SIM proteins. *Journal of Biological Chemistry*, Vol. 270, pp 26292-302.

Szollosi A, Nenquin M & Henquin JC (2007) Overnight culture unmasks glucose-induced insulin secretion in mouse islets lacking ATP-sensitive K+ channels by improving the triggering Ca2+ signal. J.Biol.Chem. 282: 14768-14776

Tomita S, Sinal CJ, Yim SH & Gonzalez FJ (2000). Conditional disruption of the aryl hydrocarbon receptor nuclear translocator (Arnt) gene leads to loss of target gene induction by the aryl hydrocarbon receptor and hypoxia-inducible factor 1alpha. *Molecular Endocrinology*, Vol. 14, pp 1674-1681.

Tripathy D, Chavez AO (2010). Defects in insulin secretion and action in the pathogenesis of type 2 diabetes mellitus. *Current Diabetes Reports.* Vol. 10, No. 3, pp 184-91.

Tsuchiya T, Kominato K & Ueda M(2002). Human Hypoxic Signal Transduction through a Signature Motif in Hepatocyte Nuclear Factor 4. *Journal of Biochemistry*, Vol. 132, pp 37-44.

Wang GL & Semenza GL (1995 a). Purification and characterization of hypoxia-inducible factor 1. *Journal of Biological Chemistry*, Vol. 270, pp 1230-1237.

Wang GL, Jiang BH, Rue EA & Semenza GL (1995 b). Hypoxia-inducible factor 1 is a basic-helix-loop-helix-PAS heterodimer regulated by cellular O2 tension. *Proceedings of the national academy of sciences,* Vol. 92, No. 12, pp 5510-5514.

Wang XL, Suzuki R, Lee K, Tran T, Gunton JE, Saha AK, Patti ME, Goldfine A, Ruderman NB, Gonzalez FJ & Kahn CR (2009). Ablation of ARNT/HIF-1beta in liver alters gluconeogenesis, lipogenic gene expression, and serum ketones. *Cell Metabolism,* Vol. 9, pp 428-439.

Wood SM, Gleadle JM, Pugh CW, Hankinson O & Ratcliffe PJ (1996). The Role of the Aryl Hydrocarbon Receptor Nuclear Translocator (ARNT) in Hypoxic Induction of Gene Expression. Studies in ARNT-Deficient Cells. *Journal of Biological Chemistry,* Vol. 271, No. 25, pp 15117–15123.

Woods SL & Whitelaw ML (2002). Differential activities of murine single minded 1 (SIM1) and SIM2 on a hypoxic response element. Cross-talk between basic helix-loop-helix/per-Arnt-Sim homology transcription factors. *Journal of Biological Chemistry,* Vol. 277, No. 12, pp 10236-10243.

Yamagata K (2003). Regulation of pancreatic beta-cell function by the HNF transcription network: lessons from maturity-onset diabetes of the young (MODY). *Endocrinology Journal,* Vol. 50, No. 5, pp 491-499.

Zehetner J, Danzer C, Collins S, Eckhardt K, Gerber PA, Ballschmieter P, Galvanovskis J, Shimomura K, Ashcroft FM, Thorens B, Rorsman P & Krek W (2008). pVHL is a regulator of glucose metabolism and insulin secretion in pancreatic beta-cells. *Genes & Development,* Vol. 22, No. 22, pp 3135–3146.

5

HIV-1 Selectively Integrates Into Host DNA *In Vitro*

Tatsuaki Tsuruyama

Department of Molecular Pathology, Graduate School of Medicine, Kyoto University
Kyoto, Kyoto Prefecture
Japan

1. Introduction

1.1 Summary

The biochemistry of retroviral integration selectivity is not fully understood. We modified the previously reported *in vitro* integration reaction protocol and developed a novel reaction system with higher efficiency. We used a DNA target composed of a repeat sequence DNA, 5'-(GTCCCTTCCCAGT)$_6$(ACTGGGAAGGGAC) $_6$-3', that was ligated into a circular plasmid. Target DNA was reacted with a pre-integration (PI) complex that was formed by incubation of the end cDNA of the HIV-1 genome and recombinant integrase. It was confirmed that integration selectively occurred in the middle segment of the repeat sequence. On the other hand, both frequency and selectivity of integration markedly decreased when target sequences were used in which CAGT bases in the middle position of the original target sequence were deleted. Moreover, upon incubation with a combination of these deleted DNAs and the original sequence, the integration efficiency and selectivity towards the original target sequence were significantly reduced, which indicated interference effects by the deleted sequence DNAs. Efficiency and selectivity were also found to vary with changes in the manganese dichloride concentration of the reaction buffer, probably due to induction of fluctuation in the secondary structure of the substrate DNA. Such fluctuation may generate structural isomers that are favorable for selective integration into the target sequence DNA. In conclusion, there is considerable selectivity in HIV-integration into the specified target sequence. The present *in vitro* integration system will therefore be useful for monitoring viral integration activity or for testing of integrase inhibitors.

1.2 Background

Retroviral integration into host DNA is a critical step in the viral life cycle. Once integrated, the proviral genome will be stably duplicated along with the host cellular DNA duplication and will be transmitted to the daughter cells. Retroviruses can thus serve as powerful tools for the integration of foreign genes into a host genome (Fig. 1), and an MLV vector can be used to examine the function of introduced genes and the development of induced pluripotent stem (iPS) cells [1]. However, integrated retroviral genomes also have the potential to cause unexpected transformation through up-regulation of target genes by retroviral promoter elements located at the long terminal repeat (LTR) following integration [2]. Although

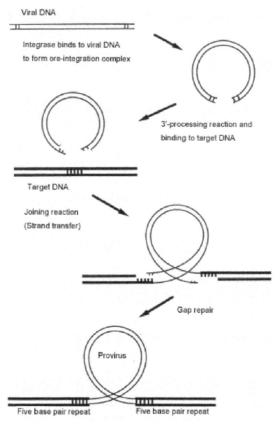

Fig. 1. Scheme of viral integration into target DNA: 3'-processing, join reaction, and gap repair. After the HIV-1 RNA genome has been transcribed into double-stranded DNA, the viral protein integrase binds to the termini of the viral DNA ends in a tetrameric fashion and the integrase creates overlapping 5'-ends by removing two nucleotides from the 3'-ends (3'end-processing). The HIV-1 DNA and the host cell DNA are ligated by synthesis of phosphodiester bonds between the terminal nucleotides of the viral 3'-ends and overlapping 5'-ends of the host chromosome. The non-homologous 5'-ends from the viral DNA are removed by integrase. Finally, the gaps are filled up by host cellular repair proteins, which recognize single strand breaks. A five base-repeat is observed in the flanking sequence after the gap repair reaction.

integration events have long been considered to be random, several recent findings have shown that integration of the murine leukemia retrovirus (MLV) and of HIV-1 is detected more frequently in actively transcribed genes [3, 4] or in promoter regions [5]. Previous statistical studies have also demonstrated that weak palindromic sequences are a common feature of the sites targeted for retroviral integration [6, 7]. Similar target preferences have also been reported for human T cell leukemia retrovirus type I (HTLV-I) integration sites [8]. Because of these findings, we investigated the biophysical mechanisms underlying *in vitro* integration. In the present study, we aimed to establish an *in vitro* integration assay using retroviral cDNA and integrase. Yoshinaga et al. previously reported the development of an *in vitro* integration assay using recombinant HIV-1 integrase, and short viral and target DNA sequences [9]. Using this method, they successfully detected retroviral cDNA-target DNA complexes *in vitro* and reported that the dinucleotide motif 5'-CA that is located at the proviral genome termini was essential for HIV-1 integration. Yoshinaga et al. called these dinucleotides the integration signal sequence. We modified Yoshinaga's method of *in vitro* integration in order to identify the precise HIV-1 integration sites using a target DNA that corresponds to an actual gene sequence.

In our previous study, we identified a common MLV integration site within the *signal transducer and activators of transcription 5*a (*Stat5a*) gene in MLV-induced spontaneous murine lymphoma in an inbred strain of mice, SL/Kh [10]. This is the first report of MLV integration target sequence. It has also been previously demonstrated that the *Stat5a* gene represents one of the common integration sites of MLV (the Mouse Retrovirus Tagged Cancer Gene Database (MRTCGD) (http://rtcgd.ncifcrf.gov/cgi-bin/mm7/easy_search.cgi) [11]. The encoded STAT5A protein is a transcription factor that is known to play an essential role in the development of myelo- and lympho-proliferative disease [12, 13]. In the current study, we modified this *Stat5a* gene sequence for use as a target for HIV-1 integration *in vitro*. This target gene consists of a 5'-CA-rich sequence, which may provide a useful clue for preparation of target DNA sequences, because terminal CA dinucleotide motifs are shared by MLV and HIV-1, as well as by HTLV-I proviruses.

2. *In vitro* integration assay used in previous studies

In many previous *in vitro* integration protocols, the double stranded DNA of the HIV-1 3'-LTR proviral end alone and the substrate DNA are mixed in an appropriate buffer containing MnCl₂ [14] (Fig. 2, 3). Subsequently, a PCR reaction using a primer set targeted to the proviral

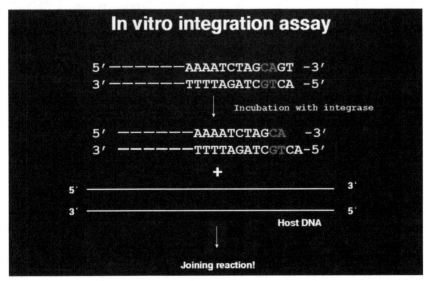

Fig. 2. Scheme of the previously described incubation with recombinant integrase

DNA and the target DNA amplifies a DNA segment that includes the viral-host DNA junction. In some protocols, even the viral DNA itself is used as the target DNA. Thus, previous studies paid little attention to the target sequence. It is commonly known that integrase binds to the proviral DNA in a regular tetrameric fashion. Indeed, some sequence motifs should be favored by an integrase oligomer, because a dimeric transcriptional factor protein has the ability to bind to palindromic sequence motifs such as E-box and GAS elements.

The 3' end of the HIV-1 LTR sequence is shown. Red letters indicate the conserved dinucleotide motif. Incubation of the 3' end of the HIV-1 LTR sequence DNA with integrase results in processing of the end of the 5'-GT dinucleotide. The resulting exposed hydroxyl group then attacks the target DNA.

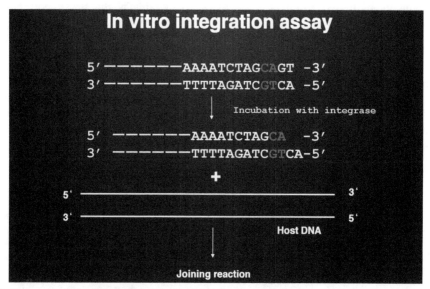

Fig. 3. Protocol of the previously described *in vitro* integration assay

A nick is introduced into the host DNA that is attached to a CA dinucleotide in the HIV-1 DNA end. Red letters indicate the conserved dinucleotide motif. Arrows indicate the PCR primer set and the direction of DNA polymerization.

2.1 Preparation of target sequences for *in vitro* integration

We recently reported the target sequence of MLV integration. We developed an inbred strain of mice suffering from spontaneous B cell lymphoma by MLV integration. MLV integration into *Stat5a* was identified in 25% of the lymphoma genome [15, 16] (Fig. 4). As depicted in Fig. 4, the hot spot of integration included a 5'-CA-rich sequence as well as a palindromic motif. Downward facing arrows in the figure indicated the MLV integration sites. The abundance of 5'-CA dinucleotides in the integration hot spots provided us with a hint for the preparation of target DNA for HIV integration, because these motifs are shared by the genome ends of MLV and HIV-1 (Fig. 5). We hypothesized that HIV-1 DNA also favors such a 5'-CA-rich sequence motif.

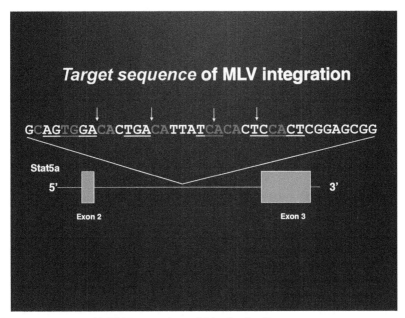

Fig. 4. Target sequence of MLV integration within the *Stat5a* gene.

Downward-facing arrows indicate the MLV integration sites (15, 16). Red letters highlight the CA-rich sequence. Blue boxes represent exons of the *Stat5a* gene. Underlines indicates the palindromic motif.

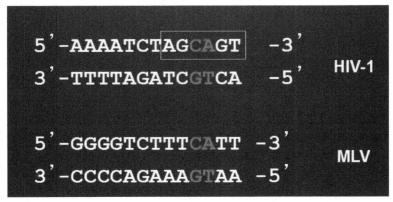

Fig. 5. Conserved 5'-CA dinucleotide in the proviral ends of HIV-1 and MLV.

A box indicate the integration signal sequence reported by Yoshinaga et al (9).

2.2 Modified *In vitro* integration assay

We then prepared a target sequence for HIV-1 integration. A repeat sequence was prepared in order to enhance integration efficiency. We used the repeat sequence, 5'-

(GTCCCTTCCC*AGT*)₆(*ACT*GGGAAGGGAC)₆-3', or a modification of this sequence, which was ligated into a circular plasmid. The sequence within parenthesis is the unit of the repeat. This target sequence includes the 5'-CA dinucleotide motif, and includes 5'-AC at the HIV-1 DNA termini (Fig. 5). 5'-CAGT and 5'-ACTG (shown in *italics* in the above sequence) in the repeat units are also present in the HIV-1 proviral genome ends. This target DNA was reacted with recombinant integrase and formed a pre-integration (PI) complex. Figures 6 and 7 show our scheme of *in vitro* integration as well as the sequences of the HIV-1 proviral 5'- and 3'-ends. Following incubation of the proviral LTR sequence DNAs with recombinant integrase, the resultant pre-integration complexes were reacted with the target DNA. PCR amplification was performed and the integration sites were analyzed by direct sequencing. Unlike previously reported protocols, we used both 5'- and 3'-LTR sequences in our protocol. Such a target sequence unit was expected to directly interact with complementary HIV-1 DNA end sequences present in the target DNA. Complementarity between HIV-1 DNA and host DNA is shown in Fig. 8.

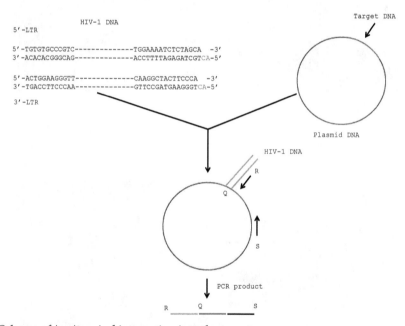

Fig. 6. Scheme of *in vitro* viral integration into the target sequence.

The red segment in the target sequence DNA that includes a circular plasmid represents the 144-bp target DNA, and the black line represents the remainder of the circular plasmid DNA used for ligation. Following incubation of the proviral LTR sequence DNAs with integrase, the resultant pre-integration complexes were reacted with the substrate DNA. Red letters in the HIV-1 cDNA represent the LTR termini. PCR amplification was performed using primers corresponding to regions in the proviral ends and a plasmid region. The integration sites were analyzed by direct sequencing. "Q" in the PCR product indicates the junction between the provirus and the target DNA and R&S represents the 3'-ends of the primer within the HIV-1 DNA and the plasmid DNA, respectively [14].

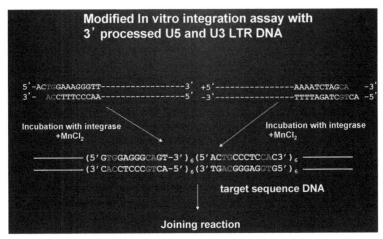

Fig. 7. Target sequence for integration.

The top sequences show the termini of the HIV-1 provirus. The bottom sequence indicates the target sequence and highlights the dinucleotide motif CA/TG (red), and the AC bases (yellow) that are also present at the HIV-1 DNA termini. We prepared a repeat sequence in order to enhance integration efficiency. The sequence shown in parenthesis is the unit of the repeat. Our protocol differs from previous protocols in that we used both 5'- and 3'-LTR sequences rather than a single 3'-LTR DNA.

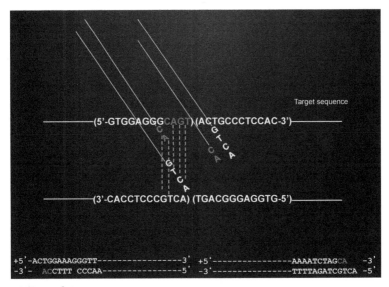

Fig. 8. Integration scheme.

The repeat sequence unit in the target sequence was expected to directly interact with HIV-1 DNA. The dotted lines indicate complementarity between target and viral DNA. The dsDNA sequence at the bottom indicates the 5'- and 3'-ends of the proviral HIV-1 DNA.

2.3 Reaction protocol

The detailed reaction protocol that we used is as follows. First, 75 ng of the U5'-LTR cDNA sequence of HIV-1.; (+) 5'-TGT GTG CCC GTC TGT TGT GTG ACT CTG GTA ACT AGA GAT CCT CAG ACC TTT TTG GTA GTG TGG AAA ATC TCT AGC A-3' and (-) 5'-ACT GCT AGA GAT TTT CCA CAC TAC CAA AAA GGG TCT GAG GGA TCT CTA GTT ACC AGA GTC ACA CAA CAG ACG GGC ACA CA-3', was incubated with 50 ng recombinant HIV-1 integrase in 10 μl of binding buffer for 1 h at 30°C. The binding buffer consisted of 1-0.1 mM $MnCl_2$, 80 mM glutamate potassium glutamate, 10 mM mercaptoethanol, 10% DMSO, and 35 mM MOPS (pH 7.2).

Similarly, the 3'-LTR cDNA sequence (75 ng) of HIV-1.; (+) 5'-ACT GGA AGG GTT AAT TTA CTC CAA GCA AAG GCA AGA TAT CC TTG ATT TGT GGG TCT ATA ACA CAC AAG GCT ACT TCC CA-3' and (-) 5'-ACTG GGA AGT AGC CTT GTG TGT TAT AGA CCC ACA AAT CAA GGA TAT CTT GCC TTT GCT TGG AGT AAA TTA ACC CTT CCAGT-3', was incubated with the recombinant retroviral integrase. After incubation, the double-stranded (ds) 5'-LTR DNA was combined with the ds 3'-LTR DNA for 1 h at 30 °C, and the LTR DNA was then further incubated with the target DNA for 1 h at 30°C. As controls, ds 5'-LTR DNA and ds 3'-LTR DNA were also individually incubated with the target DNA. For control target DNAs we synthesized four random 144-bp sequences, which were designed by a random number generator, and we ligated these sequences into circular DNA in the same manner as described below for the target DNA

In order to prevent non-specific reactions at the target DNA sequence, we ligated the target sequence DNA into circular plasmid DNA (Invitrogen pCR2.1 TOPO vector) and used this entire DNA as the target DNA for the assay (Fig. 6). The proportion of LTRs and target DNAs was optimized to prevent both non-specific reactions and integration due to an excess of LTRs. The DNA reacted in the buffer was purified using a QIA quick column (QIAGEN, GmbH, Germany). PCR amplification was then performed using retroviral primers: the HIV-1 U5'-LTR primer, 5'-GTG TGC CCG TCT GTT GTG TGA CTCTGG-3', or the HIV-1 U3'-LTR primer, 5'-CTG GGA AGT AGC CTT GTG TGT TAT AG-3', and a TOPO vector primer 5'-TCA CTC ATG GTT ATG GCA GC -3' whose first nucleotide corresponds to nucleotide position 2222 in the TOPO-pCR2.1 plasmid (Invitrogen, Carlsberg, CA). Amplicon copy number was quantified following identification of the HIV-1-substrate DNA junction [14].

3. Results

3.1 Selective viral DNA integration into the target sequence

Figures 9 and 10 show the percentage of viral DNA integration into the target sequence or into the same length random sequences. Four types of the same length random sequences were used as controls. The horizontal blue line shows the percentage integration when uniform integration into the substrate DNA, including into the target DNA plus the circular plasmid, was thought to occur. These data indicate that the percentage of integration into the target sequence was significantly higher than that into random sequences. Also, when a target sequence was used in which the middle 5'-CA and 5'-GT nucleotides were deleted, the integration efficiency was significantly decreased. Thus, local nucleotide motifs within the target sequence affect integration efficiency.

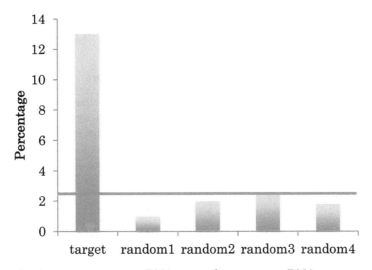

Fig. 9. Integration into target sequence DNA vs. random sequence DNAs.

The percentage of PCR product copies derived from viral DNA that had integrated into the target sequence or into random sequences is plotted vs. the total number of PCR product copies, including the PCR products that were integrated into the remainder of the DNA sequence of the plasmid. The horizontal line shows the ratio of these PCR products when integration was thought to occur in a uniform manner in the 4-kb substrate DNA.

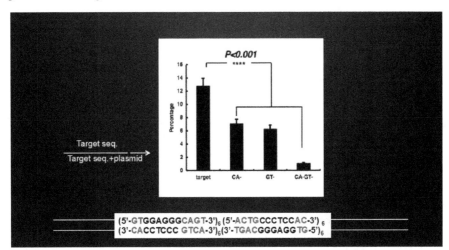

Fig. 10. A graph of the percentage integration into the target sequence or into random sequences.

The left arrows represents the percentage (~2.3%) of integration into target DNA to integration into control when integration was thought to occur in a uniform manner in the substrate DNA, the target sequence DNA plus the circular plasmid with which it was ligated. The integration efficiency was significantly decreased when CA and GT were removed from the middle region of the repeat. Thus, local nucleotide motifs affect integration efficiency (****P < 0.001).

3.2 *In vitro* integration site in the target repeat sequence DNA

Figure 11 shows the *in vitro* integration site in the target repeat sequence DNA. The entire target sequence is shown in this figure. The vertical axis indicates the percentage of PCR amplicons derived from the integration of individual LTR units. Integration efficiency was significantly higher when both the 5'- and the 3'-LTR DNA were used than when either LTR DNA was used alone. The use of both 5'- and 3'-LTR DNA is one of the unique points of our protocol, since previous protocols used a single 3'-LTR DNA (Figs. 7 and 11).

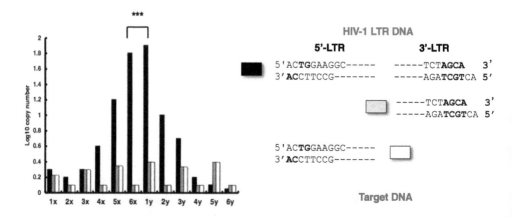

Fig. 11. *In vitro* integration site in the target repeat sequence DNA.

The vertical axis indicates the percentage of the PCR amplicons derived from proviral DNA integrating into individual units. The entire target sequence is shown. The integration efficiency was significantly higher when both 5'- and 3'-LTR DNA were used rather than when a single LTR DNA was used. The use of both 5'- and 3'-LTR DNA is one of the unique points of our protocol. x, GTGGAGGGCAGT; y, ACTGCCCCCAC. (*** P < 0.001)

Interestingly, we found that the middle segment of the target sequence was more favorable for integration, even though the same sequence units were repeated in the target sequence. To explain this observation, we considered the possibility that a structural factor may contribute to selective integration into the middle segment. Thus, if a single strand of the target DNA focally appeared by rewinding of the target double strand DNA, a long hairpin or cruciform structure may form in the target sequence site. It is probable that, if the target sequence DNA is open, or rewound, then the top of such a secondary structure would be favorable for integration. DNA folding thermodynamic analysis was performed to determine secondary structure in the target DNA and a hairpin structure was indeed predicted by this analysis (Fig. 12).

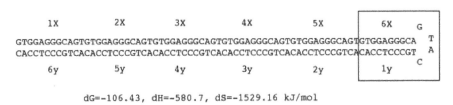

```
5'-GTGGAGGGCAGTGTGGAGGGCAGTGTGGAGGGCAGTGTGGAGGGCAGTGTGGAGGGCAGTGTGGAGGGCAGT
  ACTGCCCTCCACACTGCCCTCCACACTGCCCTCCACACTGCCCTCCACACTGCCCTCCACACTGCCCTCCAC-3'
```

1X	2X	3X	4X	5X	6X

```
GTGGAGGGCAGTGTGGAGGGCAGTGTGGAGGGCAGTGTGGAGGGCAGTGTGGAGGGCAGTGTGGAGGGCA   G
CACCTCCCGTCACACCTCCCGTCACACCTCCCGTCACACCTCCCGTCACACCTCCCGTCACACCTCCCGT   T
                                                                         A
                                                                         C
```

6y	5y	4y	3y	2y	1y

```
dG=-106.43, dH=-580.7, dS=-1529.16 kJ/mol
```

Fig. 12. A presumed secondary structure in the target sequence DNA.

The blue lines in the target DNA sequence shown at the top indicate the most frequent integration site. The presumed hairpin like structure shown at the bottom was constructed based on calculation of the target DNA sequence using the *m*-fold program (http://mfold.rna.albany.edu/?q=mfold/DNA-Folding-Form). dG, dH, and dS represent Gibbs' free energy, enthalpy, and entropy in ssDNA, respectively. A box represents the most frequent integration site.

3.2 Decoy effect of modified target sequences

We prepared two modified DNA target sequences in which the 5'-CA and 5'-GT were removed from the repeat unit at the middle site, termed modified sequence I and II, respectively (Fig. 13). PCR analysis of in vitro integration into modified sequence I or II revealed significant reductions in the number of copies of the PCR products compared to integration into the unmodified sequence. In addition, integration selectivity was not evident when using the modified DNA sequences ($P < 0.05$). We next mixed substrate DNA

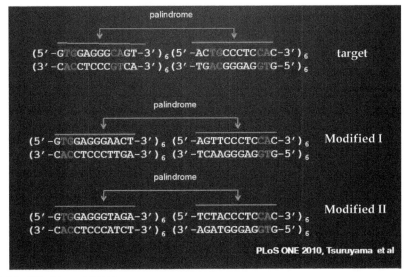

Fig. 13. Modified target DNA sequences.

containing the target sequence with substrate DNA containing modified sequence I or II in equal amounts, and examined the number of PCR product copies that originated from integration into the non-modified target sequence. Integration into the original, non-modified target sequence of the substrate DNA was significantly reduced when this DNA was mixed with the modified sequences (Fig. 14).

Two modified DNA sequences were prepared in which CA and GT were removed from the repeat unit at the middle site, termed modified sequence I and modified sequence II respectively. Red letters represent the TG/CA motifs. Yellow letters represent the GT/AC motifs that are observed in the HIV-1 proviral genome.

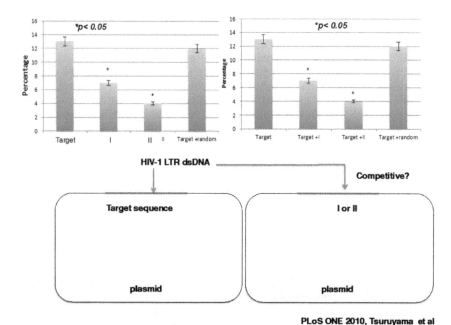

PLoS ONE 2010, Tsuruyama et al

Fig. 14. *In vitro* integration using modified sequence I or II

The result showed significant reductions in the number of copies of PCR products derived from integrated DNA. In addition, integration selectivity was evidently suppressed when using the modified DNA targets (left graph, $*P < 0.05$). Substrate DNA containing the target sequence was then mixed with substrate DNA containing modified sequence I or II in equal amounts, and the percentage of PCR product copies originating from integration into the original target sequence were determined. Integration into the original target sequence was significantly reduced when this target was mixed with the modified sequences (right graph, $*P < 0.05$).

3.3 Biochemistry of the integrase: DNA structure fluctuation enhances selective integration

We digested circular DNA with HIV-1 integrase in a buffer containing various concentrations of manganese dichloride and measured the band intensity of linearized DNA following electrophoresis. The relative band intensity increased when the concentration of $MnCl_2$ in the reaction buffer exceeded 40 mM (Fig. 15). This result raised the question of how such fluctuation in DNA structure influences the selectivity of *in vitro* HIV-1 integration. Furthermore, the percentage of integration into the target sequence DNA was found to increase significantly when the concentration of $MnCl_2$ exceeded 40 mM. The ratio of the PCR product number derived from integration into the target sequence DNA was also found to increase significantly when the $MnCl_2$ concentration exceeded 40 mM (Fig. 16). In conclusion, such fluctuation may generate a favorable conformation of target DNA for integration of the HIV-1 LTR [16, 17].

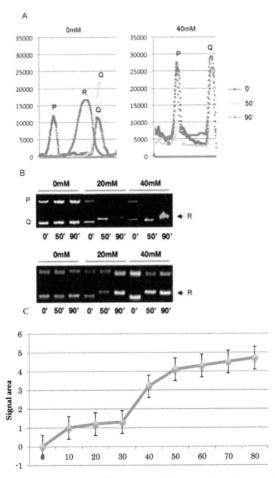

Fig. 15. Effect of $MnCl_2$ concentration and structural-fluctuation of target DNA on integration

(A) Electrophoretogram of plasmid plus target DNA (left) and plasmid DNA (right). P and Q indicate structural isomers of the circular DNA. R, a single fragment, indicates the digested DNA fragment after 90 min incubation. The vertical axis indicates relative signal intensity and the horizontal axis indicates electrophoretic mobility distance. (B) Electrophoresis of the DNA following incubation of recombinant integrase in buffer containing 0, 20, or 40 mM MnCl₂ following incubation for 0, 50, and 90 min. P, Q, and R are as described in (A). Upper and lower photographs display electrophoresis of plasmid plus target DNA and plasmid DNA, respectively. Significant fluctuation in R was observed in the electrophoresed DNA following 90 min incubation. (C) Relative signal area of the digested 4.0-kb substrate DNA corresponding to R in (B). Error bars represent standard deviation (S.D.). Fragment R area was calculated by integration of the individual curves shown in the electrophoretogram (A) with respect to electrophoretic mobility distance. Signal area significantly increased when 40 mM MnCl₂ was included in the buffer.

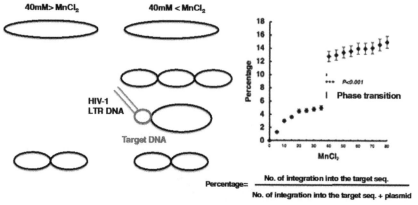

Fig. 16. DNA structural fluctuation and integration

A MnCl₂ concentration greater than 40 mM significantly increased the percentage of integration into the target sequence DNA. The percentage of copy number of PCR products derived from integration into the target sequence DNA was found to increase significantly when the concentration of MnCl₂ exceeded 40 mM. It is concluded that such fluctuation may generate a favorable conformation of target DNA (in red) for integration of HIV-1 LTR DNA.

4. Conclusion

In conclusion, selective HIV-1 integration was proved at an *in vitro* level in this study. The factors that determine this selectivity are (i) a sequence motif, including CAGT, and (ii) a structural factor that can be induced by fluctuation of a high concentration of MnCl₂ [16, 17]. The findings shown in Figs. 9 and 10 indicate that the percentage of integration into the target sequence was significantly greater than the integration rate into the random and deleted sequences. Moreover, the entire repeat sequence or secondary structure may be a target of integration.

In particular, our findings that sequences similar to the target DNA sequence interfere with integration (Fig. 14). Thus, a modified DNA can act as a decoy for the target DNA. In the present study, integration efficiency and selectivity were highly sensitive to MnCl₂ concentration in the reaction buffer. In particular, the integration efficiency and selectivity increased significantly when the MnCl₂ concentration was increased from 30 mM to 40 mM.

Fluctuations in the electrophoretic mobility of the substrate DNA also increased. These results suggest that there is a threshold concentration of $MnCl_2$ for *in vitro* integration, probably because $MnCl_2$ induces instability of secondary structure and therefore phase transition of the host DNA strand may occur. Target DNA can probably not generate the specified stable conformation under 40mM of $MnCl_2$. Based on these data as well as the data shown in Fig. 15, we propose that there are close correlations between structural changes in substrate DNA and integration selectivity and efficiency (Fig. 16, 17). We have used $MnCl_2$ for studies of *in vitro* integration because this salt is more appropriate than other salts for the generation of *in vivo* integration. However, during *in vivo* integration into the host genome, numerous DNA binding proteins and metal ions regulate the reaction in a complex manner. Therefore, the present data cannot be immediately applied to *in vivo* systems and further investigation using cell culture systems are necessary. Nevertheless, this *in vitro* integration assay is expected to facilitate understanding of the pathogenicity of HIV-1.

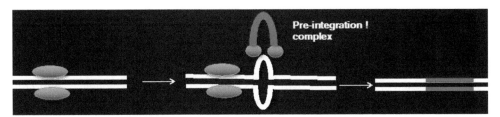

Fig. 17. A model of integration

The top of the secondary structure may be favorable for integration when the target DNA sequence is open or rewound by protein binding to the upstream of the target DNA sequence.

5. Acknowledgments

This work was supported by a Grant-in-Aid for Cancer Research from the Ministry of Education, Culture, Sports, Science, and Technology, Japan, and a Grant for Strategic Research on Cancer from the Ministry of Health, Labor, and Welfare, Japan (No. 72602-010-A03-0001) (http://www.jsps.go.jp/j-grantsinaid/index.html). The funders had no role in study design, data collection or analysis, in the decision to publish, or in preparation of the manuscript. The method described in this manuscript has been registered as "Nucleic acid having retroviral integration target-activity," patent number 4631084 in Japan (GenBank, DD323298). In relation to this patent, we also declare no conflict of interest. We are grateful to Masakazu Hatanaka and Tomokazu Yoshinaga for their helpful advice and insightful comments regarding this manuscript. In particular, we are also grateful to Dr. Tasuku Honjo and Dr. Hiroshi Hiai (Kyoto University) for their review of this study and for providing critical advice, and to Miss Hiroko Saito for her excellent technical support.

6. References

[1] Coffin, J.M.; Hughes, S.H. &Varmus, H.E. (1997). Retroviruses. New York: Cold Spring Harbor Laboratory Press.

[2] Wu, X.;Li Y.; Crise, B. & Burgess, S.M. (2003). Transcription start regions in the human genome are favored targets for MLV integration. Science 300:1749-1751.

[3] Schröder, A.R.; Shinn, P.; Chen, H.; Berry, C.; Ecker, J.R.; et al. (2002). HIV-1 integration in the human genome favors active genes and local hotspots. Cell 110: 521-529.

[4] Tsukahara, T.; Agawa, H.; Matsumoto, S.; Matsuda, M.; Ueno, S.; et al. (2006). Murine leukemia virus vector integration favors promoter regions & regional hot spots in a human T-cell line. Biochem Biophys Res Commun. 345:1099-1107.

[5] Holman, A.G. & Coffin, J.M. (2005). Symmetrical base preferences surrounding HIV-1,avian sarcoma/leukosis virus, and murine leukemia virus integration sites. Proc Natl Acad Sci U S A 102: 6103-6107.

[6] Wu, X.; Li, Y.; Crise, B.; Burgess, S.M. & Munroe, D.J. (2005). Weak palindromic consensus sequences are a common feature found at the integration target sites of many retroviruses. J Virol. 79: 5211-5214.

[7] Derse, D.; Crise, B.; Li, Y.; Princler, G.; Lum N.; et al. (2007). Human T-cell leukemia virus type 1 integration target sites in the human genome, comparison with those of other retroviruses. J Virol. 81: 6731-6741.

[8] Yoshinaga, T. & Fujiwara, T. (1995). Different roles of bases within the integration signal sequence of human immunodeficiency virus type 1 in vitro. J Virol. 69: 3233-3226.

[9] Tsuruyama, T.; Nakamura, T.; Jin, G.; Ozeki, M.; Yamada, Y.; et al. (2002). Constitutive activation of Stat5a by retrovirus integration in early pre-B lymphomas of SL/Kh strain mice. Proc Natl Acad. Sci. U. S. A. 99: 8253-8258.

[10] Akagi, K.; Suzuki, T.; Stephens, RM.; Jenkins, NA. & Copeland, NG (2004). RTCGD, retroviral tagged cancer gene database. Nucleic Acids Res. 32 (Database issue),D523-527.

[11] Nosaka, T.; Kawashima, T.; Misawa, K.; Ikuta, K.; Mui, AL.; et al. (1999). STAT5 as a molecular regulator of proliferation, differentiation, & apoptosis in hematopoieticcells. EMBO J 17: 4754-4765.

[12] Schwaller, J.; Parganas, E.; Wang, D.; Cain, D.; Aster, JC. et al. (2000). Stat5 is essential for the myelo- & lymphoproliferative disease induced by TEL/JAK2. Mol Cell 6:693-704.

[13] Tsuruyama,T.; Nakai, T.; Hiratsuka, T.; Jin, G.; Nakamura, T. & Yoshikawa, K. (2010).In vitro HIV-1 selective integration into the target sequence and decoy-effect of the modified sequence. PLoS One 5 : e13841.

[14] Tsuruyama, T.; Nakamura, T.; Jin G.; Ozeki, M.; Yamada, Y. &Hiai, H. Constitutive activation of Stat5a by retrovirus integration in early pre-B lymphomas of SL/Kh strain mice. Proc Natl Acad Sci U S A. 2002 Jun 11.;99(12):8253-825.

[15] Iwaki, T.; Makita, N. &Yoshikawa, K (2008). Folding transition of a single semiflexible polyelectrolyte chain through toroidal bundling of loop structures. Journal of Chemical Physics 129: 065103.

[16] Ueda, M.; & Yoshikawa, K. (1996). Phase transition and phase segregation in a single double-stranded DNA molecule. Physical Review Letters 77: 2133-2136.

6

Functional Genomics of Anoxygenic Green Bacteria *Chloroflexi* Species and Evolution of Photosynthesis

Kuo-Hsiang Tang

Carlson School of Chemistry and Biochemistry, and Department of Biology
Clark University, Worcester
USA

1. Introduction

In addition to the most recently reported aerobic anoxygenic phototrophic bacterium *Chloroacidobacterium thermophilium* [1], five phyla of phototrophic bacteria have been reported, including four phyla anoxygenic phototrophic bacteria (anaerobic and aerobic anoxygenic phototrophic Proteobacteria, filamentous anoxygenic phototrophs (FAPs), green sulfur bacteria and heliobacteria) and oxygenic phototrophic bacteria (cyanobacteria). According to 16S rRNA analysis, *Chloroflexi* species in FAPs are the earliest branching bacteria capable of photosynthesis [2,3] (**Fig. 1**), and the thermophilic bacterium *Chloroflexus* [Cfl.] *aurantiacus* among the *Chloroflexi* species has been long regarded as a key organism to resolve the obscurity of the origin and early evolution of photosynthesis. *Cfl. aurantiacus* can grow phototrophically under anaerobic conditions or chemotrophically under aerobic and dark conditions [4]. During phototrophic growth of *Cfl. aurantiacus*, the light energy is first absorbed by the peripheral light-harvesting complex chlorosomes, then transferred to the integral membrane B808-866 core antenna complex and finally to the reaction center (RC). *Cfl. aurantiacus* contains a chimeric photosystem that comprises some characters of green sulfur bacteria (chlorosomes) and anoxygenic phototrophic Proteobacteria (the B808-866 core antenna complex), and also has some unique electron transport proteins compared to other photosynthetic bacteria. The complete genomic sequence of *Cfl. aurantiacus* has been recently determined, analyzed and compared to the genomes of other photosynthetic bacteria [5].

Significant contributions of horizontal/lateral gene transfer among uni-cellular [6] and multi-cellular [7] organisms during the evolution, including the evolution of photosynthesis [8,9], have been recognized. Various perspectives on evolution of photosynthesis have been reported in literature [8-25], whereas our understanding of transition from anaerobic to aerobic world is still fragmentary. The recent genomic report on *Cfl. aurantiacus* [5], along with previous physiological, ecological and biochemical studies, indicate that the anoxygenic phototroph bacterium *Cfl. aurantiacus* has many interesting and certain unique features in its metabolic pathways. The *Cfl. aurantiacus* genome contains numerous aerobic/anaerobic gene pairs and oxygenic/anoxygenic metabolic pathways in the *Cfl. aurantiacus* genome [5], suggesting numerous gene adaptations/replacements in *Cfl. aurantiacus* to facilitate life under both anaerobic and aerobic growth conditions. These

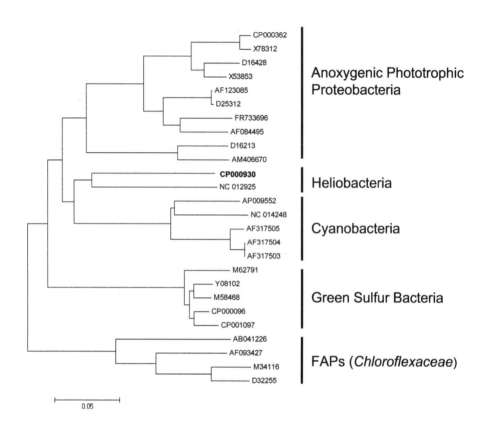

Fig. 1. Phylogenetic tree of photosynthetic bacteria.
The tree was constructed with un-rooted neighbor joining 16S rRNA dendrogram from five phyla of photosynthetic microbes, including cyanobacteria, heliobacteria, phototrophic anoxygenic Proteobacteria, green sulfur bacteria and filamentous anoxygenic phototrophs (FAPs). Bacterial names and accession numbers of 16S rRNA genes: (1) Phototrophic anoxygenic Proteobacteria: *Roseobacter denitrificans* OCh114 (CP000362), *Roseobacter litoralis* (X78312), *Rhodobacter capsulatus* (D16428), *Rhodobacter sphaeroides* 2.4.1 (X53853), *Rhodopseudomonas faecalis* strain gc (AF123085), *Rhodopseudomonas palustris* (D25312), *Rhodopseudomonas acidophila* (FR733696), *Rhodopseudomonas viridis* DSM 133 (AF084495), *Rubrivivax gelatinosus* (D16213); (2) heliobacteria: *Heliobacterium gestii* (AB100837), *Heliobacterium modesticaldum* (CP000930); (3) cyanobacteria: *Oscillatoria amphigranulata* str. 19-2 (AF317504), *Oscillatoria amphigranulata* str. 11-3 (AF317503), *Oscillatoria amphigranulata* str. 23-3 (AF317505), *Microcystis aeruginosa* NIES-843 (AP009552), *Nostoc azollae* 0708 (NC_014248); (4) green sulfur bacteria: *Chlorobaculum thiosulfatiphilum* DSM 249 (Y08102), *Pelodictyon luteolum* DSM 273 (CP000096), *Chlorobium limicola* DSM 245 (CP001097), *Chlorobaculum tepidum* TLS (M58468), *Chlorobium vibrioforme* DSM 260 (M62791); and (5) FAPs: *Chloroflexus aurantiacus* J-10-fl (M34116), *Chloroflexus aggregans* (D32255), *Oscillochloris trichoides* (AF093427), *Roseiflexus castenholzii* DSM 13941 (AB041226)

include duplicate genes and gene clusters for the alternative complex III (ACIII) [26,27], auracyanin (a type I blue copper protein) [28,29] and NADH:quinone oxidoreductase (complex I); and several aerobic/anaerobic enzyme pairs in central carbon metabolism (pyruvate metabolism and the tricarboxylic acid (TCA) cycle) and tetrapyrroles and nucleic acids biosynthesis [5]. Overall, genomic information is consistent with a high tolerance for oxygen that has been reported in the growth of *Cfl. aurantiacus*.

Phylogenetic analyses on the photosystems and comparisons to the genome and reports of other photosynthetic bacteria suggest lateral or horizontal gene transfers between *Cfl. aurantiacus* and other photosynthetic bacteria [3,30,31]. The *Cfl. aurantiacus* genome suggests possible evolutionary connections of photosynthesis. Here we probe some proposed lateral gene transfers using the phylogenetic analyses on important proteins/enzymes on chlorophyll biosynthesis, photosynthetic electron transport chain, and central carbon metabolism. Further, we also discuss the evolutionary perspectives on assembling photosynthetic machinery, autotrophic carbon assimilation and unique components on the electron transport chains of *Cfl. aurantiacus* and other phototrophic and non-phototrophic bacteria.

2. Results and discussion

a.   Photosynthetic components

The photosystem of *Cfl. aurantiacus* is a chimeric system with contains a peripheral light harvesting complex chlorosomes and an integral membrane B808-866-type II RC (quinone-type) core complex. Chlorosomes are typically found in type I (Fe-S type) RC phototrophic organisms, such as green sulfur bacteria (GSBs) [32] and the recently discovered aerobic anoxygenic bacterium *Chloroacidobacterium thermophilium* [1], whereas the B808-866-RC core complex is arranged similarly to the LH-RC core complex in phototrophic Proteobacteria [33]. Thus, the *Cfl. aurantiacus* photosystem indicates little correlation between the RC type and light-harvesting antenna complexes in the assembly of the photosystem of anoxygenic phototrophic bacteria [8,34]. Two hypotheses, which are selective loss and fusion, for evolutionary of photosynthetic RCs have been proposed [8,35]. The phylogenic analyses and evolutionary perspectives of the integral membrane-RC core complex in *Cfl. aurantiacus* and other phyla of phototrophic bacteria are presented in several reports [8,36,37] for readers who are interested in further information. It is possible that during the evolution of photosynthesis chlorosomes were transferred between *Cfl. aurantiacus* and GSBs, which have larger chlorosomes and more genes encoding chlorosome proteins [38,39], and that the integral membrane core antenna complex and a type II RC in *Cfl. aurantiacus* were possibly transferred either to or from photosynthetic anoxygenic Proteobacteria.

b.   Electron transfer complexes

Four copies of auracyanin genes have been identified in the *Cfl. aurantiacus* genome and two aurancyanin proteins have been characterized biochemically and structurally [28]. Auracyanin has also been biochemically characterized in *Roseiflexus castenholzii* [40], which only has one copy of aurancyanin gene in the genome [5]. The gene encoding a putative auracyanin has been identified in the genome of the non-photosynthetic aerobic thermophilic bacterium *Thermomicrobium roseum* DSM 5159, which is evolutionarily related to *Cfl. aurantiacus* [41]. Genes encoding auracyanin may have been transferred either to or from

Thermomicrobium roseum. Further, higher plants, green algae and cyanobacteria operate the photosynthetic electron transport via a water-soluble mobile type I blue copper protein plastocyanin. Auracyanin may have evolved from or to plastocyanin in cyanobacteria.

Most of phototrophic bacteria use the cytochrome bc_1 or b_6/f complex for transferring electrons during phototrophic growth, whereas *Chloroflexi* species operate photosynthetic electron transport using a unique complex, namely alternative complex III (ACIII) [1,26,27]. Two sets of ACIII gene clusters, one containing seven genes and the other containing thirteen genes, have been identified in the *Cfl. aurantiacus* genome [5]. The seven subunit complex has been characterized biochemically [27]. In contrast, *Roseiflexus castenholzii*, which is a member of a familia *Chloroflexaceae* and phylogenetically closely related to *Cfl. aurantiacus* [42], contains only one copy of the ACIII operon with a six-gene cluster (Rcas_1462-1467) [5]. In addition to *Cfl. aurantiacus* and other members of *Chloroflexaceae*, genes encoding ACIII, which contains seven subunits [27], have also been identified in the *Chloroacidobacterium thermophilium* genome [1]. ACIII has also been identified in non-phototrophic bacterium *Rhodothermus marinus* [43] and suggested to wide-spread in prokayrotes [44]. Genes encoding ACIII may have been transferred either from or to evolved from or to *Chloroacidobacterium thermophilium* (and/or *Rhodothermus marinus*). Further, ACIII may have evolved from or to the cytochrome bc_1 or b_6/f complex.

NADH:quinone oxidoreductase (Complex I, EC 1.6.5.3) is known to be responsible for the electron transport in the respiratory chain. Two sets of the Complex I genes, one of which forms a gene cluster, have been identified in the *Cfl. aurantiacus* genome [5]. Two Complex I gene clusters have also been identified in some anaerobic anoxygenic phototrophic Proteobacteria (AnAPs), such as *Rhodobacter* [Rba.] *sphaeroides* and *Rhodopseudomonas* [Rps.] *palustris*, and gene expression profile in *Rba. sphaeroides* suggests that one of the gene clusters is responsible for photosynthetic electron transport during phototrophic and anaerobic growth and the other is required for the respiratory chain during aerobic and dark growth [45]. **Fig. 2** shows the phylogenetic trees constructed based on the amino acid sequences of the subunit F of Complex I (encoded by the *nuoF* gene) in phototrophic bacteria. The subunit F protein in **Fig. 2A** is encoded by the gene locus Caur_2901 in the gene cluster (Caur_2896 to Caur_2909), and the subunit F protein in **Fig. 2B** is encoded by the gene locus Caur_1185. No Complex I genes have been identified in the green sulfur bacteria, which cannot respire or grow in darkness. Note that one subunit F protein in *Cfl. aurantiacus* is more related to the protein in anoxygenic phototrophic Proteobacteria than to the protein in heliobacteria and cyanobacteria (**Fig. 2A**) and the other *Cfl. aurantiacus* subunit F protein is more related to the protein in heliobacteria and cyanobacteria than to the protein in anoxygenic Proteobacteria (**Fig. 2B**), suggesting different biological functions for two NADH:quinone oxidoreductase complexes found in the *Cfl. aurantiacus* genome.

c.   (Bacterio)chlorophyll biosynthesis

AcsF (aerobic cyclase) and BchE (anaerobic cyclase) are suggested to be responsible for biosynthesis of the isocyclic ring of (bacterio)chlorophylls and conversion of Mg-protoporphyrin monomethyl ester (MgPMMe) to Mg-divinyl-protochlorophyllide *a* (PChlide) under aerobic and anaerobic growth conditions, respectively [46-51] (**Fig. 3A**). Both MgPMMe and PChlide are suggested to be photosensitizers of higher plants and green algae that produce reactive oxygen species in response to the excess light [52]. Both *acsF* (Caur_2590) and *bchE* (Caur_3676) are detected in the *Cfl. aurantiacus* genome [5]. AcsF has

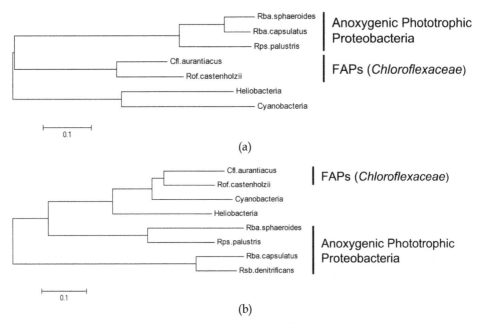

Fig. 2. Phylogenetic tree of the NADH:quinine oxidoreductase (Complex I) in phototrophic bacteria.
The subunit F proteins of *Cfl. aurantiacus*, *Roseiflexus* [Rof.] *castenholzii* (FAPs), *Rhodobacter* [Rba.] *sphaeroides* and *Rhodopseudomonas* [Rps.] *palustris* (anoxygenic Proteobacteria) in **Fig. 2A** and **2B** are encoded by different *nuoF* genes. Two Complex I are identified in *Cfl. aurantiacus*, *Rof. castenholzii*, *Rba. sphaeroides* and *Rps. palustris*, and one Complex I gene cluster is found in heliobacteria, cyanobacteria and some phototrophic anoxygenic Proteobacteria (e.g., *Rba. capsulatus* and *Roseobacter* [Rsb.] *denitrificans*). The trees are constructed based on amino acid sequences using the phylogenetic software MEGA5 [65] with un-rooted neighbor jointing method.

not been identified in any strictly anaerobic phototrophic bacteria (e.g., green sulfur bacteria and heliobacteria). In addition to Proteobacteria (including aerobic and anaerobic anoxygenic phototrophic Proteobacteria) and cyanobacteria, several non-phototrophic α-Proteobacteria also contain the *acsF* gene, including several facultative methotrophic bacteria (e.g., *Methylocella silvestris*, *Methylobacterium* [Mtb.] sp. 4-46, *Mtb. populi*, *Mtb. chloromethanicum*, *Mtb. radiotolerans* and *Mtb. extorquens*) and the environmental bacterium *Brevundimonas subvibrioides* (**Fig. 3B**). Roles of the gene encoding the putative AcsF in these non-phototrophic bacteria are unclear. AcsF has also been characterized for *Cfl. aurantiacus* grown under anaerobic conditions [50]. Together, the role of AcsF remains to be further understood. BchE is widely spread in all phyla of anoxygenic phototrophic bacteria (e.g., anoxygenic phototrophic Proteobacteria, green sulfur bacteria, heliobacteria and FAPs) and some facultative methyltrophic bacteria and cynaobacteria also contain the gene encoding the putative BchE (**Fig. 3C**). Experimental evidence indicates that the *bchE* genes in the cyanobacterium *Synechocystis* sp. PCC 6803 are important but do not contribute to the formation of the isocyclic ring of chlorophylls [47].

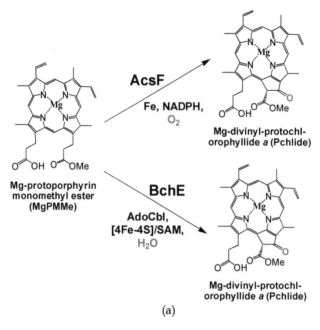

Mg-protoporphyrin
monomethyl ester
(MgPMMe)

AcsF

Fe, NADPH,
O_2

Mg-divinyl-protochl-
orophyllide *a* (Pchlide)

BchE

AdoCbl,
[4Fe-4S]/SAM,
H_2O

Mg-divinyl-protochl-
orophyllide *a* (Pchlide)

(a)

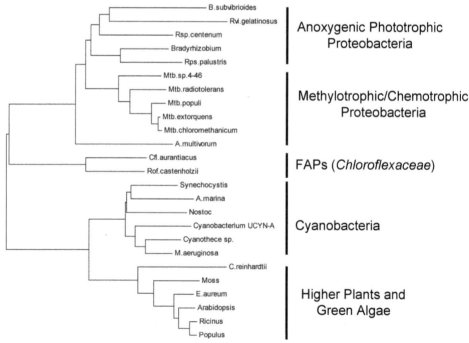

Anoxygenic Phototrophic
Proteobacteria

Methylotrophic/Chemotrophic
Proteobacteria

FAPs (*Chloroflexaceae*)

Cyanobacteria

Higher Plants and
Green Algae

(b)

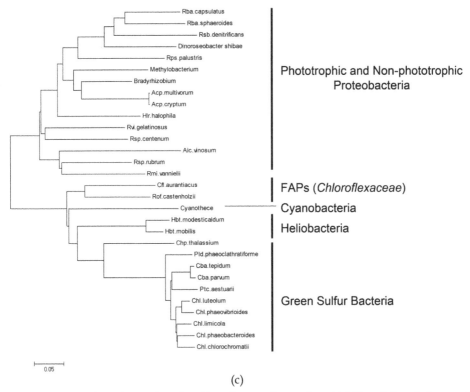

(c)

Fig. 3. Reactions of aerobic cyclase (AcsF) and anaerobic cyclase (BchE) and the phylogenetic trees.
Conversion of MgPMMe into PChlide is suggested to be catalyzed by AcsF and BchE under aerobic and anaerobic conditions, respectively (A). The phylogenetic relationships of AcsF (B) and BchE (C) are shown. The trees are constructed based on amino acid sequences using the phylogenetic software MEGA5 [65] with un-rooted neighbor jointing method.

Phylogenetic analyses suggest that the *acsF* gene in *Cfl. aurantiacus* and other *Chloroflexaceae* species are more evolutionarily related to the genes in anoxygenic phototrophic Proteobacteria than to the genes in oxygenic phototrophs (cyanobacteria, green algae and higher plants) (**Fig. 3B**), and that the *bchE* gene in *Cfl. aurantiacus* is more evolutionarily related to the genes in strictly anaerobic phototrophs (green sulfur bacteria and heliobacteria) than to the genes in phototrophic and non-phototrophic Proteobacteria (**Fig. 3C**). It is possible that the *Cfl. aurantiacus acsF* gene was transferred either to or from Proteobacteria, and the *Cfl. aurantiacus bchE* gene was transferred either to or from heliobacteria and green sulfur bacteria. The phylogenetic analyses of AcsF and BchE in **Fig. 3** likely suggest horizontal gene transfers among phototrophic bacteria and also between phototrophic and non-phototrophic bacteria.

d.   Central carbon metabolism

Here we analyze enzymes/gene products for pyruvate metabolism, which takes place in every living organism, and the TCA cycle. In contrast to other phyla of phototrophic bacteria, *Cfl. aurantiacus* and other members of *Chloroflexaceae* are only bacteria containing both anaerobic and aerobic gene pairs for pyruvate and α-ketoglutarate metabolism: pyruvate/α-ketoglutarate dehydrogenase (aerobic enzymes) and pyruvate/α-ketoglutarate synthase (or pyruvate/α-keto-glutarate:ferredoxin oxidoreductase (PFOR/KFOR)) (anaerobic enzymes).

Fig. 4A shows the phylogenetic analyses of the E1 protein of α-ketoglutarate dehydrogenase (encoded by *sucA*) from FAPs and anoxygenic phototrophic Proteobacteria. Note that the *Cfl. aurantiacus* α-ketoglutarate dehydrogenase has higher sequence identities to many gram-(+) non-phototrophic *Bacillus* strains (~50%) than phototrophic anoxygenic Proteobacteria (~40%). Similar results also find in the sequence alignments of the E1 protein of pyruvate dehydrogenase, and the *Cfl. aurantiacus* enzyme has ~51-55% identities with *Thermobifida fusca*, *Streptomyces cattleya*, *Acidothermus cellulolyticus*, *Saccharopolyspora erythraea*, and *Sanguibacter keddieii* and ~38-44% or lower identities with the phosynthetic Proteobacteria and cyanobacteria (data not shown). These results support the horizontal gene transfer between microbial genomes. **Fig. 4B** shows the phylogenetic tree of the E1 protein of pyruvate dehydrogenase. The *Cfl. aurantiacus* enzyme is less related to cyanobacteria and anoxygenic phototrophic Proteobacteria.

Fig. 4C suggests that α-ketoglutarate synthase in *Cfl. aurantiacus* are more closely related to the enzyme in heliobacteria than in green sulfur bacteria. While the biochemical studies of the *Cfl. aurantiacus* α-ketoglutarate synthase have not been reported, the phylogenetic analyses of α-ketoglutarate synthase are consistent with the central carbon flow in these three phyla of photosynthetic bacteria: the green sulfur bacteria operate the reductive (reverse) TCA cycle, and *Cfl. aurantiacus* and heliobacteria have strong carbon flow via either a complete or a partial oxidative (forward) TCA cycle [34].

Fig. 4D suggests that pyruvate synthase in heliobcteria evolved prior to the enzymes in other phyla of photosynthetic bacteria, and that the enzyme in *Cfl. auranticus* is remotely related to the enzymes in GSBs and cyanobacteria, which are likely from the same origins, similar to the tree of the E1 protein of pyruvate dehydrogenase (**Fig. 4B**). Together, the phylogenetic analyses suggest pyruvate metabolism of anoxygenic phototrophic Proteobacteria is more related to cyanobacteria than to *Cfl. aurantiacus* (and perhaps FAPs). Compared to the experimental data, acetate can support the growth of *Cfl. aurantiacus* during anaerobic growth in the light and during aerobic growth in darkness [53], and acetate excretion has been reported during the pyruvate-grown heliobacteria [54,55] but not on other phyla of photosynthetic bacteria. *Cfl. aurantiacus* likely uses pyruvate synthase for assimilate acetyl-CoA. Since heliobacteria do not have pyruvate dehydrogenase, their pyruvate synthase is supposed to convert pyruvate to acetyl-CoA, which is then converted to acetate. Further, pyruvate synthase is essential for the growth of green sulfur bacteria because it is required to convert acetyl-CoA generated from the reductive TCA cycle to pyruvate, whereas the role of pyruvate synthase in oxygenic phototrophic bacteria (cyanobacteria) is not clear, as pyruvate synthase is sensitive to oxygen during biochemical characterization *in vitro*.

e.   Autotrophic carbon assimilation

Cfl. aurantiacus can grow photoautotrophically and uses the 3-hydroxypropionate (3HOP) bi-cycle to assimilate inorganic carbon [5,56-58]. Both 3HOP bi-cycle and the widely distributed Calvin-Benson cycle can operate in both aerobic and anaerobic conditions.

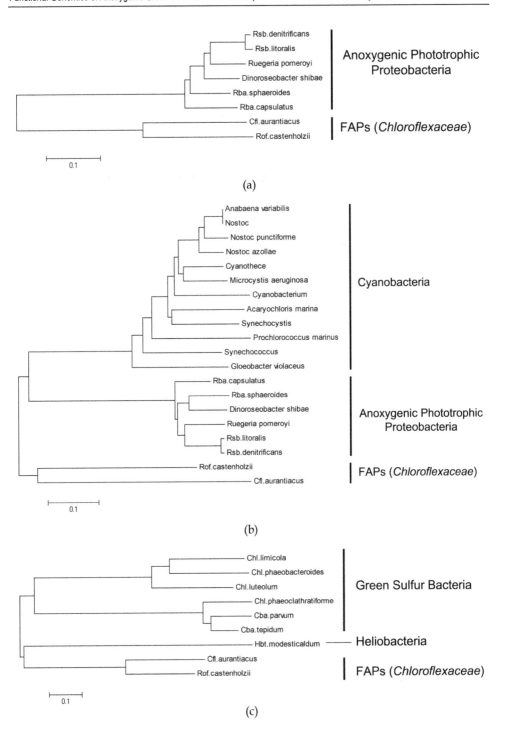

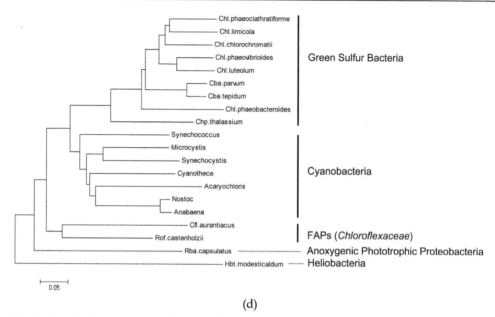

(d)

Fig. 4. The phylogenetic trees of α-ketoglutarate dehydrogenase (A), pyruvate dehydrogenase (B), α-ketoglutarate synthase (C) and pyruvate synthase (D).
The trees are constructed based on amino acid sequences using the phylogenetic software MEGA5 [65] with un-rooted neighbor jointing method.

However, one significant problem leading to low photosynthesis efficiency of higher plants and oxygenic phototrophs is photorespiration and energy waste resulting from the interactions of oxygen with RuBisCO (ribulose 1,5-bisphosphate carboxylase/oxygenase) [12], the carboxylase in the Calvin-Benson cycle. Different from the Calvin-Benson cycle, the 3HOP bi-cycle assimilates bicarbonate instead of CO_2 (**Fig. 5A**). The 3HOP bi-cycle, which operates in *Cfl. aurantiacus* and most likely in other members of *Chloroflexaceae* [57], is similar to 3-hydroxypropionate/4-hydroxybutyrate (3HOP/4HOB) cycle reported in several archaea [59,60] (**Fig. 5B**). Several enzymes operate in both 3HOP bi-cycle and 3HOP/4HOB cycle, including enzymes for assimilating inorganic carbon: acetyl-CoA carboxylase and propionyl-CoA carboxylase. 16S rRNA analyses suggest that Archaea developed earlier than the bacteria capable of using light as the energy sources [3], so the 3HOP bi-cycle may have evolved from the 3HOP/4HOB cycle.

Other horizontal gene transfers can be also found in the autotrophic carbon assimilation on other members of *Chloroflexales*. For example, several strains in the family of *Oscillochloridaceae* assimilate inorganic carbon via the Calvin-Benson cycle and have an incomplete TCA cycle [61]. In addition to oxygenic phototrophs, anaerobic anoxygenic phototrophic Proteobacteria (AnAPs) also operate the Calvin-Benson cycle. In contrast to oxygenic phototrophs, poor substrate specificity of RuBisCO should not be a serious concern for anoxygenic phototrophs like AnAPs and *Oscillochloridaceae*. It is possible that the genes in the Calvin-Benson cycle in may transfer between *Oscillochloridaceae*, AnAPs and cyanobacteria. Furthermore, *Dehalococcoides ethanogenes* strain 195, a Gram-positive non-phototrophic bacteria in the subphylum 2 of *Chloroflexi* [62], uses (*Re*)-citrate synthase [63] and has a branched TCA cycle [63,64]. Together, three members of the phylum *Chloroflexi*, *Cfl. aurantiacus*, *Oscillochloridaceae* and *Dehalococcoides ethanogenes* have distinct central carbon metabolic pathways.

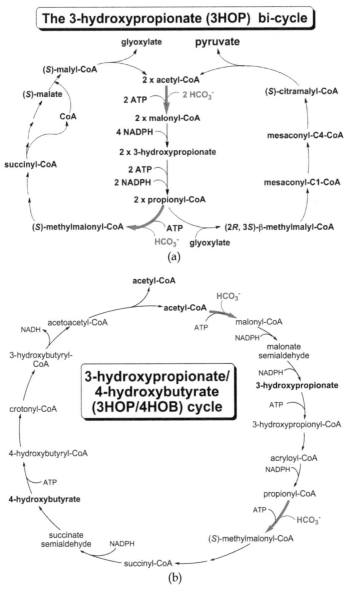

Fig. 5. Autotrophic carbon assimilations through 3-hydroxypropionate bi-cycle (A) and 3-hydroxypropionate/4-hydroxybutyrate cycle.
Several enzymes, including acetyl-CoA carboxylase and propionyl-CoA carboxylase, operate in both 3HOP bi-cycle and 3HOP/4HOB cycle.

3. Conclusions

Previous physiological, ecological and biochemical studies [4] as well as genomic analyses [5], indicate that the anoxygenic phototroph bacterium *Cfl. aurantiacus* has many interesting and certain unique features in its metabolic pathways. The evolutionary links of *Cfl. aurantiacus* and other phototrophic bacteria suggested from this report are summarized in **Table 1**. It has been recognized that the type II RCs were transferred between the *Chloroflexi* species (or FAPs) and the anoxygenic phototrophic Proteobacteria. Sequence alignments and phylogenetic analyses illustrated in this report suggest: (i) Some *Cfl. aurantiacus* enzymes in essential metabolic pathways are more related to the anoxygenic phototrophic Proteobacteria than other phototrophic bacteria, whereas other enzymes are more related to other phototrophic bacteria than anoxygenic phototrophic Proteobacteria; and (ii) some *Cfl. aurantiacus* enzymes in essential carbon metabolic pathways are more related to non-photosynthetic microbes than other phyla of phototrophic bacteria. Together, our studies support lateral/horizontal gene transfers among microbes, and suggest that photosynthesis is likely an adaption to the environments [9].

	Anoxygenic Phototrophic Proteobacteria	Green Sulfur Bacteria	Heliobacteria	Cyanobacteria
Photosynthetic Machinery				
Chlorosomes		√		
B808-866 core antenna complex	√			
Reaction center	√			
Electron Transport Chain				
Complex I -A[a]	more related		less related	less related
Complex I -B[b]	less related		related	more related
Auracyanin				√
Pigment Biosynthesis				
Aerobic cyclase (AcsF)	more related			less related
Anaerobic cyclase (BchE)	less related	more related	more related	?[c]
Central Carbon Metabolism				
Pyruvate dehydrogenase	√			√
Pyruvate synthase	related	more related	less related	more related
α-ketoglutarate dehydrogenase	√			
α-ketoglutarate synthase		less related	more related	

[a] The *Cfl. aurantiacus* complex I with clustered genes (**Fig. 2A**)
[b] The *Cfl. aurantiacus* complex I without clustered genes (**Fig. 2B**)
[c] The putative bchE genes in some cyanobacteria have been reported not to function as anaerobic cyclase.

Table 1. Evolutionary perspectives of selective proteins/enzymes/cellular complexes in *Cfl. aurantiacus* versus other phyla of phototrophic bacteria

4. Abbreviations

AnAPs:	anaerobic anoxygenic phototrophic Proteobacteria
AcsF:	aerobic cyclase
ACIII:	alternative complex III
BchE:	anaerobic cyclase
FAPs:	filamentous anoxygenic phototrophs (or green non-sulfur bacteria or green gliding bacteria)
GSBs:	green sulfur bacteria
3HOP bi-cycle:	3-hydroxypropionate bi-cycle
3HOP/4HOB cycle:	3-hydroxypropionate/4-hydroxybutyrate cycle
RC:	reaction center
RuBisCO:	ribulose 1,5-bisphosphate carboxylase/oxygenase
TCA cycle:	tricarboxylic acid cycle

5. Acknowledgements

The author thanks Dr. Robert E. Blankenship for introducing the author into the fields of photosynthesis and the financial support of start-up fund from Clark University.

6. References

[1] Bryant, D.A. et al. (2007). *Candidatus* Chloracidobacterium thermophilum: an aerobic phototrophic Acidobacterium. Science 317, 523-6.

[2] Giovannoni, S.J. and Stingl, U. (2005). Molecular diversity and ecology of microbial plankton. Nature 437, 343-8.

[3] Pace, N.R. (1997). A molecular view of microbial diversity and the biosphere. Science 276, 734-40.

[4] Hanada, S. and Pierson, B.K. (2006). The family chloroflexaceae. *The Prokaryotes*, 3rd Ed., Vol. 7, Springer, New York. pp. 815 – 842.

[5] Tang, K.H. et al. (2011). Complete Genome Sequence of the Filamentous Anoxygenic Phototrophic Bacterium *Chloroflexus aurantiacus*. BMC Genomics 12, 334.

[6] Jain, R., Rivera, M.C., Moore, J.E. and Lake, J.A. (2002). Horizontal gene transfer in microbial genome evolution. Theor Popul Biol 61, 489-95.

[7] Pierce, S.K., Massey, S.E., Hanten, J.J. and Curtis, N.E. (2003). Horizontal transfer of functional nuclear genes between multicellular organisms. Biol Bull 204, 237-40.

[8] Hohmann-Marriott, M.F. and Blankenship, R.E. (2011). Evolution of photosynthesis. Annu Rev Plant Biol 62, 515-48.

[9] Leslie, M. (2009). Origins. On the origin of photosynthesis. Science 323, 1286-7.

[10] Blankenship, R.E. (1992). Origin and early evolution of photosynthesis. Photosynth Res 33, 91-111.

[11] Blankenship, R.E. (2001). Molecular evidence for the evolution of photosynthesis. Trends Plant Sci 6, 4-6.

[12] Blankenship, R.E. (2002). *Molecular Mechanisms of Photosynthesis*. Blackwell Science Ltd, Oxford.

[13] Blankenship, R.E. and Hartman, H. (1998). The origin and evolution of oxygenic photosynthesis. Trends Biochem Sci 23, 94-7.

[14] Allen, J.F. (2005). A redox switch hypothesis for the origin of two light reactions in photosynthesis. FEBS Lett 579, 963-8.
[15] Allen, J.F. and Martin, W. (2007). Evolutionary biology: out of thin air. Nature 445, 610-2.
[16] Bjorn, L.O. and Govindjee. (2009). The evolution of photosynthesis and chloroplasts. Curr Sci 96, 1466-1474.
[17] Buick, R. (2006). When did oxygenic photosynthesis evolve? Phil Trans R Soc B 363, 2731-2743.
[18] De Marais, D.J. (2000). Evolution. When did photosynthesis emerge on Earth? Science 289, 1703-5.
[19] Gupta, R.S. (2010). Molecular signatures for the main phyla of photosynthetic bacteria and their subgroups. Photosynth Res 104, 357-72.
[20] Hengeveld, R. (2007). Two approaches to the study of the origin of life. Acta Biotheor 55, 97-131.
[21] Mulkidjanian, A.Y. (2009). On the origin of life in the zinc world: 1. Photosynthesizing, porous edifices built of hydrothermally precipitated zinc sulfide as cradles of life on Earth. Biol Direct 4, 26.
[22] Olson, J.M. and Blankenship, R.E. (2004). Thinking about the evolution of photosynthesis. Photosynth Res 80, 373-86.
[23] Olson, J.M. and Pierson, B.K. (1987). Evolution of reaction centers in photosynthetic prokaryotes. Int Rev Cytol 108, 209-48.
[24] Raven, J. (2009). Functional evolution of photochemical energy transformation in oxygen-producing organisms. Funct Plant Biol 36, 505-515.
[25] Shi, T. and Falkowski, P.G. (2008). Genome evolution in cyanobacteria: the stable core and the variable shell. Proc Natl Acad Sci U S A 105, 2510-5.
[26] Yanyushin, M.F., del Rosario, M.C., Brune, D.C. and Blankenship, R.E. (2005). New class of bacterial membrane oxidoreductases. Biochemistry 44, 10037-45.
[27] Gao, X., Xin, Y., Bell, P.D., Wen, J. and Blankenship, R.E. (2010). Structural analysis of alternative complex III in the photosynthetic electron transfer chain of Chloroflexus aurantiacus. Biochemistry 49, 6670-9.
[28] Lee, M., del Rosario, M.C., Harris, H.H., Blankenship, R.E., Guss, J.M. and Freeman, H.C. (2009). The crystal structure of auracyanin A at 1.85 A resolution: the structures and functions of auracyanins A and B, two almost identical "blue" copper proteins, in the photosynthetic bacterium Chloroflexus aurantiacus. J Biol Inorg Chem 14, 329-45.
[29] McManus, J.D., Brune, D.C., Han, J., Sanders-Loehr, J., Meyer, T.E., Cusanovich, M.A., Tollin, G. and Blankenship, R.E. (1992). Isolation, characterization, and amino acid sequences of auracyanins, blue copper proteins from the green photosynthetic bacterium Chloroflexus aurantiacus. J Biol Chem 267, 6531-40.
[30] Oyaizu, H., Debrunner-Vossbrinck, B., Mandelco, L., Studier, J.A. and Woese, C.R. (1987). The green non-sulfur bacteria: a deep branching in the eubacterial line of descent. Syst Appl Microbiol 9, 47-53.
[31] Woese, C.R. (1987). Bacterial evolution. Microbiol Rev 51, 221-71.
[32] Frigaard, N.-U. and Bryant, D.A. (2006). Chlorosomes: antenna organelles in photosynthetic green bacteria. In Shively JM (ed) Complex intracellular structures in prokaryotes. Springer, Berlin, pp 79–114.

[33] Tang, K.H., Urban, V.S., Wen, J., Xin, Y. and Blankenship, R.E. (2010). SANS investigation of the photosynthetic machinery of *Chloroflexus aurantiacus*. Biophys J 99, 2398-407.

[34] Tang, K.H., Tang, Y.J. and Blankenship, R.E. (2011). Carbon metabolic pathways in phototrophic bacteria and their broader evolutionary implications. Front Microbio 2, 165.

[35] Mathis, P. (1990). Compared structure of plant and bacterial photosynthetic reaction centers. Evolutionary implications. Biochim Biophys Acta 1018, 163–7.

[36] Blankenship, R.E. (2010). Early evolution of photosynthesis. Plant Physiol 154, 434-8.

[37] Mix, L.J., Haig, D. and Cavanaugh, C.M. (2005). Phylogenetic analyses of the core antenna domain: investigating the origin of photosystem I. J Mol Evol 60, 153-63.

[38] Bryant, D.A. et al. (2011). Comparative and functional genomics of anoxygenic green bacteria from the taxa *Chlorobi*, *Chloroflexi*, and *Acidobacteria*. *In* Burnap RL, and Vermaas W, (eds.), *Advances in Photosynthesis and Respiration, Functional Genomics and Evolution of Photosynthetic Systems*. Vol. 33, Springer, Dordrecht. in press.

[39] Frigaard, N.U., Chew, A.G.M., Maresca, J.A. and Bryant, D.A. (2006). Bacteriochlorophyll Biosynthesis in Green Bacteria. *In* Grimm B, Porra RJ, Rudiger W, and Scheer H (eds), *Chlorophylls and Bacteriochlorophylls*. Springer Academic Publishers, Dordrecht. pp 201-221

[40] Tsukatani, Y., Nakayama, N., Shimada, K., Mino, H., Itoh, S., Matsuura, K., Hanada, S. and Nagashima, K.V. (2009). Characterization of a blue-copper protein, auracyanin, of the filamentous anoxygenic phototrophic bacterium *Roseiflexus castenholzii*. Arch Biochem Biophys 490, 57-62.

[41] Wu, D. et al. (2009). Complete genome sequence of the aerobic CO-oxidizing thermophile *Thermomicrobium roseum*. PLoS One 4, e4207.

[42] Hanada, S., Takaichi, S., Matsuura, K. and Nakamura, K. (2002). *Roseiflexus castenholzii* gen. nov., sp. nov., a thermophilic, filamentous, photosynthetic bacterium that lacks chlorosomes. Int J Syst Evol Microbiol 52, 187-93.

[43] Refojo, P.N., Teixeira, M. and Pereira, M.M. (2010). The alternative complex III of *Rhodothermus marinus* and its structural and functional association with caa3 oxygen reductase. Biochim Biophys Acta 1797, 1477-82.

[44] Refojo, P.N., Sousa, F.L., Teixeira, M. and Pereira, M.M. (2010). The alternative complex III: a different architecture using known building modules. Biochim Biophys Acta 1797, 1869-76.

[45] Pappas, C.T. et al. (2004). Construction and validation of the *Rhodobacter sphaeroides* 2.4.1 DNA microarray: transcriptome flexibility at diverse growth modes. J Bacteriol 186, 4748-58.

[46] Gough, S.P., Petersen, B.O. and Duus, J.O. (2000). Anaerobic chlorophyll isocyclic ring formation in Rhodobacter capsulatus requires a cobalamin cofactor. Proc Natl Acad Sci U S A 97, 6908-13.

[47] Minamizaki, K., Mizoguchi, T., Goto, T., Tamiaki, H. and Fujita, Y. (2008). Identification of two homologous genes, chlAI and chlAII, that are differentially involved in isocyclic ring formation of chlorophyll a in the cyanobacterium Synechocystis sp. PCC 6803. J Biol Chem 283, 2684-92.

[48] Ouchane, S., Steunou, A.S., Picaud, M. and Astier, C. (2004). Aerobic and anaerobic Mg-protoporphyrin monomethyl ester cyclases in purple bacteria: a strategy adopted to bypass the repressive oxygen control system. J Biol Chem 279, 6385-94.

[49] Pinta, V., Picaud, M., Reiss-Husson, F. and Astier, C. (2002). Rubrivivax gelatinosus acsF (previously orf358) codes for a conserved, putative binuclear-iron-cluster-containing protein involved in aerobic oxidative cyclization of Mg-protoporphyrin IX monomethylester. J Bacteriol 184, 746-53.

[50] Tang, K.H., Wen, J., Li, X. and Blankenship, R.E. (2009). Role of the AcsF protein in Chloroflexus aurantiacus. J Bacteriol 191, 3580-7.

[51] Tottey, S., Block, M.A., Allen, M., Westergren, T., Albrieux, C., Scheller, H.V., Merchant, S. and Jensen, P.E. (2003). Arabidopsis CHL27, located in both envelope and thylakoid membranes, is required for the synthesis of protochlorophyllide. Proc Natl Acad Sci U S A 100, 16119-24.

[52] Li, Z., Wakao, S., Fischer, B.B. and Niyogi, K.K. (2009). Sensing and responding to excess light. Annu Rev Plant Biol 60, 239-60.

[53] Madigan, M.T., Petersen, S.R. and Brock, T.D. (1974). Nutritional studies on Chloroflexus, a filamentous photosynthetic, gliding bacterium. Arch Microbiol 100, 97-103.

[54] Pickett, M.W., Williamson, M.P. and Kelly, D.J. (1994). An enzyme and 13C-NMR of carbon metabolism in heliobacteria. Photosynth. Res. 41, 75-88.

[55] Tang, K.H., Yue, H. and Blankenship, R.E. (2010). Energy metabolism of Heliobacterium modesticaldum during phototrophic and chemotrophic growth. BMC Microbiol 10, 150.

[56] Holo, H. (1989). Chloroflexus aurantiacus secretes 3-hydroxypropionate, a possible intermediate in the assimilation of CO_2 and acetate. Arch Microbiol 151, 252-256.

[57] Zarzycki, J., Brecht, V., Muller, M. and Fuchs, G. (2009). Identifying the missing steps of the autotrophic 3-hydroxypropionate CO_2 fixation cycle in Chloroflexus aurantiacus. Proc Natl Acad Sci U S A 106, 21317-22.

[58] Strauss, G. and Fuchs, G. (1993). Enzymes of a novel autotrophic CO2 fixation pathway in the phototrophic bacterium Chloroflexus aurantiacus, the 3-hydroxypropionate cycle. Eur J Biochem 215, 633-43.

[59] Hugler, M. and Sievert, S.M. (2011). Beyond the Calvin cycle: autotrophic carbon fixation in the ocean. Ann Rev Mar Sci 3, 261-89.

[60] Berg, I.A., Kockelkorn, D., Buckel, W. and Fuchs, G. (2007). A 3-hydroxypropionate/4-hydroxybutyrate autotrophic carbon dioxide assimilation pathway in Archaea. Science 318, 1782-6.

[61] Berg, I.A., Keppen, O.I., Krasil'nikova, E.N., Ugol'kova, N.V. and Ivanovskii, R.N. (2005). Carbon metabolism of filamentous anoxygenic phototrophic bacteria of the family Oscillochloridaceae. Microbiology 74, 258-264.

[62] Seshadri, R. et al. (2005). Genome sequence of the PCE-dechlorinating bacterium Dehalococcoides ethenogenes. Science 307, 105-8.

[63] Tang, Y.J., Yi, S., Zhuang, W.Q., Zinder, S.H., Keasling, J.D. and Alvarez-Cohen, L. (2009). Investigation of carbon metabolism in Dehalococcoides ethenogenes strain 195 by use of isotopomer and transcriptomic analyses. J Bacteriol 191, 5224-31.

[64] West, K.A. et al. (2008). Comparative genomics of Dehalococcoides ethenogenes 195 and an enrichment culture containing unsequenced Dehalococcoides strains. Appl Environ Microbiol 74, 3533-40.

[65] Tamura, K., Peterson, D., Peterson, N., Stecher, G., Nei, M. and Kumar, S. (2011). MEGA5: Molecular Evolutionary Genetics Analysis using Maximum Likelihood, Evolutionary Distance, and Maximum Parsimony Methods. Mol Biol Evol, in press.

7

Modulation of EAAC1-Mediated Glutamate Uptake by Addicsin

Mitsushi J. Ikemoto and Taku Arano
Biomedical Research Institute
National Institute of Advanced Industrial Science and Technology (AIST)
Graduate School of Science, Toho University
Japan

1. Introduction

Glutamate is the major excitatory neurotransmitter in the mammalian central nervous system (CNS). In addition to functioning as a neurotransmitter at the majority of brain synapses, it is the substrate for synthesis of the major inhibitory transmitter γ-aminobutyric acid (GABA). However, glutamate is also a neurotoxin, and a number of molecular control mechanisms are responsible for maintaining extracellular glutamate below excitotoxic levels. Na^+-dependent excitatory amino acid transporters (EAATs) are crucial regulators of extracellular glutamate and also act to control the dynamics of excitatory transmission in the CNS (Danbolt, 2001). The Na^+-dependent excitatory amino acid carrier 1 (EAAC1) is expressed in the somata and dendrites of many neuronal types, including pyramidal cells of the hippocampal formation and cortex, and many subtypes of GABAergic inhibitory neurons (Rothstein et al., 1994). The physiological significance of EAAC1 is unclear because the subcellular distribution and kinetic properties of this transporter would not allow for a substantial contribution to glutamate clearance from the synaptic cleft; rather, these functions are mediated by glial EAATs (EAAT1 and EAAT2) located in the perisynaptic region. Recent studies have demonstrated multiple functions for EAAC1 distinct from clearance of glutamate from CNS synapses (Kiryu-Seo et al., 2006; Levenson et al., 2002; Peghini et al., 1997; Sepkuty et al., 2002). For example, decreased EAAC1 expression in the CNS impairs neuronal glutathione (GSH) synthesis, leading to oxidative stress and age-dependent neurodegeneration (Aoyama et al., 2006), suggesting that aberrant EAAC1 expression contributes to the pathogenesis of neurodegenerative diseases.

Studies conducted over the past decade on the kinetics of EAAC1 and regulation of transporter expression and function have lead to a greater appreciation of the physiological and pathophysiological relevance of EAAC1 (Aoyama et al., 2008b; Danbolt, 2001; Kanai & Hediger, 2004; Nieoullon et al., 2006), but there are many issues to be resolved for a thorough understanding of the significance of EAAC1 in normal brain function and disease. In particular, the regulatory mechanisms of EAAC1-mediated glutamate uptake are largely unknown. The recent discovery of addicsin (glutamate transporter-associated protein 3-18, GTRAP3-18) as an EAAC1 binding protein has contributed greatly to our understanding of the regulatory mechanisms of EAAC1 activity (Lin et al., 2001). Furthermore, we recently proposed a regulatory model of EAAC1-mediated glutamate uptake by addicsin complexes (Akiduki & Ikemoto, 2008). In this chapter, we describe the regulation of EAAC1-mediated glutamate uptake based on our recent results. To better understand this regulatory mechanism, we first explain three key molecules involved in this regulatory pathway — EAAC1, addicsin, and ADP-ribosylation factor-like 6 interacting protein 1 (Arl6ip1).

1.1 EAAC1

The EAAC1 protein was first identified as a Na^+-dependent high-affinity glutamate transporter by expression cloning in *Xenopus* oocytes (Kanai & Hediger, 1992). Stoichiometric analysis demonstrates that EAAC1 transports L-glutamate, L-aspartate, and D-aspartate, accompanied by the cotransport of 3 Na^+ and 1 H^+, and the countertransport of 1 K^+ (Kanai & Hediger, 2003). In mammalian tissues, there are five different subtypes of EAATs — EAAT1 (glutamate/aspartate transporter, GLAST), EAAT2 (glutamate transporter 1, GLT-1), EAAT3 (EAAC1), EAAT4, and EAAT5 (Danbolt, 2001). These EAATs are structurally similar; all have eight transmembrane domains and a pore loop between the seventh and eighth domain. Most EAATs play an important role in removing extracellular glutamate from the synaptic and extrasynaptic space (Kanai & Hediger, 2003), particularly GLAST and GLT-1. These two isoforms are primarily expressed in glial cells and play a major role in protecting neurons from glutamate-induced toxicity (Rothstein et al., 1994) as well as terminating glutamatergic transmission (Rothstein et al., 1993; Tong & Jahr, 1994). In contrast, EAAC1 is diffusely localized to the cell bodies and dendrites of neurons and is enriched in cortical and hippocampal pyramidal cells as well as in some inhibitory neurons (Conti et al., 1998; Rothstein et al., 1994). This subcellular localization and restricted distribution indicate that EAAC1 does not play a major role in glutamate clearance from the synaptic cleft (Rothstein et al., 1996). Recent studies suggest that EAAC1 contributes to multiple physiological functions distinct from glutamate clearance. Indeed, EAAC1 transport provides cysteine as a substrate of GSH synthesis (Y. Chen & Swanson, 2003; Himi et al., 2003; Watabe et al., 2008; Zerangue & Kavanaugh, 1996). Neurons cannot transport extracellular GSH and therefore must transport cysteine from the extracellular space for *de novo* GSH synthesis from cysteine (Aoyama et al., 2008b). In the CNS, the depletion of GSH is associated with neurodegenerative disorders, including Alzheimer's and Parkinson's diseases (Ramassamy et al., 2000; Sian et al., 1994). Consistent with these results, EAAC1 knockout mice show oxidative stress in neurons and age-dependent neurodegeneration, pathologies that are rescued by *N*-acetylcysteine, a membrane-permeable cysteine precursor (Aoyama et al., 2006). These mice also show alteration of zinc homeostasis and increased neural damage after transient cerebral ischemia (Won et al., 2010). Furthermore, in a knockin mouse model of Huntington's disease, in which human *huntingtin* exon 1 with 140 CAG repeats was inserted into the wild-type low CGA repeat mouse *huntingtin* gene, oxidative stress and cell death were caused by abnormal Rab11-dependent EAAC1 trafficking to the cell surface (X. Li et al., 2010). In addition, 1-methyl-4-phenyl-1,2,3,6-tetrahydropyridine-treated mice, an animal model of Parkinson's disease, show reduced EAAC1-mediated neuronal cysteine uptake, impaired GSH synthesis, and motor dysfunction (Aoyama et al., 2008a). These results indicate that dysfunctional EAAC1-mediated cysteine transport increases neural vulnerability to oxidative stress and could contribute to the pathogenesis of neurodegenerative diseases.

In addition to cysteine transport, EAAC1 has several other functions unrelated to removal of extracellular glutamate. For instance, EAAC1 promotes GABA synthesis by supplying the substrate glutamate (Mathews & Diamond, 2003; Sepkuty et al., 2002). Therefore, EAAC1 can strengthen inhibitory synapses in response to elevations in extracellular glutamate and contribute indirectly to GABA release (Mathews & Diamond, 2003). Indeed, a loss of EAAC1 function leads to epilepsy (Sepkuty et al., 2002), underscoring the importance of EAAC1 in GABAergic transmission. Furthermore, EAAC1 plays a crucial role in preventing neuronal death by suppressing glutamate excitotoxicity (Kiryu et al., 1995; Murphy et al., 1989) and has a mitochondria-mediated anti-apoptotic function in injured motor neurons (Kiryu-Seo et al., 2006). These studies and those discussed in Section 3.4 strongly suggest that EAAC1 contributes to multiple functions in the CNS distinct from glutamate clearance.

The regulatory mechanisms of EAAC1 have been widely investigated *in vitro*. Cumulative evidence demonstrates that glutamate uptake by EAAC1 is facilitated by cell signaling molecules and accessory proteins that promote the redistribution of EAAC1 from the endoplasmic reticulum (ER) to the plasma membrane. First, several reports demonstrate that several kinase signaling cascades regulate EAAC1 activity. In C6BU-1 glioma cells and primary neuronal cultures, phorbol 12-myristate 13-acetate (PMA), a protein kinase C (PKC) activator, rapidly increases EAAC1-mediated glutamate uptake (Dowd & Robinson, 1996). This effect is regulated by mechanisms that are independent of *de novo* synthesis of new transporters but is related to the redistribution of EAAC1 from subcellular compartments to the plasma membrane (Davis et al., 1998; Fournier et al., 2004; Sims et al., 2000). Pharmacological analyses demonstrate that PKCα regulates EAAC1 translocation from intracellular compartments to the cell surface, and that PKCε increases EAAC1 functional activity (Gonzalez et al., 2002). PKCα interacts with EAAC1 in a PKC-dependent manner and phosphorylates EAAC1 (Gonzalez et al., 2003). Platelet-derived growth factor (PDGF) increases the delivery of EAAC1 to the cell surface through phosphatidylinositol 3-kinase (PI3K) activity (Fournier et al., 2004; Sheldon et al., 2006; Sims et al., 2000). Consistent with this result, wortmannin, a PI3K inhibitor, decreases cell surface expression of EAAC1 and inhibits EAAC1-mediated glutamate uptake (Davis et al., 1998). In addition, PKC and PDGF have different effects on trafficking and internalization of EAAC1; PMA, but not PDGF, reduces internalization of EAAC1 (Fournier et al., 2004). Thus, EAAC1 trafficking is regulated by two independent signaling pathways. In contrast, PKC negatively regulates EAAC1-mediated glutamate uptake in *Xenopus* oocytes (Trotti et al., 2001) and in Madin–Darby canine kidney (MDCK) cells (Padovano et al., 2009) by inhibiting cell surface expression through calcineurin-mediated internalization (Padovano et al., 2009; Trotti et al., 2001), suggesting that the regulatory mechanisms of EAAC1 surface expression and function by PKC are specific to cell type and depend on specific PKC isozymes. Second, accessory proteins regulate EAAC1 activity. For instance, δ opiod receptor interacts with EAAC1 and inhibits EAAC1-mediated glutamate uptake in *Xenopus* oocytes and rat hippocampal neurons (Xia et al., 2006). In addition, N-methyl-D-aspartate receptors containing NR1, NR2A, and/or NR2B interact with EAAC1 and facilitate the cell surface expression of EAAC1 in C6BU-1 cells and rat hippocampal neurons (Waxman et al., 2007). Moreover, the cell surface expression of EAAC1 is controlled by interactions with Na^+/H^+-exchanger regulatory factor 3 (NHERF-3, also called PDZK1) and adaptor protein 2 (AP-2). While NHERF-3 promotes the delivery of EAAC1 to the plasma membrane, AP-2 regulates constitutive endocytosis of EAAC1 in MDCK cells (D'Amico et al., 2010). Furthermore, reticulon 2B (RTN2B) interacts with EAAC1 and addicsin/GTRAP3-18, and promotes intracellular trafficking of EAAC1 in HEK293 cells and cultured cortical neurons (Liu et al., 2008). Addicsin/GTRAP3-18 interacts with EAAC1 and inhibits EAAC1 trafficking in HEK293 cells (Ruggiero et al., 2008). Thus, multiple regulatory mechanisms control EAAC1 trafficking and membrane expression, but the molecular details are generally unclear. In this study, we focus on the regulation of EAAC1 trafficking by addicsin.

1.2 Addicsin

In many papers, human addicsin and rat addicsin are called JWA and GTRAP3-18, respectively. Addicsin, GTRAP3-18, and JWA have been independently identified by several research groups (Ikemoto et al., 2002; Lin et al., 2001; Zhou et al., GeneBank, AF070523, unpublished observations). We first identified *addicsin* as a novel mRNA encoding a 22-kDa hydrophobic protein that is highly expressed in the basomedial nucleus of the mouse amygdala following repeated morphine administration (Ikemoto et al., 2002). Meanwhile, *GTRAP3-18* cDNA was identified as encoding an EAAC1 binding protein by yeast two-hybrid screening of a rat brain cDNA library using the C-terminal intracellular domain of EAAC1 as bait (Lin et al., 2001). The *JWA* gene was identified as an all-*trans* retinoic acid (RA)-responsive factor from human tracheobronchial epithelial cells (Zhou et al., GeneBank, AF070523, unpublished observations). Bioinformatic analysis demonstrates that JWA has a prenylated Rab acceptor 1 (PRA1) domain and 62% similarity with Jena-Muenchen 4 (JM4), a protein recently identified as PRA1 domain family member 2 (PRAF2) (Schweneker et al., 2005). Proteins containing a large PRA1 domain form a new family of PRA1 domain family proteins (PRAFs) that regulate intracellular protein trafficking. Thus, addicsin is a new member of the PRAF family, PRAF3.

The *addicsin* cDNA is approximately 1.4 kbp and consists of a 564-bp single open reading frame (Ikemoto et al., 2002). The *addicsin* gene contains three exons separated by two introns, and the sequence is highly conserved among vertebrates (Butchbach et al., 2002). Furthermore, *addicsin* is located on mouse chromosome 6, a location corresponding to human chromosome 3p (Butchbach et al., 2002; Ikemoto et al., 2002).

Mouse addicsin is a 22-kDa protein of 188 amino acids with putative transmembrane segments (Butchbach et al., 2002; Ikemoto et al., 2002). Mouse addicsin is 98% identical to rat GTRAP3-18 and 95% similar to human JWA (Butchbach et al., 2002; Ikemoto et al., 2002). Moreover, addicsin has two putative PKC phosphorylation motifs (amino acids 18–20 and 138–140) as well as two putative cAMP-dependent protein kinase and calcium/calmodulin-dependent protein kinase II phosphorylation motifs (amino acids 27–31 and 35–39) (Butchbach et al., 2002; Ikemoto et al., 2002) (Fig. 1). However, there is no evidence that these phosphorylation sites are phosphorylated by protein kinases *in vitro* and *in vivo*.

Expression profiles of addicsin and *addicsin* mRNA were investigated in the developing and mature brain. In the developing rat brain, the expression levels of addicsin decrease significantly from embryonic day 17 to post-natal day 0 (Maier et al., 2009). Meanwhile, *addicsin* mRNA levels increase gradually during early maturation, peaking around post-natal day 5, and then declining by about 50% by post-natal day 14 (Inoue et al., 2005). This developmental expression pattern corresponds to periods of elevated synaptogenesis, suggesting that addicsin is involved in synapse formation. Indeed, later in this chapter, we discuss evidence that addicsin participates in intracellular protein trafficking of neurotransmitter receptors. Addicsin is widely distributed in the brain (Akiduki et al., 2007; Butchbach et al., 2002). In the mature CNS, addicsin is expressed in the cerebral cortex, amygdala, striatum, hippocampus (CA1–3 fields), dentate gyrus, and cerebellum. Addicsin is expressed in the somata of glutamatergic and GABAergic neurons and exhibits presynaptic localization in restricted regions such as CA3 stratum lucidum (Akiduki et al., 2007). In situ hybridization analysis reveals that addicsin mRNA is widely distributed in the brain, predominantly expressed in principal neurons, including glutamatergic and GABAergic neurons in the mature CNS (Inoue et al., 2005). However, the precise subcellular localization of addicsin remains controversial. Recent reports found that addicsin is an integral ER membrane protein that prevents EAAC1 maturation and function by inhibiting ER trafficking (Ruggiero et al., 2008). However, our protein fractionation analysis using mouse whole brain lysates prepared in PBS, NaCl, or Na_2CO_3 buffer, all indicate that addicsin is predominantly present in the S1 soluble fraction, while the ER transmembrane protein calnexin is present in the P2 pellet fraction (Ikemoto et al., 2002). Our subcellular fractionation analysis with highly purified synaptic fractions prepared from mouse forebrain also support the notion that addicsin is present in the cytoplasmic and presynaptic membrane fractions (Akiduki et al., 2007). Furthermore, immunocytochemical studies reveal that addicsin is present in both the plasma membrane and the intracellular compartments, including the ER (Ikemoto et al., 2002; Watabe et al., 2007, 2008). Consistent with these findings, bioinformatic analysis demonstrates that the α-helix is not long enough for a transmembrane domain; nevertheless, addicsin is predicted to be a hydrophobic protein composed of 62% α-helix and 8% β-sheet (Butchbach et al., 2002), suggesting that it is membrane-associated. Further investigations are needed to clarify the subcellular localization of addicsin, but it is apparent that this protein can exist in both soluble and membrane-associated forms.

Addicsin easily forms homo- and heteromultimers (Ikemoto et al., 2002; Lin et al., 2001) and many reports demonstrate that addicsin can associate with a multitude of proteins (Akiduki & Ikemoto, 2008), including Arl6ip1 (Akiduki & Ikemoto, 2008), ARL6 (Ingley et al., 1999), δ opioid receptor (Wu et al., 2011), EAAC1 (Lin et al., 2001), Rab1 (Maier et al., 2009), and RTN2B (Liu et al., 2008). Moreover, recent studies using the yeast two-hybrid system revealed many potential addicsin-binding proteins (M.J. Ikemoto et al., unpublished data), strongly suggesting that addicsin exerts multiple physiological functions by forming various molecular complexes. It is vital to catalog these interacting proteins and to determine the presence and location of these molecular complexes.

These potential functions remain largely speculative, but molecular studies have provided several intriguing candidates (Fig. 2). First, addicsin is involved in apoptosis induced by 12-O-tetradecanoylphorbol-13-acetate, all-trans RA, N-(4-hydroxyphenyl) retinamide, arsenic trioxide, and cadmium (Mao et al., 2006; Zhou et al., 2008). Knockdown of addicsin attenuates all-trans RA-induced and arsenic trioxide-induced apoptosis (Mao et al., 2006; Zhou et al., 2008). Therefore, addicsin serves as a pro-apoptotic molecule. Second, addicsin acts as an environmental stress sensor to protect cells from oxidative stress and subsequent genomic damage. Addicsin is also involved in cellular responses to environmental stresses, including oxidative stress and heat shock, and in the differentiation of leukemia cells under nonphysiological conditions (Cao et al., 2007; Huang et al., 2006a, 2006b; T. Zhu et al., 2005). Addicsin is upregulated after exposure to the pro-oxidants benzo[α]pyrene and hydrogen peroxide through activation of the nuclear transcription factor I (NFI) (R. Chen et al., 2007). Addicsin facilitates DNA repair by interacting with X-ray cross-complementing group 1 protein, a regulator of the DNA base excision repair processes that translocates to the nucleus in response to oxidative stress (R. Chen et al., 2007; Wang et al., 2009). Thus, NFI-mediated addicsin upregulation protects against DNA damage induced by benzo[α]pyrene and hydrogen peroxide. Third, addicsin also inhibits cancer cell migration as was observed in HeLa, B16, and HCCLM3 cancer cells. (H. Chen et al., 2007). Addicsin has an important role in maintaining the stability of F-actin and in the initiation of actin cytoskeletal rearrangements. Moreover, knockdown of addicsin results in the inactivation of the MEK–ERK signaling cascade. Thus, addicsin inhibits cell migration by activating the mitogen-activated protein kinase (MAPK) cascade and regulating the rearrangement of the F-actin cytoskeleton (H. Chen et al., 2007). Fourth, addicsin participates in the regulation of GSH synthesis; the association of addicsin with EAAC1 at the plasma membrane inhibits the uptake of cysteine for GSH synthesis and thus determines the intracellular GSH content *in vitro* and *in vivo* (Watabe et al., 2007, 2008). This suggests that addicsin is a therapeutic target for enhancing GSH levels in patients with neurodegenerative disorders, such as Alzheimer's and Parkinson's diseases, associated with oxidative stress. Fifth, addicsin significantly inhibits neurite growth in differentiated CAD cells by inactivating Rab1, a positive regulator of ER-to-Golgi trafficking (Maier et al., 2009). Finally, addicsin participates in the regulation of EAAC1-mediated glutamate uptake (Akiduki & Ikemoto, 2008) and ER protein trafficking (Liu et al., 2008; Ruggiero et al., 2008). We discuss these latter two physiological functions in detail (Section 2).

1.3 Arl6ip1

The "ADP-ribosylation factor-like 6 interacting protein 1 (Arl6ip1)" is the new name assigned to three independently described factors: the original Arl6ip, apoptotic regulator in the membrane of the ER (ARMER), and protein KIAA0069. The Arl6ip1 protein was first identified by yeast two-hybrid screening using mouse ARL6 as bait (Ingley et al., 1999) and as a negative regulatory factor during myeloid differentiation by differential display (Pettersson et al., 2000). Moreover, a novel protein, designated ARMER, initially discovered as a false-positive clone by yeast two-hybrid screening using Bcl-xL as bait, is also Arl6ip1 (Lui et al., 2003). In addition, Arl6ip1 has more than 96% homology with the human protein KIAA0069, the product of a cDNA isolated from the human myeloblast cell line KG-1 during a systematic effort to characterize complete cDNAs (Nomura et al., 1994). Amino acid analysis of Arl6ip1 demonstrates that it is composed of 203 amino acids and encodes a 23-kDa protein with four putative transmembrane segments (Pettersson et al., 2000). Several studies indicate that Arl6ip1 is an integral membrane protein localized to the ER (Lui et al., 2003; Pettersson et al., 2000). Furthermore, computational analysis of the topology of Arl6ip1 demonstrates that the N- and C-terminal ends are both exposed to the cytoplasm (Lui et al., 2003). Consistent with these results, Arl6ip1 has two putative casein kinase II phosphorylation motifs (amino acids 18–21 and 128–131), three putative PKC phosphorylation motifs (amino acids 94–96, 115–117, and 128–130), a N-glycosylation motif (amino acids 6–9), a prenyl group-binding motif (amino acids 72–75), and an ER retention signal in the C-terminal cytoplasmic region (amino acids 200–203) (Akiduki &

Ikemoto, 2008; Lui et al., 2003) (Fig. 1). Thus, Arl6ip1 function may be controlled by diverse intracellular cell signals, but it is unknown whether these motifs are physiologically functional.

The functions of Arl6ip1 remain largely unknown, but culture studies have provided several intriguing possibilities. For example, Arl6ip1 protects HT1080 fibrosarcoma cells from apoptosis induced by serum starvation, doxorubicin, UV irradiation, tumor necrosis factor α, and ER stressors by inhibiting caspase-9 activity (Lui et al., 2003). In addition, Arl6ip1 suppresses cisplatin-induced apoptosis in CaSki human cervical cancer cells by regulating the expression of apoptosis-related proteins caspase-3, caspase-9, p53, NF-κB, MAPK, Bcl-2, Bcl-xL, and Bax (Guo et al., 2010a). Furthermore, Arl6ip1 is involved in cell growth, cell cycle progression, and invasion of cancer cells. Downregulation of Arl6ip1 suppresses cell proliferation and colony formation, arrests cell cycling at the G0/G1 phase, and inhibits migration of CaSki human cervical cancer cells (Guo et al., 2010b). Most relevant to the present discussion, Arl6ip1 is involved in the regulation of EAAC1. Recently, we demonstrated that Arl6ip1 is a novel addicsin-associating factor that indirectly promotes PKC-dependent EAAC1-mediated glutamate uptake by decreasing the number of addicsin molecules available for suppression of EAAC1 (Akiduki & Ikemoto, 2008).

2. Regulation of EAAC1 function by addicsin

The mechanisms by which addicsin regulates EAAC1 activity have not been definitively established. However, the discovery of addicsin/GTRAP3-18 has contributed greatly to our understanding of EAAC1 function. Recent evidence demonstrates two major mechanisms of addicsin-mediated regulation of EAAC1 activity. One regulatory pathway is dependent on the dynamic competition for free addicsin molecules by other addicsin molecules to form the homocomplex and by Arl6ip1 to form a heterocomplex. This addicsin–Arl6ip1 complex sequesters addicsin molecules and blocks the interaction of addicsin with EAAC1 in the plasma membrane, thereby reducing the inhibitory effect of addicsin on EAAC1-mediated glutamate uptake (Akiduki & Ikemoto, 2008; Lin et al., 2001) (Fig. 3). Second, addicsin functions as a negative regulator of EAAC1 trafficking through the ER and inhibits the cell surface expression of EAAC1 (Liu et al., 2008; Ruggiero et al., 2008). In this section, we discuss these two mechanisms in detail.

2.1 Modulation of EAAC1-mediated glutamate uptake by addicsin

As an introduction to addicsin/GTRAP3-18-mediated regulation of EAAC1 activity, we discuss two early papers in detail. Lin et al. demonstrated that addicsin/GTRAP3-18 binds to EAAC1 and inhibits EAAC1-mediated glutamate uptake by this direct interaction (Lin et al., 2001). The second is our study showing that addicsin inhibits EAAC1-mediated glutamate uptake in a PKC activity-dependent manner while Arl6ip1 promotes glutamate uptake (also in a PKC activity-dependent manner) by inhibiting the interaction of addicsin with EAAC1 (Akiduki & Ikemoto, 2008). Lin et al. first identified addicsin/GTRAP3-18 as an EAAC1-interacting protein by yeast two-hybrid screening of a rat brain cDNA library. To evaluate whether addicsin/GTRAP3-18 modulates EAAC1 function, they examined the effect of increasing addicsin/GTRAP3-18 expression on EAAC1-mediated glutamate uptake *in vitro* and *in vivo*. First, they showed that glutamate uptake decreased progressively with increasing expression of addicsin/GTRAP3-18 in HEK293 cells. Subsequent kinetic analyses in HEK293, C6BU-1, and COS7 cells revealed that elevated expression of addicsin decreased the glutamate affinity of EAAC1 without altering the maximal transport velocity (correlated with expression). Furthermore, HEK293 cells coexpressing addicsin/GTRAP3-18 and a truncated EAAC1 missing the addicsin/GTRAP3-18 association region showed higher glutamate uptake than cells expressing wild-type EAAC1. In addition, this truncated EAAC1 had a higher affinity for glutamate, suggesting that addicsin/GTRAP3-18 normally reduces EAAC1-mediated glutamate uptake by binding to this association region and reducing transporter glutamate affinity. Next, they evaluated the effect of intraventricular injection of an addicsin/GTRAP3-18 antisense mRNA on EAAC1-mediated glutamate uptake *in vivo*. The antisense treatment resulted in reduced addicsin/GTRAP3-18 expression, a significant increase in cortical EAAC1-mediated glutamate uptake, and an increase in glutamate affinity compared to saline-treated or sense mRNA-treated control animals. In conclusion, addicsin/GTRAP3-18 can negatively modulate EAAC1-mediated glutamate uptake by a direct interaction with EAAC1.

We first isolated addicsin as a novel protein richly expressed in the amygdala of mice under chronic morphine treatment. Addicsin has a tendency to form the multimeric complex *in vitro* (Ikemoto et al., 2002; Lin et al., 2001). The initial discovery of addicsin prompted us to perform yeast two-hybrid screening of an amygdala cDNA library constructed from chronic morphine-administered mice. From this screen, we identified Arl6ip1 as a candidate addicsin-interacting protein. As described in section 1.3, Arl6ip1 is an anti-apoptotic protein located in the ER. As previously described, addicsin inhibits EAAC1-mediated glutamate uptake by direct association at the plasma membrane (Lin et al., 2001), so we speculated that Arl6ip1 upregulates EAAC1-mediated glutamate transport by inhibiting the interaction between addicsin and EAAC1 (Fig. 3).

As a first step to verify this hypothesis, we investigated whether addicsin could bind Arl6ip1 *in vitro* and *in vivo*. To eliminate the possibility of false-positive clones, reconfirmation tests using a full length mouse Arl6ip1 as prey or bait were performed. This tests revealed the specific interaction with addicsin in the yeast AH109 strain. We next examined the reproducibility of this screening result by yeast two-hybrid screening using a different cDNA library prepared from whole brains of 7-week-old mice. We obtained 20 positive clones that clearly displayed α-galactosidase activity (the gene driven by the protein–protein interaction in the two-hybrid screen). Among these positive clones, 11 were identical to *Arl6ip1* cDNA (M.J. Ikemoto et al., unpublished data), confirming the interaction with addicsin and Arl6ip1 in the yeast AH109 strain. We then performed immunoprecipitation analysis, glycerol gradient analysis, and immunocytochemical analysis to directly test the interaction between Arl6ip1 and addicsin *in vitro*. For this purpose, we prepared cell lysates from NG108-15 cells expressing FLAG-tagged Arl6ip1 (Arl6ip1-FLAG), Myc-tagged addicsin (addicsin-myc), or both. Immunoprecipitation analysis of these cell lysates demonstrated that Arl6ip1-FLAG specifically interacted with addicsin-myc in the cell lysates prepared from coexpressing cells, but not from cells expressing Arl6ip1-FLAG or addicsin-myc alone. Glycerol gradient analysis revealed that the elution profile of Arl6ip1-FLAG was similar to that of addicsin-myc. The elution peaks of both proteins were observed in the fraction with a deduced molecular mass of 24 kDa. Moreover, the elution peak of the addicsin homodimer was present in the 44-kDa fraction, suggesting that addicsin forms Arl6ip1–addicsin heterodimers and addicsin–addicsin homodimers *in vitro*. Immunocytochemical analysis in NG108-15 cells overexpressing Arl6ip1-FLAG and addicsin-myc demonstrated subcellular colocalization (M.J. Ikemoto et al., unpublished data). To examine the interaction of both proteins *in vivo*, we performed *in vivo* immunoprecipitation assays of whole brain lysates using an anti-Arl6ip1 polyclonal antibody (generated from a synthetic peptide spanning amino acids 185–199 of mouse Arl6ip1) that again revealed a specific interaction between Arl6ip1 and addicsin. Western blot analysis demonstrated that Arl6ip1 was widely expressed in the mature brain and showed substantial regional overlap with addicsin. In addition, immunohistochemical staining confirmed that Arl6ip1 was widely expressed in the mature brain and localized in neuron-like cells. The neural expression pattern of Arl6ip1 was the same as addicsin, suggesting that Arl6ip1 is colocalized with addicsin in the mature CNS. We concluded that addicsin specifically interacted with Arl6ip1 *in vitro* and *in vivo*.

As a second step, we then determined the Arl6ip1- and addicsin-binding regions on addicsin. If Arl6ip1 does regulate EAAC1 activity by competitively binding to addicsin molecules and thus preventing the formation of addicsin homodimers that downregulate EAAC1 activity, the Arl6ip1- and addicsin-binding regions on addicsin should be located close enough for such a competitive interaction. Immunoprecipitation assays using several addicsin truncation mutants indicated that Arl6ip1 associated with full length addicsin (wt), a truncation lacking the C-terminal region at amino acids 145–188 (d1), a deletion mutant of the N-terminal domain at amino acids 1–102 (d2), and a mutant missing the region containing the C-terminal phosphorylation motif at amino acids 136–144 (d3). However, Arl6ip1 could not interact with a mutant lacking a portion of the hydrophobic region at amino acids 103–117 (d4). As expected, addicsin was able to associate with the wt, d1, d2, or d3 mutant, but not the d4 truncation mutant, indicating that the hydrophobic region at amino acids 103–117 of addicsin is a crucial domain for the formation of addicsin–addicsin homodimers and addicsin-Arl6ip1 heterodimers (Fig. 1). These results strongly support our hypothesis that Arl6ip1 antagonizes addicsin-mediated downregulation of EAAC1 activity by sequestering free addicsin.

As a third step, we investigated whether Arl6ip1 had a positive effect on EAAC1-mediated glutamate uptake. For this purpose, we selected C6BU-1 glioma cells that expressed EAAC1 as the principal or only EAAT (Palos et al., 1996). We created two stably expressing C6BU-1 cell lines, designated C6BU-1-pSw-addicsin and C6BU-1-pSw-Arl6ip1. In these cell lines, we could strictly control the expression levels of V5-tagged addicsin (addicsin-V5) or V5-tagged Arl6ip1 (Arl6ip1-V5) by exposure to 10 nM mifepristone (11β-[4-dimethylamino]phenyl-17β-hydroxy-17-[1-propynyl]estra-4,9-dien-3-one), a synthetic 19-norsteroid. In addition, a cell viability assay demonstrated that upregulation of Arl6ip1-V5 or addicsin-V5 by exposure to 10 nM mifepristone was not cytotoxic, making these cell lines excellent models to evaluate the effects of changing Arl6ip1 and addicsin expression on the functional activity of EAAC1. Compared to control cells untreated with mifepristone or the PKC agonist PMA, the upregulation of Arl6ip-V5 or addicsin-V5 by 10 nM mifepristone alone did not change EAAC1-mediated glutamate uptake. When these cells were stimulated with 100 nM PMA alone, the glutamate uptake activity in C6BU-1-pSw-addicsin cells and C6BU-1-pSw-Arl6ip1 cells increased about two-fold compared to untreated controls. EAAC1-mediated glutamate uptake was significantly lower in C6BU-1-pSw-addicsin cells stimulated with both mifepristone and PMA compared to C6BU-1-pSw-addicsin cells treated with PMA alone, indicating that activation of addicsin expression inhibited PKC-dependent EAAC1 activity. In contrast, C6BU-1-pSw-Arl6ip1 cells treated with PMA and mifepristone exhibited a three-fold increase in glutamate uptake compared to the same line treated with PMA alone, indicating that Arl6ip1 overexpression enhanced PKC-dependent EAAC1 activity. On the other hand, the nonstimulating PMA analog 4α phorbol did not increase glutamate uptake relative to controls.

To further support these conclusions, we performed a knockdown experiment by transient transfection of double-stranded siRNAs into C6BU-1-pSw-Arl6ip1 cells to investigate the effect of decreased addicsin expression on EAAC1-mediated glutamate uptake. As expected, cells transfected with either of two alternative addicsin siRNAs showed about a two-fold increase in glutamate uptake in response to PMA exposure compared to cells treated with control scrambled siRNA. The elevated glutamate uptake concomitant with addicsin knockdown strongly supported the proposed mechanism for EAAC1 regulation by addicsin and Arl6ip1.

To investigate the molecular mechanisms for altered EAAC1-mediated glutamate uptake in C6BU-1-pSw-Arl6ip1 cells, we performed kinetic analysis of glutamate flux across C6BU-1-pSw-Arl6ip1 cell membranes. When Arl6ip1 was conditionally overexpressed using mifepristone, PMA treatment increased the glutamate affinity but not the maximal velocity compared to vehicle-treated controls (PMA: K_m = 647 µM, V_{max} = 1.5 × 10^3 pmole/mg/min; vehicle: K_m = 824 µM, V_{max} = 1.5 × 10^3 pmole/mg/min) with no change of addicsin expression levels. Thus, Arl6ip1 promoted EAAC1-mediated glutamate uptake by increasing the catalytic efficacy of EAAC1. Specifically, Arl6ip1 blocked the addicsin-mediated reduction in EAAC1 glutamate affinity.

As a fourth step, we then examined the subcellular localization of Arl6ip1 in C6BU-1-pSw-Arl6ip1 cells. Western blot analysis revealed that Arl6ip1-V5 expression levels were unaffected by 100 nM PMA exposure. Immunocytochemical analysis demonstrated that Arl6ip1-V5 was predominantly localized to cytoplasmic structures such as the ER and that this subcellular expression pattern was not changed by PMA. Furthermore, cell biotinylation analysis indicated that Arl6ip1 did not interact with the plasma membrane, consistent with our previous result that Arl6ip1 failed to interact with EAAC1 by immunoprecipitation. Therefore, Arl6ip1 was localized to the ER under all conditions tested and acted to "trap" addicsin molecules in Arl6ip1–addicsin heterodimers, thus preventing the direct interaction of addicsin with EAAC1. To confirm our hypothesis, we produced an addicsin mutant that lacked interaction with Arl6ip1 but not with other addicsin molecules. Fine mutational analysis was used to separate the Arl6ip1- and addicsin-binding regions within the addicsin d4 region. We compared addicsin sequences among various species and noted that two amino acids at positions 110 and 112 of mouse addicsin were completely conserved from fruit fly to human. We created a double-mutated form of addicsin that substituted both the native tyrosine at amino acid 110 and the leucine at amino acid 112 with alanine. The mutant, designated addicsin Y110A/L112A (or addicsinYL), showed markedly less binding to Arl6ip1 (40% of wild-type addicsin) but normal wild-type binding to addicsin, as revealed by immunoprecipitation. In addition, a cell biotinylation assay indicated that addicsinYL was unable to localize to the plasma membrane, suggesting that addicsinYL lost EAAC1-binding activity. To evaluate the effect of addicsinYL on EAAC1-mediated glutamate uptake, we created a conditional C6BU-1 cell line, designated C6BU-1-pSw-addicsinYL. This cell line exhibited mifepristone-dependent upregulation of V5-tagged addicsinYL and increased glutamate uptake in response to PMA that was unchanged by mifepristone-induced upregulation of addicsinYL. That is, glutamate uptake was not reduced by induced addicsinYL expression. These data strongly suggest that addicsin is a key negative regulator of EAAC1 in the plasma membrane and that Arl6ip1 is a negative regulator of addicsin.

As a final step, we examined the effect of addicsin PKC phosphorylation sites on EAAC1-mediated glutamate uptake in C6BU-1 cells. Addicsin has putative PKC phosphorylation motifs at amino acids 18-20 and 138-140, and PKC activation increases EAAC1-mediated glutamate uptake. We established conditional C6BU-1 cell lines, designated C6BU-1-pSw-addicsinS18A and C6BU-1-pSw-addicsinS138A. C6BU-1-pSw-addicsinS18A cells expressed a V5-tagged addicsin point mutant that substituted native serine 18 for alanine in the N-terminal motif in response to mifepristone, while C6BU-1-pSw-addicsinS138A cells expressed a V5-tagged addicsin point mutant that substituted native serine 138 for alanine in the C-terminal motif. These cells showed no cytotoxicity in response to 10 nM mifepristone. In contrast to cells expressing wild-type addicsin, expression of addicsinS18A did not suppress the PMA-induced increase in EAAC1-mediated glutamate uptake. Moreover, increased expression of addicsinS18A caused a significant increase in glutamate uptake even without PMA stimulation by a dominant negative effect. Similarly, addicsinS138A expression did not suppress the PMA-induced increase in EAAC1-mediated glutamate uptake. Thus, these mutations abolished the inhibitory effect of addicsin. However, in contrast to addicsinS18A, addicsinS138A expression had no influence on EAAC1-mediated glutamate uptake activity in the absence of PMA stimulation. Both serine 18 and serine 138 within the putative PKC phosphorylation motifs are critical for the negative regulation of EAAC1-mediated glutamate uptake and suggest that the PKC phosphorylation site at serine 138 is functional under physiological conditions.

Based on these data, we proposed the regulatory model of EAAC1-mediated glutamate uptake illustrated in Fig. 3. If addicsin expression is high enough relative to Arl6ip1 to form many more addicsin homodimers than addicsin–Arl6ip1 heterodimers, EAAC1-mediated glutamate uptake is reduced. Furthermore, activation of the PKC isozyme that phosphorylates addicsin at S18 or S138 may further potentiate this negative regulation. On the other hand, if addicsin expression is low enough or Arl6ip1 expression high enough that formation of heterodimers predominates, fewer addicsin homodimers are available to suppress EAAC1 activity. The resulting decrease in addicsin–EAAC1 binding will enhance the catalytic efficacy of EAAC1, in a PKC-activity dependent manner. In sum, Arl6ip1 acts as a positive regulator of EAAC1-mediated glutamate uptake (Fig. 3) and may therefore possess significant neuroprotective efficacy against neurodegenerative diseases linked to excitotoxicity and oxidative stress.

2.2 Modulation of ER protein trafficking by addicsin

Addicsin is a member of the PRAF protein family with homology to PRA1 and PRAF2 (JM4) (Schweneker et al., 2005). PRA1 is associated with the Golgi membrane and interacts with Rab, a member of the Ras superfamily of small GTP-binding proteins, which regulates intracellular protein trafficking (Bucci et al., 1999; Liang & Li, 2000; Martincic et al., 1997).

Immunocytochemical studies reveal that mature addicsin is present in both the plasma membrane and the intracellular compartment, including the ER (Ikemoto et al., 2002; Watabe et al., 2007, 2008). Thus, addicsin may also be involved in intracellular protein trafficking. To investigate this possibility, we examined EAAC1 oligosaccharide residues under conditions of varying addicsin expression. The oligosaccharide residues on EAAC1 are an excellent indicator of the extent of ER-to-Golgi trafficking and plasma membrane localization because the newly synthesized EAAC1 is N-glycosylated with high mannose oligosaccharide chains that are subsequently processed into more complex sugar chains by resident Golgi enzymes (Yang & Kilberg, 2002). In HEK293T cells coexpressing addicsin, EAAC1 is predominantly modified by high mannose oligosaccharides, suggesting that EAAC1 proteins are largely confined to the ER. Furthermore, addicsin delays oligosaccharide maturation of EAAC1 but does not induce EAAC1 degradation (Ruggiero et al., 2008). These data suggest that addicsin delays ER-to-Golgi trafficking of EAAC1. Moreover, addicsin inhibits ER-to-Golgi trafficking of dopamine transporter, GABA transporter 1, and several G-protein-coupled receptors, including β_2-adrenergic receptor, α_1-β receptor, and D_2 receptor (Ruggiero et al., 2008). Furthermore, addicsin inhibits the function of RTN2B, a member of the reticulon protein family localized in the ER, which enhances ER-to-Golgi trafficking of EAAC1 (Liu et al., 2008). As addicsin, RTN2B, and EAAC1 are coexpressed in neurons, they may interact in one complex. Indeed, addicsin and EAAC1 can interact with RTN2B by binding to different regions of the protein. In addition, coexpression of RTN2B and EAAC1 in HEK293 cells increases EAAC1 cell surface expression, while increasing addicsin expression blocks this effect. Thus, EAAC1 trafficking is inhibited by addicsin and facilitated by RTN2B (Liu et al., 2008). Based on these data, Liu et al. proposed a model in which the regulation of ER trafficking governs the activity and density of EAAC1 at the plasma membrane. Under normal conditions, RTN2B facilitates EAAC1 trafficking from the ER because basal expression of addicsin is too low to have an inhibitory effect. Under stressful conditions, such as oxidative and chemical stress, addicsin expression is upregulated and the inhibitory effect on EAAC1 trafficking predominates over the facilitating effect of RTN2B (Liu et al., 2008). Addicsin can delay ER-to-Golgi trafficking of structurally and functionally distinct proteins in addition to EAAC1. Thus, addicsin is a stress-induced multifunctional protein that participates in various physiological and pathological functions by regulating ER trafficking of many membrane effector proteins, including receptors and transporters.

3. Addicsin & neurological disorders

Recent studies have also linked addicsin to the pathophysiology of several neurological diseases, including drug addiction, schizophrenia, and epilepsy. In this section, we focus on these diseases and review the putative pathophysiological functions of addicsin in the mammalian CNS.

3.1 Drug abuse

Several studies demonstrate that addicsin is involved in drug abuse, the development of morphine dependence (Ikemoto et al., 2002; Wu et al., 2011), and ethanol tolerance (C. Li et al., 2008). In an effort to clarify the molecular mechanism of opiate addiction, we performed subtractive hybridization of mRNA expressed in the amygdala of mice treated with repeated doses of morphine and identified *addicsin* mRNA as a factor selectively upregulated relative to drug-naïve mice (Ikemoto et al., 2000, 2002). Upregulation of *addicsin* mRNA was specifically induced by chronic, but not acute, morphine administration and was completely inhibited by coadministration of naloxone, an opiate receptor antagonist (Ikemoto et al., 2002). In that study, we used a morphine administration protocol that had been previously shown to induce morphine dependence and tolerance (Kaneto et al., 1973). Thus, our data strongly suggested that addicsin was involved in the development of morphine dependence in this animal model. Later reports have confirmed our findings by directly demonstrating that addicsin is directly involved in the development of morphine dependence (Wu et al., 2011). Chronic morphine treatment upregulated addicsin in prefrontal cortex, nucleus accumbens, and amygdala, which are regions known to be critical for the development of morphine dependence and other addictive behaviors. Furthermore, addicsin knockdown by infusion of addicsin antisense nucleotides into the cerebral ventricles significantly decreased withdrawal behaviors following chronic morphine treatment in rats (Wu et al., 2011). Addicsin knockdown suppressed the upregulation of δ opioid receptors, the activation of the dopamine- and cAMP-regulated phosphoprotein of 32 kDa (DARPP-32), and MAPK activation normally induced by chronic morphine treatment. Furthermore, addicsin knockdown enhanced the degradation of δ opioid receptors through the ubiquitin-proteasome pathway (Wu et al., 2011). These data suggest that addicsin directly contributes to the regulation of δ opioid receptor stability and the development of morphine dependence by suppressing δ opioid receptor expression and the activation of DARPP-32 and MAPK. The δ opioid receptor knockout mice do not develop analgesic tolerance to morphine without affecting the development of physical dependence (Kieffer & Gaveriaux-Ruff, 2002; Nitsche et al., 2002; Y. Zhu et al., 1999). Thus, further investigations are needed to clarify whether addicsin is involved in analgesic tolerance.

Ethanol-induced cellular responses are analogous to those elicited by heat shock stresses (Piper, 1995; Wilke et al., 1994). Similarly, addicsin expression is enhanced in response to various environmental stressors, such as oxidative stress and heat shock stress (R. Chen et al., 2007). Furthermore, our study demonstrated that addicsin plays an important role in the development of morphine dependence and tolerance (Ikemoto et al., 2002). In the light of these observations, addicsin is considered to be essential for the development of ethanol tolerance. To address this issue, addicsin knockdown flies were generated. To estimate ethanol tolerance objectively, the inebriation test was performed (Bellen, 1998). Flies were exposed to ethanol vapor, and the mean elution time (MET) was measured three times after inebriation. The addicsin knockdown flies showed no difference between the first MET and third MET, while wild-type flies exhibited a significant higher third MET (C. Li, et al., 2008), indicating that addicsin knockdown flies failed to acquire ethanol tolerance.

3.2 Schizophrenia

Glutamatergic neurotransmission and plasticity are disrupted in patients with schizophrenia (Javitt, 2010; Kantrowitz & Javitt, 2010; Paz et al., 2008). This has led some researchers to speculate that EAATs and EAAT-interacting proteins that regulate glutamate transport efficacy or transporter expression may be abnormal in patients with schizophrenia (Bauer et al., 2008; Huerta et al., 2006). Indeed, addicsin/JWA transcripts were overexpressed in the thalamus (Huerta et al., 2006) and the anterior cingulate cortex of schizophrenics as shown by *in situ* hybridization (Bauer et al., 2008). In these studies, the protein expression levels of addicsin/JWA were not determined. In addition, expression of EAAT3, the human homolog of EAAC1, was also upregulated in the anterior cingulate cortex of schizophrenic patients (Bauer et al., 2008). Furthermore, a microarray study of multiple human brain regions demonstrates that the anterior cingulate cortex is more vulnerable to these aberrant gene expression patterns (Katsel et al., 2005), and hypofrontality is a key feature of schizophrenia. Addicsin is thus a promising target for further research focusing on the role of glutamate transporters in schizophrenia. Moreover, addicsin regulates trafficking of a plethora of other membrane proteins, including dopamine receptors, suggesting another pathway through which addicsin participates in the pathogenesis of schizophrenia.

3.3 Epilepsy

Anatomical analysis of EAAT expression reveals that EAAC1 is enriched in neurons and particularly localized to inhibitory GABAergic neurons (Conti et al., 1998; He et al., 2000; Rothstein et al., 1994). Cerebroventricular injection of EAAC1 antisense oligonucleotides caused no elevation of extracellular glutamate in the rat striatum but did produce mild neurotoxicity and epileptiform activity (Rothstein et al., 1996). Furthermore, epilepsy in EAAC1 knockdown rats is caused by decreased GABA synthesis (Sepkuty et al., 2002). Glutamate is a precursor for GABA synthesis, so molecules that alter the intracellular availability of glutamate in GABAergic interneurons, including addicsin/GTRAP3-18, may have an important role in epileptogenesis or ictogenesis. In a recent study of the antiepileptic drug levetiracetam (LEV), changes in the expression of addicsin/GTRAP3-18, glutamate transporters, and GABA transporters were examined in a rat post-traumatic epilepsy model induced by $FeCl_3$ injection into the amygdala. Administration of LEV increased expression of EAAC1 and GABA transporter 3 (GAT-3) but decreased expression of addicsin/GTRAP3-18 in the rat hippocampal formation (Ueda et al., 2007). These results suggest that both the suppression of glutamatergic excitation and the enhancement of GABAergic inhibition induced by chronic LEV administration are due to the upregulation of EAAC1 and GAT-3 subsequent to downregulation of addicsin/GTRAP3-18. A long-lasting suppression of addicsin/GTRAP3-18 expression was observed in the rat pentylenetetrazole (PTZ)-induced kindling model of epilepsy (Ueda et al., 2006). Similarly, antisense-mediated knockdown of addicsin/GTRAP3-18 decreases seizure threshold and promotes PTZ kindling. In addition, addicsin/GTRAP3-18 knockdown increases basal release of glutamate and GABA in the rat hippocampal formation, indicating that knockdown of addicsin/GTRAP3-18 promotes GABA synthesis (Ueda et al., 2006). These studies, demonstrating that addicsin can increase GABA synthesis by increasing the substrate (i.e., glutamate) supply, define addicsin as a novel therapeutic target in epilepsy.

3.4 Other neurological disorders

Addicsin directly modulates glutamate and cysteine uptake by EAAC1, suggesting that addicsin participates in the pathogenesis of neurological disorders associated with excitotoxicity and oxidative stress. Here we briefly discuss some representative EAAC1 functions relevant to CNS pathology. A recent study demonstrated that EAAC1-deficient mice developed age-dependent brain atrophy and behavioral abnormalities in the cognitive and motivational domains. In addition, EAAC1 knockout mice displayed impaired GSH homeostasis and age-dependent neurodegeneration, and these pathologies were rescued by treatment with the membrane permeable cysteine precursor N-acetylcysteine (Aoyama et al., 2006). These EAAC1 knockout mice also display dicarboxylic aminoaciduria and significant motor impairments (Peghini et al., 1997). These results indicate that EAAC1 functions as a cysteine transporter in neurons and sustains intracellular GSH to ameliorate oxidative stress *in vivo*. Furthermore, neuronal glutamate uptake can also regulate memory formation (Levenson et al., 2000; Maleszka et al., 2000). The increase of EAAC1-mediated neuronal glutamate uptake is associated with the induction and expression of early phase long-term potentiation (LTP) in the CA1 area of the hippocampal formation and with contextual fear conditioning, a form of hippocampus-dependent memory thought to depend on induction of LTP (Levenson et al., 2002). These results suggest that regulation of glutamate uptake by EAAC1 is a physiologically important mechanism for the modulation of synaptic strength during long-term changes in synaptic efficacy (plasticity). Thus, dysfunction of EAAC1 induced by aberrant addicsin expression may lead to neurodegeneration and cognitive decline. Of particular interest is the role of addicsin in the pathogenesis of neurodegenerative diseases such as Alzheimer's and Parkinson's diseases. These questions warrant further research.

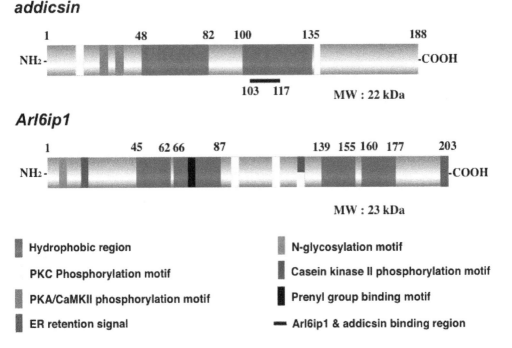

Fig. 1. A schematic presentation of addicsin and Arl6ip1

4. Future research perspective

Despite these advances, our understanding of the regulatory mechanisms of addicsin expression and the range of addicsin functions is far from complete. The elucidation of the regulatory mechanism of addicsin expression under basal and pathological conditions is essential for understanding the physiological and pathological roles of addicsin. For instance, while addicsin has consensus PKC phosphorylation sequences, it is unclear whether PKC actually phosphorylates addicsin and controls addicsin functions *in vivo*. It is also unknown whether or how PKC phosphorylation affects the interaction between addicsin and Arl6ip1. To overcome these challenges, it is crucial to clarify whether PKC phosphorylation sites of addicsin are physiologically controlled by PKC signaling and by which PKC isoforms. Furthermore, it remains controversial whether addicsin is an integral membrane protein. Our results strongly support the notion that addicsin is a membrane-associating protein with a soluble and membrane-localized form. Thus, it is important to clarify the different molecular features and functions of the soluble and membrane-localized forms of addicsin.

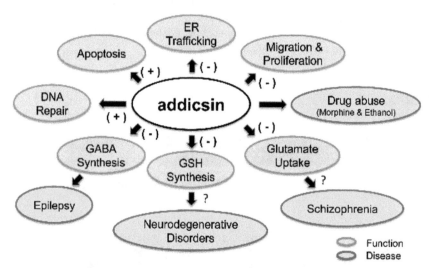

Fig. 2. A scheme of the proposed physiological functions of addicsin

Second, *in vivo* functional studies are still needed to clarify the physiological and pathological functions of addicsin. Accumulating evidence suggests that addicsin participates in various physiological and pathological processes *in vivo*, but the molecular mechanisms controlling the selective interaction of addicsin with multiple targets, including receptors and transporters, are unknown. Furthermore, many reports demonstrate that the physiological and pathological roles of addicsin are observed when expression of addicsin is increased by various stresses, including oxidative and chemical stress. Thus, the production of animal models that overexpressed addicsin in a tissue- or region-specific manner may be useful to analyze addicsin functions in various tissues, including the brain. At present, no studies have been undertaken in tissues outside the brain, although addicsin is ubiquitously expressed in kidney, heart, and liver (Butchbach et al., 2002; Ikemoto et al., 2002).

We believe that studies using transgenic or conditional knockin/knockout animal models will lead to novel insights into addicsin function. Of particular interest is whether dysfunctional addicsin expression or function can lead to neurodegenerative diseases through dysregulation of EAAC1 or other proteins. Finally, we hope that studies on addicsin will continue to advance our understanding of the role of addicsin in the pathogenesis of diseases, such as drug abuse, and lead to the development of curative therapies.

5. Conclusion

In this chapter, we argued that Arl6ip1 is a novel addicsin-interacting protein that indirectly promotes PKC-dependent, EAAC1-mediated glutamate uptake by inhibiting the interaction of addicsin with EAAC1 at the plasma membrane. Based on these findings, we proposed the regulatory model of EAAC1-mediated glutamate uptake illustrated in Fig. 3. In this model, EAAC1-mediated glutamate uptake activity can be negatively and positively regulated by PKC activity depending on dynamic modulation by addicsin complexes. Thus, the cellular dynamics of addicsin is a key element regulating EAAC1-mediated glutamate uptake. The study of addicsin is still in its infancy, but future findings on the physiological and pathophysiological functions of addicsin could greatly clarify the role of EAAC1 (and other proteins regulated by addicsin) in health and disease.

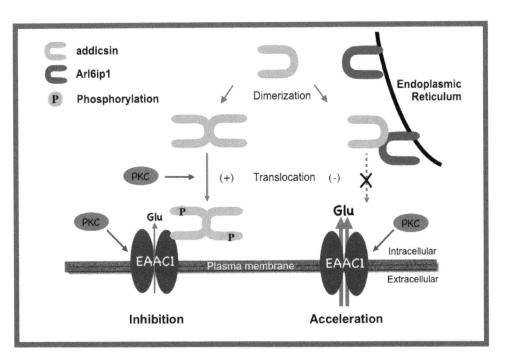

Fig. 3. A regulatory model of EAAC1-mediated glutamate uptake in C6BU-1 cells

6. Acknowledgment

This work was supported by a research grant from the National Institute of Advanced Industrial Science and Technology (AIST) of Japan. We thank Dr. S. Akiduki, M.Sc. M. Takumori, Dr. M. Ohtomi, Dr. K. Inoue, and Dr. T. Ochiishi for helpful discussions, and Ms. K. Nemoto for her excellent technical assistance.

7. References

Akiduki, S. & Ikemoto, M. J. (2008). Modulation of the Neural Glutamate Transporter EAAC1 by the Addicsin-Interacting Protein Arl6ip1, *The Journal of Biological Chemistry* Vol.283, No.46, (November 2008), pp. 31323-31332, ISSN 0021-9258

Akiduki, S., Ochiishi, T. & Ikemoto, M. J. (2007). Neural Localization of Addicsin in Mouse Brain, *Neuroscience Letters* Vol.426, No.3, (Octbor 2007), pp. 149-154

Aoyama, K., Suh, S. W., Hamby, A. M., Liu, J., Chan, W. Y., Chen, Y. & Swanson, R. A. (2006). Neuronal Glutathione Deficiency and Age-Dependent Neurodegeneration in the EAAC1 Deficient Mouse, *Nature Neuroscience* Vol.9, No.1, (January 2006), pp. 119-126

Aoyama, K., Matsumura, N., Watabe, M. & Nakaki, T. (2008a). Oxidative Stress on EAAC1 Is Involved in MPTP-Induced Glutathione Depletion and Motor Dysfunction, *The European Journal of Neuroscience* Vol.27, No.1, (January 2008), pp. 20-30, ISSN 1460-9568 (Electronic), 0953-816X (Linking)

Aoyama, K., Watabe, M. & Nakaki, T. (2008b). Regulation of Neuronal Glutathione Synthesis, *Journal of Pharmacological Sciences* Vol.108, No.3, (November 2008), pp. 227-238, ISSN 1347-8613

Bauer, D., Gupta, D., Harotunian, V., Meador-Woodruff, J. H. & McCullumsmith, R. E. (2008). Abnormal Expression of Glutamate Transporter and Transporter Interacting Molecules in Prefrontal Cortex in Elderly Patients with Schizophrenia, *Schizophrenia Research* Vol.104, No.1-3, (September 2008), pp. 108-120, ISSN 1573-2509

Bellen, H. J. (1998). The Fruit Fly: A Model Organism to Study the Genetics of Alcohol Abuse and Addiction?, *Cell* Vol.93, No.6, (June 1998), pp. 909-912, ISSN 0092-8674

Bucci, C., Chiariello, M., Lattero, D., Maiorano, M. & Bruni, C. B. (1999). Interaction Cloning and Characterization of the cDNA Encoding the Human Prenylated Rab Acceptor (PRA1), *Biochemical and Biophysical Research Communications* Vol.258, No.3, (May 1999), pp. 657-662, ISSN 0006-291X

Butchbach, M. E., Lai, L. & Lin, C. L. (2002). Molecular Cloning, Gene Structure, Expression Profile and Functional Characterization of the Mouse Glutamate Transporter (EAAT3) Interacting Protein GTRAP3-18, *Gene* Vol.292, No.1-2, (June 2002), pp. 81-90

Cao, X. J., Chen, R., Li, A. P. & Zhou, J. W. (2007). JWA Gene Is Involved in Cadmium-Induced Growth Inhibition and Apoptosis in HEK-293T Cells, *Journal of Toxicology and Environmental Health A* Vol.70, No.11, (June 2007), pp. 931-937

Chen, H., Bai, J., Ye, J., Liu, Z., Chen, R., Mao, W., Li, A. & Zhou, J. (2007). JWA as a Functional Molecule to Regulate Cancer Cells Migration Via MAPK Cascades and F-Actin Cytoskeleton, *Cellular Signalling* Vol.19, No.6, (June 2007), pp. 1315-1327

Chen, R., Qiu, W., Liu, Z., Cao, X., Zhu, T., Li, A., Wei, Q. & Zhou, J. (2007). Identification of JWA as a Novel Functional Gene Responsive to Environmental Oxidative Stress Induced by Benzo[a]Pyrene and Hydrogen Peroxide, *Free Radical Biology and Medicine* Vol.42, No.11, (June 2007), pp. 1704-1714

Chen, Y. & Swanson, R. A. (2003). The Glutamate Transporters EAAT2 and EAAT3 Mediate Cysteine Uptake in Cortical Neuron Cultures, *Journal of Neurochemistry* Vol.84, No.6, (March 2003), pp. 1332-1339, ISSN 0022-3042

Conti, F., DeBiasi, S., Minelli, A., Rothstein, J. D. & Melone, M. (1998). EAAC1, a High-Affinity Glutamate Tranporter, Is Localized to Astrocytes and GABAergic Neurons Besides Pyramidal Cells in the Rat Cerebral Cortex, *Cerebral Cortex* Vol.8, No.2, (March 1998), pp. 108-116, ISSN 1047-3211

D'Amico, A., Soragna, A., Di Cairano, E., Panzeri, N., Anzai, N., Vellea Sacchi, F. & Perego, C. (2010). The Surface Density of the Glutamate Transporter EAAC1 Is Controlled by Interactions with PDZK1 and AP2 Adaptor Complexes, *Traffic* Vol.11, No.11, (November 2010), pp. 1455-1470, ISSN 1600-0854 (Electronic), 1398-9219 (Linking)

Danbolt, N. C. (2001). Glutamate Uptake, *Progress in Neurobiology* Vol.65, No.1, (September 2001), pp. 1-105

Davis, K. E., Straff, D. J., Weinstein, E. A., Bannerman, P. G., Correale, D. M., Rothstein, J. D. & Robinson, M. B. (1998). Multiple Signaling Pathways Regulate Cell Surface Expression and Activity of the Excitatory Amino Acid Carrier 1 Subtype of Glu Transporter in C6 Glioma, *The Journal of Neuroscience* Vol.18, No.7, (April 1998), pp. 2475-2485, ISSN 0270-6474

Dowd, L. A. & Robinson, M. B. (1996). Rapid Stimulation of EAAC1-Mediated Na+-Dependent L-Glutamate Transport Activity in C6 Glioma Cells by Phorbol Ester, *Journal of Neurochemistry* Vol.67, No.2, (August 1996), pp. 508-516, ISSN 0022-3042

Fournier, K. M., Gonzalez, M. I. & Robinson, M. B. (2004). Rapid Trafficking of the Neuronal Glutamate Transporter, EAAC1: Evidence for Distinct Trafficking Pathways Differentially Regulated by Protein Kinase C and Platelet-Derived Growth Factor, *The Journal of Biological Chemistry* Vol.279, No.33, (August 2004), pp. 34505-34513, 0021-9258

Gonzalez, M. I., Kazanietz, M. G. & Robinson, M. B. (2002). Regulation of the Neuronal Glutamate Transporter Excitatory Amino Acid Carrier-1 (EAAC1) by Different Protein Kinase C Subtypes, *Molecular Pharmacology* Vol.62, No.4, (October 2002), pp. 901-910, ISSN 0026-895X

Gonzalez, M. I., Bannerman, P. G. & Robinson, M. B. (2003). Phorbol Myristate Acetate-Dependent Interaction of Protein Kinase C alpha and the Neuronal Glutamate Transporter EAAC1, *The Journal of Neuroscience* Vol.23, No.13, (July 2003), pp. 5589-5593, ISSN 1529-2401 (Electronic), 0270-6474 (Linking)

Guo, F., Li, Y., Liu, Y., Wang, J. & Li, G. (2010a). Arl6ip1 Mediates Cisplatin-Induced Apoptosis in Caski Cervical Cancer Cells, *Oncology Reports* Vol.23, No.5, (May 2010), pp. 1449-1455, ISSN 1791-2431 (Electronic), 1021-335X (Linking)

Guo, F., Liu, Y., Li, Y. & Li, G. (2010b). Inhibition of ADP-Ribosylation Factor-Like 6 Interacting Protein 1 Suppresses Proliferation and Reduces Tumor Cell Invasion in Caski Human Cervical Cancer Cells, *Molecular Biology Reports* Vol.37, No.8, (December 2010), pp. 3819-3825, ISSN 1573-4978 (Electronic), 0301-4851 (Linking)

He, Y., Janssen, W. G., Rothstein, J. D. & Morrison, J. H. (2000). Differential Synaptic Localization of the Glutamate Transporter EAAC1 and Glutamate Receptor Subunit GluR2 in the Rat Hippocampus, *The Journal of Comparative Neurology* Vol.418, No.3, (March 2000), pp. 255-269, ISSN 0021-9967

Himi, T., Ikeda, M., Yasuhara, T., Nishida, M. & Morita, I. (2003). Role of Neuronal Glutamate Transporter in the Cysteine Uptake and Intracellular Glutathione Levels in Cultured Cortical Neurons, *Journal of Neural Transmission* Vol.110, No.12, (December 2003), pp. 1337-1348, ISSN 0300-9564

Huang, S., Shen, Q., Mao, W. G., Li, A. P., Ye, J., Liu, Q. Z., Zou, C. P. & Zhou, J. W. (2006a).
JWA, a Novel Signaling Molecule, Involved in All-*Trans* Retinoic Acid Induced
Differentiation of HL-60 Cells, *Journal of Biomedical Science* Vol.13, No.3, (May 2006),
pp. 357-371

Huang, S., Shen, Q., Mao, W. G., Li, A. P., Ye, J., Liu, Q. Z., Zou, C. P. & Zhou, J. W. (2006b).
JWA, a Novel Signaling Molecule, Involved in the Induction of Differentiation of
Human Myeloid Leukemia Cells, *Biochemical and Biophysical Research
Communications* Vol.341, No.2, (March 2006), pp. 440-450

Huerta, I., McCullumsmith, R. E., Haroutunian, V., Gimenez-Amaya, J. M. & Meador-
Woodruff, J. H. (2006). Expression of Excitatory Amino Acid Transporter
Interacting Protein Transcripts in the Thalamus in Schizophrenia, *Synapse* Vol.59,
No.7, (June 2006), pp. 394-402, ISSN 0887-4476

Ikemoto, M., Takita, M., Imamura, T. & Inoue, K. (2000). Increased Sensitivity to the
Stimulant Effects of Morphine Conferred by Anti-Adhesive Glycoprotein SPARC in
Amygdala, *Nature Medicine* Vol.6, No.8, (August 2000), pp. 910-915, ISSN 1078-8956

Ikemoto, M. J., Inoue, K., Akiduki, S., Osugi, T., Imamura, T., Ishida, N. & Ohtomi, M.
(2002). Identification of Addicsin/GTRAP3-18 as a Chronic Morphine-Augmented
Gene in Amygdala, *Neuroreport* Vol.13, No.16, (November 2002), pp. 2079-2084

Ingley, E., Williams, J. H., Walker, C. E., Tsai, S., Colley, S., Sayer, M. S., Tilbrook, P. A.,
Sarna, M., Beaumont, J. G. & Klinken, S. P. (1999). A Novel ADP-Ribosylation Like
Factor (ARL-6), Interacts with the Protein-Conducting Channel SEC61beta Subunit,
FEBS Letters Vol.459, No.1, (October 1999), pp. 69-74, ISSN 0014-5793

Inoue, K., Akiduki, S. & Ikemoto, M. J. (2005). Expression Profile of Addicsin/GTRAP3-18
Mrna in Mouse Brain, *Neuroscience Letters* Vol.386, No.3, (October 2005), pp. 184-188

Javitt, D. C. (2010). Glutamatergic Theories of Schizophrenia, *The Israel Journal of Psychiatry
and Related Sciences* Vol.47, No.1, pp. 4-16, ISSN 0333-7308

Kanai, Y. & Hediger, M. A. (1992). Primary Structure and Functional Characterization of a
High-Affinity Glutamate Transporter, *Nature* Vol.360, No.6403, (December 1992),
pp. 467-471

Kanai, Y. & Hediger, M. A. (2003). The Glutamate and Neutral Amino Acid Transporter
Family: Physiological and Pharmacological Implications, *European Journal of
Pharmacology* Vol.479, No.1-3, (October 2003), pp. 237-247, ISSN 0014-2999

Kanai, Y. & Hediger, M. A. (2004). The Glutamate/Neutral Amino Acid Transporter Family
SLC1: Molecular, Physiological and Pharmacological Aspects, *Pflugers Archiv
European Journal of Physiology* Vol.447, No.5, (Feb 2004), pp. 469-479, ISSN 0031-6768
(Print),0031-6768 (Linking)

Kaneto, H., Koida, M., Nakanishi, H. & Sasano, H. (1973). A Scoring System for Abstinence
Syndrome in Morphine Dependent Mice and Application to Evaluate Morphine
Type Dependence Liability of Drugs, *The Japanese Journal of Pharmacology* Vol.23,
No.5, (October 1973), pp. 701-707, ISSN 0021-5198

Kantrowitz, J. T. & Javitt, D. C. (2010). Thinking Glutamatergically: Changing Concepts of
Schizophrenia Based Upon Changing Neurochemical Models, *Clinical Schizophrenia
& Related Psychoses* Vol.4, No.3, (October 2010), pp. 189-200, ISSN 1935-1232

Katsel, P., Davis, K. L., Gorman, J. M. & Haroutunian, V. (2005). Variations in Differential
Gene Expression Patterns across Multiple Brain Regions in Schizophrenia,
Schizophrenia Research Vol.77, No.2-3, (September 2005), pp. 241-252, ISSN 0920-9964

Kieffer, B. L. & Gaveriaux-Ruff, C. (2002). Exploring the Opioid System by Gene Knockout,
Progress in Neurobiology Vol.66, No.5, (Apr 2002), pp. 285-306, ISSN 0301-0082
(Print), 0301-0082 (Linking)

Kiryu, S., Yao, G. L., Morita, N., Kato, H. & Kiyama, H. (1995). Nerve Injury Enhances Rat Neuronal Glutamate Transporter Expression: Identification by Differential Display PCR, *The Journal of Neuroscience* Vol.15, No.12, (December 1995), pp. 7872-7878, ISSN 0270-6474

Kiryu-Seo, S., Gamo, K., Tachibana, T., Tanaka, K. & Kiyama, H. (2006). Unique Anti-Apoptotic Activity of EAAC1 in Injured Motor Neurons, *The EMBO Journal* Vol.25, No.14, (July 2006), pp. 3411-3421

Levenson, J., Endo, S., Kategaya, L. S., Fernandez, R. I., Brabham, D. G., Chin, J., Byrne, J. H. & Eskin, A. (2000). Long-Term Regulation of Neuronal High-Affinity Glutamate and Glutamine Uptake in *Aplysia*, *Proceedings of the National Academy of Sciences of the United States of America* Vol.97, No.23, (November 2000), pp. 12858-12863, ISSN 0027-8424

Levenson, J., Weeber, E., Selcher, J. C., Kategaya, L. S., Sweatt, J. D. & Eskin, A. (2002). Long-Term Potentiation and Contextual Fear Conditioning Increase Neuronal Glutamate Uptake, *Nature Neuroscience* Vol.5, No.2, (February 2002), pp. 155-161

Li, C., Zhao, X., Cao, X., Chu, D., Chen, J. & Zhou, J. (2008). The Drosophila Homolog of JWA Is Required for Ethanol Tolerance, *Alcohol and Alcoholism* Vol.43, No.5, (October 2008), pp. 529-536, ISSN 1464-3502 (Electronic), 0735-0414 (Linking)

Li, X., Valencia, A., Sapp, E., Masso, N., Alexander, J., Reeves, P., Kegel, K. B., Aronin, N. & Difiglia, M. (2010). Aberrant Rab11-Dependent Trafficking of the Neuronal Glutamate Transporter EAAC1 Causes Oxidative Stress and Cell Death in Huntington's Disease, *The Journal of Neuroscience* Vol.30, No.13, (March 2010), pp. 4552-4561, ISSN 1529-2401 (Electronic), 0270-6474 (Linking)

Liang, Z. & Li, G. (2000). Mouse Prenylated Rab Acceptor Is a Novel Golgi Membrane Protein, *Biochemical and Biophysical Research Communications* Vol.275, No.2, (August 2000), pp. 509-516, ISSN 0006-291X

Lin, C. I., Orlov, I., Ruggiero, A. M., Dykes-Hoberg, M., Lee, A., Jackson, M. & Rothstein, J. D. (2001). Modulation of the Neuronal Glutamate Transporter EAAC1 by the Interacting Protein GTRAP3-18, *Nature* Vol.410, No.6824, (March 2001), pp. 84-88

Liu, Y., Vidensky, S., Ruggiero, A. M., Maier, S., Sitte, H. H. & Rothstein, J. D. (2008). Reticulon RTN2b Regulates Trafficking and Function of Neuronal Glutamate Transporter EAAC1, *The Journal of Biological Chemistry* Vol.283, No.10, (March 2008), pp. 6561-6571, ISSN 0021-9258

Lui, H. M., Chen, J., Wang, L. & Naumovski, L. (2003). ARMER, Apoptotic Regulator in the Membrane of the Endoplasmic Reticulum, a Novel Inhibitor of Apoptosis, *Molecular Cancer Research* Vol.1, No.7, (May 2003), pp. 508-518, ISSN 1541-7786

Maier, S., Reiterer, V., Ruggiero, A. M., Rothstein, J. D., Thomas, S., Dahm, R., Sitte, H. H. & Farhan, H. (2009). GTRAP3-18 Serves as a Negative Regulator of Rab1 in Protein Transport and Neuronal Differentiation, *Journal of Cellular and Molecular Medicine* Vol.13, No.1, (January 2009), pp. 114-124, ISSN 1582-4934 (Electronic), 1582-1838 (Linking)

Maleszka, R., Helliwell, P. & Kucharski, R. (2000). Pharmacological Interference with Glutamate Re-Uptake Impairs Long-Term Memory in the Honeybee, *Apis Mellifera*, *Behavioural Brain Research* Vol.115, No.1, (October 2000), pp. 49-53, ISSN 0166-4328

Mao, W. G., Liu, Z. L., Chen, R., Li, A. P. & Zhou, J. W. (2006). JWA Is Required for the Antiproliferative and Pro-Apoptotic Effects of All-*Trans* Retinoic Acid in HeLa Cells, *Clinical and Experimental Pharmacology & Physiology* Vol.33, No.9, (September 2006), pp. 816-824, ISSN 0305-1870

Martincic, I., Peralta, M. E. & Ngsee, J. K. (1997). Isolation and Characterization of a Dual Prenylated Rab and VAMP2 Receptor, *The Journal of Biological Chemistry* Vol.272, No.43, (October 1997), pp. 26991-26998, ISSN 0021-9258

Mathews, G. C. & Diamond, J. S. (2003). Neuronal Glutamate Uptake Contributes to GABA Synthesis and Inhibitory Synaptic Strength, *The Journal of Neuroscience* Vol.23, No.6, (March 2003), pp. 2040-2048, ISSN 1529-2401 (Electronic), 0270-6474 (Linking)

Murphy, T. H., Miyamoto, M., Sastre, A., Schnaar, R. L. & Coyle, J. T. (1989). Glutamate Toxicity in a Neuronal Cell Line Involves Inhibition of Cystine Transport Leading to Oxidative Stress, *Neuron* Vol.2, No.6, (Jun 1989), pp. 1547-1558, ISSN 0896-6273 (Print), 0896-6273 (Linking)

Nieoullon, A., Canolle, B., Masmejean, F., Guillet, B., Pisano, P. & Lortet, S. (2006). The Neuronal Excitatory Amino Acid Transporter EAAC1/EAAT3: Does It Represent a Major Actor at the Brain Excitatory Synapse?, *Journal of Neurochemistry* Vol.98, No.4, (Aug 2006), pp. 1007-1018, ISSN 0022-3042 (Print), 0022-3042 (Linking)

Nitsche, J. F., Schuller, A. G., King, M. A., Zengh, M., Pasternak, G. W. & Pintar, J. E. (2002). Genetic Dissociation of Opiate Tolerance and Physical Dependence in Delta-Opioid Receptor-1 and Preproenkephalin Knock-out Mice, *The Journal of Neuroscience* Vol.22, No.24, (Dec 2002), pp. 10906-10913, ISSN 1529-2401 (Electronic), 0270-6474 (Linking)

Nomura, N., Nagase, T., Miyajima, N., Sazuka, T., Tanaka, A., Sato, S., Seki, N., Kawarabayasi, Y., Ishikawa, K. & Tabata, S. (1994). Prediction of the Coding Sequences of Unidentified Human Genes. II. The Coding Sequences of 40 New Genes (KIAA0041-KIAA0080) Deduced by Analysis of cDNA Clones from Human Cell Line KG-1, *DNA Research* Vol.1, No.5, pp. 223-229, ISSN 1340-2838

Padovano, V., Massari, S., Mazzucchelli, S. & Pietrini, G. (2009). PKC Induces Internalization and Retention of the EAAC1 Glutamate Transporter in Recycling Endosomes of MDCK Cells, *American Journal of Physiology. Cell Physiology* Vol.297, No.4, (October 2009), pp. C835-844, ISSN 1522-1563 (Electronic), 0363-6143 (Linking)

Palos, T. P., Ramachandran, B., Boado, R. & Howard, B. D. (1996). Rat C6 and Human Astrocytic Tumor Cells Express a Neuronal Type of Glutamate Transporter, *Brain Research. Molecular Brain Research* Vol.37, No.1-2, (April 1996), pp. 297-303, ISSN 0169-328X

Paz, R. D., Tardito, S., Atzori, M. & Tseng, K. Y. (2008). Glutamatergic Dysfunction in Schizophrenia: From Basic Neuroscience to Clinical Psychopharmacology, *European Neuropsychopharmacology* Vol.18, No.11, (November 2008), pp. 773-786, ISSN 0924-977X

Peghini, P., Janzen, J. & Stoffel, W. (1997). Glutamate Transporter EAAC-1-Deficient Mice Develop Dicarboxylic Aminoaciduria and Behavioral Abnormalities but No Neurodegeneration, *The EMBO Journal* Vol.16, No.13, (July 1997), pp. 3822-3832

Pettersson, M., Bessonova, M., Gu, H. F., Groop, L. C. & Jonsson, J. I. (2000). Characterization, Chromosomal Localization, and Expression During Hematopoietic Differentiation of the Gene Encoding Arl6ip, ADP-Ribosylation-Like Factor-6 Interacting Protein (ARL6), *Genomics* Vol.68, No.3, (September 2000), pp. 351-354, ISSN 0888-7543

Piper, P. W. (1995). The Heat Shock and Ethanol Stress Responses of Yeast Exhibit Extensive Similarity and Functional Overlap, *FEMS Microbiology Letters* Vol.134, No.2-3, (December 1995), pp. 121-127, ISSN 0378-1097

Ramassamy, C., Averill, D., Beffert, U., Theroux, L., Lussier-Cacan, S., Cohn, J. S., Christen, Y., Schoofs, A., Davignon, J. & Poirier, J. (2000). Oxidative Insults Are Associated with Apolipoprotein E Genotype in Alzheimer's Disease Brain, *Neurobiology of Disease* Vol.7, No.1, (February 2000), pp. 23-37, ISSN 0969-9961

Rothstein, J. D., Jin, L., Dykes-Hoberg, M. & Kuncl, R. W. (1993). Chronic Inhibition of Glutamate Uptake Produces a Model of Slow Neurotoxicity, *Proceedings of the National Academy of Sciences of the United States of America* Vol.90, No.14, (July 1993), pp. 6591-6595

Rothstein, J. D., Martin, L., Levey, A. I., Dykes-Hoberg, M., Jin, L., Wu, D., Nash, N. & Kuncl, R. W. (1994). Localization of Neuronal and Glial Glutamate Transporters, *Neuron* Vol.13, No.3, (September 1994), pp. 713-725, ISSN 0896-6273

Rothstein, J. D., Dykes-Hoberg, M., Pardo, C. A., Bristol, L. A., Jin, L., Kuncl, R. W., Kanai, Y., Hediger, M. A., Wang, Y., Schielke, J. P. & Welty, D. F. (1996). Knockout of Glutamate Transporters Reveals a Major Role for Astroglial Transport in Excitotoxicity and Clearance of Glutamate, *Neuron* Vol.16, No.3, (March 1996), pp. 675-686, ISSN 0896-6273

Ruggiero, A. M., Liu, Y., Vidensky, S., Maier, S., Jung, E., Farhan, H., Robinson, M. B., Sitte, H. H. & Rothstein, J. D. (2008). The Endoplasmic Reticulum Exit of Glutamate Transporter Is Regulated by the Inducible Mammalian Yip6b/GTRAP3-18 Protein, *The Journal of Biological Chemistry* Vol.283, No.10, (March 2008), pp. 6175-6183

Schweneker, M., Bachmann, A. S. & Moelling, K. (2005). JM4 Is a Four-Transmembrane Protein Binding to the CCR5 Receptor, *FEBS Letters* Vol.579, No.7, (March 2005), pp. 1751-1758, ISSN 0014-5793

Sepkuty, J. P., Cohen, A. S., Eccles, C., Rafiq, A., Behar, K., Ganel, R., Coulter, D. A. & Rothstein, J. D. (2002). A Neuronal Glutamate Transporter Contributes to Neurotransmitter GABA Synthesis and Epilepsy, *The Journal of Neuroscience* Vol.22, No.15, (August 2002), pp. 6372-6379

Sheldon, A. L., Gonzalez, M. I. & Robinson, M. B. (2006). A Carboxyl-Terminal Determinant of the Neuronal Glutamate Transporter, EAAC1, Is Required for Platelet-Derived Growth Factor-Dependent Trafficking, *The Journal of Biological Chemistry* Vol.281, No.8, (February 2006), pp. 4876-4886, ISSN 0021-9258

Sian, J., Dexter, D. T., Lees, A. J., Daniel, S., Agid, Y., Javoy-Agid, F., Jenner, P. & Marsden, C. D. (1994). Alterations in Glutathione Levels in Parkinson's Disease and Other Neurodegenerative Disorders Affecting Basal Ganglia, *Annals of Neurology* Vol.36, No.3, (September 1994), pp. 348-355, ISSN 0364-5134

Sims, K. D., Straff, D. J. & Robinson, M. B. (2000). Platelet-Derived Growth Factor Rapidly Increases Activity and Cell Surface Expression of the EAAC1 Subtype of Glutamate Transporter through Activation of Phosphatidylinositol 3-Kinase, *The Journal of Biological Chemistry* Vol.275, No.7, (February 2000), pp. 5228-5237, ISSN 0021-9258

Tong, G. & Jahr, C. E. (1994). Block of Glutamate Transporters Potentiates Postsynaptic Excitation, *Neuron* Vol.13, No.5, (November 1994), pp. 1195-1203

Trotti, D., Peng, J. B., Dunlop, J. & Hediger, M. A. (2001). Inhibition of the Glutamate Transporter EAAC1 Expressed in *Xenopus* Oocytes by Phorbol Esters, *Brain Research* Vol.914, No.1-2, (September 2001), pp. 196-203, ISSN 0006-8993

Ueda, Y., Doi, T., Nakajima, A., Tokumaru, J., Tsuru, N. & Ishida, Y. (2006). The Functional Role of Glutamate Transporter Associated Protein (GTRAP3-18) in the Epileptogenesis Induced by PTZ-Kindling, *The Annual Report of The Japan Epilepsy Research Foundation* Vol.17, pp. 33-40, Available from: http:hdl.handle.net/10458/2116

Ueda, Y., Doi, T., Nagatomo, K., Tokumaru, J., Takaki, M. & Willmore, L. J. (2007). Effect of Levetiracetam on Molecular Regulation of Hippocampal Glutamate and GABA Transporters in Rats with Chronic Seizures Induced by Amygdalar $FeCl_3$ Injection, *Brain Research* Vol.1151, (June 2007), pp. 55-61

Wang, S., Gong, Z., Chen, R., Liu, Y., Li, A., Li, G. & Zhou, J. (2009). JWA Regulates XRCC1 and Functions as a Novel Base Excision Repair Protein in Oxidative-Stress-Induced DNA Single-Strand Breaks, *Nucleic Acids Research* Vol.37, No.6, (April 2009), pp. 1936-1950

Watabe, M., Aoyama, K. & Nakaki, T. (2007). Regulation of Glutathione Synthesis Via Interaction between Glutamate Transport-Associated Protein 3-18 (GTRAP3-18) and Excitatory Amino Acid Carrier-1 (EAAC1) at Plasma Membrane, *Molecular Pharmacology* Vol.72, No.5, (November 2007), pp. 1103-1110

Watabe, M., Aoyama, K. & Nakaki, T. (2008). A Dominant Role of GTRAP3-18 in Neuronal Glutathione Synthesis, *The Journal of Neuroscience* Vol.28, No.38, (September 2008), pp. 9404-9413

Waxman, E. A., Baconguis, I., Lynch, D. R. & Robinson, M. B. (2007). N-Methyl-D-Aspartate Receptor-Dependent Regulation of the Glutamate Transporter Excitatory Amino Acid Carrier 1, *The Journal of Biological Chemistry* Vol.282, No.24, (June 2007), pp. 17594-17607, ISSN 0021-9258

Wilke, N., Sganga, M., Barhite, S. & Miles, M. F. (1994). Effects of Alcohol on Gene Expression in Neural Cells, *Experientia. Supplementum* Vol.71, pp. 49-59, ISSN 1023-294X

Won, S. J., Yoo, B. H., Brennan, A. M., Shin, B. S., Kauppinen, T. M., Berman, A. E., Swanson, R. A. & Suh, S. W. (2010). EAAC1 Gene Deletion Alters Zinc Homeostasis and Exacerbates Neuronal Injury after Transient Cerebral Ischemia, *The Journal of Neuroscience* Vol.30, No.46, (November 2010), pp. 15409-15418, ISSN 1529-2401 (Electronic), 0270-6474 (Linking)

Wu, Y., Chen, R., Zhao, X., Li, A., Li, G. & Zhou, J. (2011). JWA Regulates Chronic Morphine Dependence Via the Delta Opioid Receptor, *Biochemical and Biophysical Research Communications* Vol.409, No.3, (June 2011), pp. 520-525, ISSN 1090-2104 (Electronic), 0006-291X (Linking)

Xia, P., Pei, G. & Schwarz, W. (2006). Regulation of the Glutamate Transporter EAAC1 by Expression and Activation of Delta-Opioid Receptor, *The European Journal of Neuroscience* Vol.24, No.1, (July 2006), pp. 87-93, ISSN 0953-816X

Yang, W. & Kilberg, M. S. (2002). Biosynthesis, Intracellular Targeting, and Degradation of the EAAC1 Glutamate/Aspartate Transporter in C6 Glioma Cells, *The Journal of Biological Chemistry* Vol.277, No.41, (October 2002), pp. 38350-38357, ISSN 0021-9258

Zerangue, N. & Kavanaugh, M. P. (1996). Interaction of L-Cysteine with a Human Excitatory Amino Acid Transporter, *The Journal of Physiology* Vol.493 (Pt 2), (June 1996), pp. 419-423, ISSN 0022-3751

Zhou, J., Ye, J., Zhao, X. & Li, A. (2008). JWA Is Required for Arsenic Trioxide Induced Apoptosis in HeLa and MCF-7 Cells Via Reactive Oxygen Species and Mitochondria Linked Signal Pathway, *Toxicology and Applied Pharmacology* Vol.230, No.1, (July 2008), pp. 33-40, ISSN 0041-008X

Zhu, T., Chen, R., Li, A. P., Liu, J., Liu, Q. Z., Chang, H. C. & Zhou, J. W. (2005). Regulation of a Novel Cell Differentiation-Associated Gene, JWA During Oxidative Damage in K562 and MCF-7 Cells, *Journal of Biomedical Science* Vol.12, No.1, pp. 219-227

Zhu, Y., King, M. A., Schuller, A. G., Nitsche, J. F., Reidl, M., Elde, R. P., Unterwald, E., Pasternak, G. W. & Pintar, J. E. (1999). Retention of Supraspinal Delta-Like Analgesia and Loss of Morphine Tolerance in Delta Opioid Receptor Knockout Mice, *Neuron* Vol.24, No.1, (Sep 1999), pp. 243-252, ISSN 0896-6273 (Print), 0896-6273 (Linking)

8

Mechanism of Cargo Recognition During Selective Autophagy

Yasunori Watanabe[1,2] and Nobuo N. Noda[2]
[1]Graduate School of Life Science, Hokkaido University
[2]Institute of Microbial Chemistry, Tokyo
Japan

1. Introduction

Autophagy is an intracellular bulk degradation system conserved among eukaryotes from yeast to mammals. It is responsible for the degradation of cytosolic components and organelles in response to nutrient deprivation. There are three main types of autophagy: macroautophagy, microautophagy and chaperone-mediated autophagy (CMA). Microautophagy sequesters cytoplasmic components and delivers them for degradation by direct invagination or protrusion/septation of the lysosomal or vacuolar membrane (Mijaljica et al., 2011; Uttenweiler and Mayer, 2008). CMA targets specific cytosolic proteins that are trapped by the heat shock cognate protein of 70 kDa (hsc70) and, through interaction with lysosome-associated membrane protein type 2A (LAMP-2A), they are then translocated into the lysosomal lumen for rapid degradation (Orenstein and Cuervo, 2010). Macroautophagy, hereafter referred to as autophagy, is the most well characterized process of the three. During autophagy, double membrane structures called autophagosomes sequester a portion of the cytoplasm and fuse with the lysosome (or vacuole in the case of yeast and plants) to deliver their inner contents into the organelle lumen (Mizushima, 2007; Mizushima et al., 2010). Analyses of autophagy-related (Atg) proteins have unveiled dynamic and diverse aspects of mechanisms that underlie membrane formation during autophagy (Mizushima et al., 2010; Nakatogawa et al., 2009). As the contents of autophagosomes are indistinguishable from their surrounding cytoplasm (Baba et al., 1994), autophagy has long been considered a nonselective catabolic pathway. Recent studies, however, have provided evidence for the selective degradation of various targets by autophagy. In autophagy-deficient neuronal cells, intracellular protein aggregates accumulate and eventually lead to neurodegeneration, suggesting that autophagy selectively degrades harmful protein aggregates (Hara et al., 2006; Komatsu et al., 2006). Damaged or superfluous organelles, such as mitochondria and peroxisomes, and even intracellular infectious pathogens are also selectively degraded by autophagy (Goldman et al., 2010; Gutierrez et al., 2004; Manjithaya et al., 2010; Nakagawa et al., 2004; Noda and Yoshimori, 2009). In the budding yeast *Saccharomyces cerevisiae*, α-mannosidase and aminopeptidase I are selectively transported to the vacuole through autophagic pathways (Baba et al., 1997; Hutchins and Klionsky, 2001).

Although the precise molecular mechanisms of cargo selection by autophagy are yet to be established, an increasing number of autophagic receptors that are responsible for recognition of specific cargoes have been identified. These include Atg19 and Atg34 in the selective transport of vacuolar enzymes to the vacuole through autophagy (Leber et al., 2001; Scott et al., 2001; Suzuki et al., 2010), p62 and neighbor of BRCA1 gene 1 (NBR1) in the autophagic degradation of ubiquitinated protein aggregates (Bjorkoy et al., 2005; Kirkin et al., 2009), PpAtg30 in pexophagy (autophagic degradation of peroxisome) (Farre et al., 2008), and Atg32 and Nix1 in mitophagy (autophagic degradation of mitochondria) (Kanki et al., 2009; Novak et al., 2010; Okamoto et al., 2009). Most of these receptors interact directly with Atg8-family proteins, which are crucial factors in autophagosome biogenesis.

We have been studying the mechanisms of specific cargo recognition during autophagy, especially those of the selective delivery of vacuolar enzymes into the vacuole in yeast. We summarize here the current knowledge of such mechanisms as revealed by biochemical and structural studies.

2. Recognition of vacuolar enzymes by Atg19 and Atg34

2.1 Selective transport of vacuolar enzymes by autophagic pathways

In the budding yeast *S. cerevisiae*, α-mannosidase (Ams1) and a precursor form of aminopeptidase I (prApe1) are selectively delivered into the vacuole through the cytoplasm to vacuole targeting (Cvt) pathway under vegetative conditions, and via autophagy under starvation conditions. The Cvt pathway is topologically and mechanistically similar to autophagy (Lynch-Day and Klionsky, 2010); therefore, studies on the molecular mechanisms of cargo recognition in the Cvt pathway will provide insight into the basic mechanism of selective autophagy. prApe1, the primary Cvt cargo, is synthesized in the cytosol as a precursor form with a cleavable propeptide consisting of 45 amino acid residues at the N terminus (Klionsky et al., 1992) and assembles into a dodecamer. The prApe1 dodecamer further self-assembles into a higher order structure called the Ape1 complex. The existence of a specific receptor for prApe1 was proposed when it was observed that prApe1 transport to the vacuole by the Cvt pathway is both specific and saturable.

Two groups simultaneously discovered that Atg19 has all of the characteristics needed to be a receptor for prApe1 in Cvt transport (Leber et al., 2001; Scott et al., 2001). Characterization of the protein revealed that Atg19 is needed for the stabilization of prApe1 binding to the Cvt vesicle membrane, and that in *atg19Δ* cells, prApe1 maturation is inhibited while autophagy is not affected (Suzuki et al., 2002). In addition, Atg19 binds to prApe1 in a propeptide-dependent manner, suggesting that the propeptide region is responsible for the recognition of prApe1 by the Cvt pathway machinery (Shintani et al., 2002). A secondary-structure prediction suggested that the prApe1 propeptide forms a helix-turn-helix structure and that the first helix exhibits the characteristics of an amphipathic α helix (Martinez et al., 1997). Our previous study revealed that the region containing the first helix of the prApe1 propeptide (residues 1-20) is sufficient for interaction with Atg19 (Watanabe et al., 2010). This is consistent with a previous report showing that the first helix of the prApe1 propeptide is critical for prApe1processing (Oda et al., 1996). *In vitro* pull-down assays showed that the coiled coil domain of Atg19 (residues 124-253), which contains a predicted coiled coil between amino acids 160 and 187, directly interacts with the prApe1 propeptide. This is consistent with a previous report showing that the prApe1-binding site of Atg19 is located in the region between amino acid residues 153 and 191 (Shintani et al., 2002).

Ams1, another Cvt cargo, oligomerizes after synthesis and associates with the Ape1 complex through the action of Atg19. Atg19 has two stable domains, the N-terminal domain (residues 1-123) and the Ams1 binding domain (ABD; residues 254-367, see below for further details). Ams1 associates with Atg19 via the ABD that is distinct from the prApe1 binding site and therefore Atg19 can simultaneously interact with both prApe1 and Ams1. prApe1, Ams1, and Atg19 assemble into a large complex called the Cvt complex, which was identified as an electron-dense structure localized close to the vacuole by electron microscopy (Baba et al., 1997). Atg11 interacts with Atg19 to recruit the Cvt complex to the preautophagosomal structure (PAS), which plays a central role in autophagosome formation near the vacuole (Shintani et al., 2002; Suzuki and Ohsumi, 2010). Atg19 further interacts with Atg8, which is localized at the PAS and involved in the elongation of autophagosomes, using the Atg8 family-interacting motif (AIM; 412-WEEL-415) to induce formation of the Cvt vesicle (Noda et al., 2008). Atg8 is conjugated to phosphatidylethanolamine (PE) and associates with autophagosomes or the Cvt vesicle (Ichimura et al., 2000). This explains why the vesicle selectively surrounds only the cargo. After transport to the vacuole, the prApe1 propeptide is removed via a proteinase B-dependent reaction to generate mature Ape1 (mApe1), and the Ape1 complex disassembles back into dodecamers. Atg34, an Atg19 paralog, functions as an additional receptor protein for Ams1 but not prApe1 only under starvation conditions (Suzuki et al., 2010). Although Atg34, similar to Atg19, has the predicted coiled coil (residues 130-157), Atg34 is not capable of interacting with prApe1.

Recently, two cargoes that are selectively delivered to the vacuole have been identified: leucine aminopeptidase III (Lap3) (Kageyama et al., 2009) and aspartyl aminopeptidase (Ape4) (Yuga et al., 2011). Lap3 is transported to the vacuole for degradation only when it is overproduced under nitrogen starvation conditions. Lap3 forms a homohexameric complex of ~220 kDa, which further forms an aggregate independently of prApe1. Although this transport is partially mediated by Atg19, it remains to be determined whether Lap3 can interact with Atg19. Ape4 is the third Cvt cargo, which is similar in primary structure and subunit organization to Ape1. Ape4 lacks the N-terminal propeptide that is used by prApe1 for binding to Atg19. As the Ape4-binding site in Atg19 is located between the prApe1- and Ams1-binding sites (residues 204-247), these enzymes are unlikely to compete with each other for binding to Atg19. As Atg34 did not interact with Ape4, it might not be involved in Ape4 transport. More recently, Suzuki *et al.* elucidated that selective autophagy downregulates Ty1 transposition by eliminating Ty1 virus-like particles (VLPs) from the cytoplasm under nutrient-limited conditions (Suzuki et al., 2011). Although Ty1 VLPs are not vacuolar enzymes, they are targeted to autophagosomes by an interaction with Atg19. The N-terminal domain of Atg19 is specifically required for selective transport of Ty1 VLPs to the vacuole, though Atg19 is able to interact with Ty1 Gag without the N-terminal domain. Selective autophagy might safeguard genome integrity against excessive insertional mutagenesis caused during nutrient starvation by transposable elements in eukaryotic cells.

2.2 Structural basis for Ams1 recognition by Ag19 and Atg34

Scott *et al.* suggested that Ams1 is delivered to the vacuole in an Atg19-dependent manner (Scott et al., 2001). Ams1 was found to associate with Atg19, and a defect in the Ape1-Atg19 complex formation was shown to severely affect the import of Ams1 into the vacuole, whereas Ams1 was dispensable for transport of the Ape1-Atg19 complex. This suggests that Ams1 might exploit the prApe1 import system to achieve its own effective transport to the vacuole and that it is tethered to the Ape1-Atg19 complex through interaction with Atg19. In our recent study, we identified the Ams1 binding domain (ABD) in Atg19 and Atg34 by limited proteolysis of full-length Atg19, an *in vitro* pull-down assay as well as sequence alignment (Watanabe et al., 2010). In *atg19Δ* cells expressing Atg19$^{\Delta ABD}$, Ams1 transport to the vacuole was inhibited, suggesting that the Atg19 ABD is required for Ams1 transport to the vacuole through the Cvt pathway. In such cells, prApe1 transport to the vacuole is the normal process. These results indicate that the Atg19 ABD is specifically responsible for the transport of Ams1, but not prApe1, to the vacuole through the Cvt pathway.

The Atg19 and Atg34 ABD structures were determined in solution using NMR spectroscopy (Figure 1A and B) (Watanabe et al., 2010). Both ABDs comprise eight β-strands (A-H), of which A, B, E, and H form an antiparallel β-sheet; the surface of this sheet faces a second antiparallel β-sheet comprising C, D, F, and G, thus forming a typical immunoglobulin-like β-sandwich fold. The Atg19 and Atg34 ABD structures are similar to each other with a root mean square difference of 2.1 Å for 102 residues (Z-score calculated by the Dalilite program (Holm and Park, 2000) is 12.8). There are relatively large structural differences between the Atg19 and Atg34 ABDs in the loops located at the bottom of the immunoglobulin fold (the loop connecting strands A and B (AB loop), CD, EF, and GH loops). In contrast, the loops located at the top of the immunoglobulin fold (the BC, DE, and FG loops) have a similar conformation. Furthermore, the residues comprising the top loops, especially those of the DE loop, are more strongly conserved between the Atg19 and Atg34 ABDs than those comprising the bottom loops (Figure 1). In the DE loop, His-310/296, Glu-311/297, Ile-314/300, and Lys-315/301 of Atg19/Atg34 are exposed. Among these exposed residues, His-310/296 and/or Glu-311/297 of the Atg19/Atg34 ABD are essential for Ams1 recognition. Further analysis showed that in *atg19Δatg34Δ* cells expressing Atg19^{H310A} (substitution of His-310 with alanine) but not Atg19^{E311A}, transport of Ams1-GFP to the vacuole under autophagy-inducing conditions is inhibited. This indicates that the conserved His residue in the DE loop of the Atg19 ABD plays a critical role in Ams1 recognition and that Ams1 binding of the Atg19 ABD is essential for Ams1 transportation to the vacuole. Similar experiments using Atg34 mutants showed that His-296 of Atg34 ABD, which corresponds to His-310 of Atg19 ABD, also plays a critical role in Ams1 recognition.

The ABDs in Atg19 and Atg34 have a β-sandwich fold that is observed in a variety of immunoglobulins and immunoglobulin-like domains responsible for recognizing various proteins. Because antibodies generally recognize antigens using the hypervariable loops from both the VH and VL regions, their manner of antigen binding should differ from that of monomeric ABDs with Ams1. Interestingly, however, the ABD-Ams1 interaction resembles that observed between camelid antibody fragments and their antigens, as camelid antibodies lack a light chain and function as a monomer where hypervariable loops of the VH are responsible for antigen binding (Muyldermans, 2001). It also mimics the interaction of monobodies (artificially designed proteins that use a fibronectin type III domain as a scaffold) and their targets, as monobodies interact with their targets using similar loops in a monomeric immunoglobulin fold. Camelid antibody fragments and monobodies interact with their target proteins using loops clustered at one side of their immunoglobulin fold; these loops are topologically equivalent to the BC, DE, and FG loops of the Atg19 and Atg34 ABDs, one of which was shown to be crucial for Ams1 recognition as mentioned above. Therefore, they might recognize Ams1 using these loops in a similar manner with camelid antibodies and monobodies. In order to further elucidate the recognition mechanism of Ams1 by the ABD, structural determination of the Ams1-ABD complex by X-ray crystallography is needed. We have already succeeded in overexpressing *S. cerevisiae* Ams1 in *Pichia pastoris* and purifying it on a large scale (Watanabe et al., 2009). Crystallization and structural determination of the Ams1-ABD complex are now in progress.

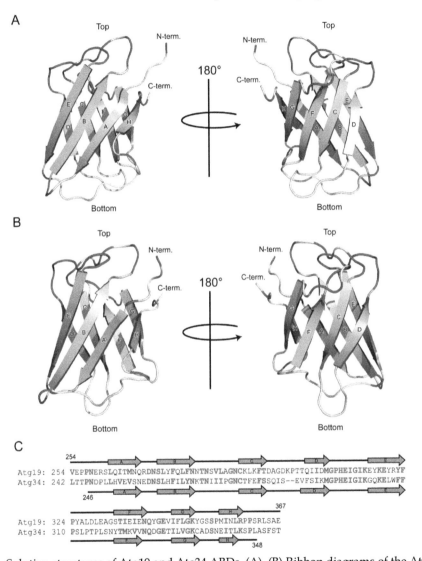

Fig. 1. Solution structures of Atg19 and Atg34 ABDs. (A), (B) Ribbon diagrams of the Atg19 ABD and Atg34 ABD structures, respectively. Strands are colored *light blue* and labeled. Loop residues conserved between Atg19 and Atg34 are colored *red*. *Left* and *right* are related by a 180° rotation along the *vertical axis*. (C) Sequence alignment between Atg19 and Atg34 ABDs. *Gaps* are introduced to maximize the similarity. Conserved or type-conserved residues are colored *red*. Secondary structure elements of the Atg19 and Atg34 ABDs are shown above and below the sequence, respectively.

3. Receptor proteins required for the selective degradation of organelles by autophagy

Damaged or superfluous organelles, such as mitochondria and peroxisomes, are also selectively degraded by autophagy. To date, several receptor proteins which function in these selective types of autophagy have been identified.

3.1 Receptor proteins in mitophagy

The mitochondrion is an organelle that produces energy through oxidative phosphorylation and simultaneously generates reactive oxygen species (ROS), causing oxidative damage to mitochondrial DNA, protein and lipids, and often inducing cell death. Therefore, an appropriate quality control of mitochondria is important to maintain proper cellular homeostasis. Selective degradation of mitochondria via autophagy, known as mitophagy, is the primary mechanism for mitochondrial quality control. Two groups independently identified Atg32 as a receptor protein for mitophagy in yeast (Kanki et al., 2009; Okamoto et al., 2009). Atg32 is a single-pass mitochondrial outer membrane protein, and its N- and C-terminal domains are oriented towards the cytoplasm and the intermembrane space, respectively. Mitochondria-anchored Atg32 binds Atg11 during mitophagy to recruit mitochondria to the PAS. When mitophagy is induced, Atg32 is phosphorylated, for which Ser-114 and Ser-119 of Atg32 are required. , The phosphorylation of Atg32 is required for Atg32-Atg11 interaction and mitophagy (Aoki et al., 2011). By controlling the activity and/or localization of the kinase that phosphorylates Atg32, cells may regulate the amount of mitochondria or remove damaged or aged mitochondria through mitophagy. Similarly to other receptor proteins, Atg32 binds to Atg8 using the AIM sequence, 86-WQAI-89, and this binding is required for the efficient sequestration of mitochondria by the autophagosome. The Atg32-Atg8 interaction may restrict autophagosome formation to the neighborhood of the targeted mitochondrion by gathering surrounding Atg proteins. Although no homolog of Atg32 has been identified in mammals, Nix was recently shown to be a mammalian mitophagy receptor protein (Novak et al., 2010). Similarly to Atg32, Nix has a transmembrane domain in its C terminus that can target the protein to the mitochondrial outer membrane and has a functional AIM, 35-WVEL-38, that directly interacts with mammalian Atg8 homologs. Thus Nix may fulfill the function of Atg32 in mammals.

3.2 Receptor protein in pexophagy

Peroxisomes have diverse functions, including the decomposition of hydrogen peroxide and the oxidation of fatty acids. The specific degradation of peroxisomes by autophagy, pexophagy, is conserved from yeast to humans and is triggered physiologically to allow cells to clear an excess of peroxisomes. The study of methylotrophic yeasts, particularly *P. pastoris* and *Hansenula polymorpha*, has led to the current understanding of the molecular mechanism governing pexophagy. Farre *et al.* identified Atg30 as a receptor protein for pexophagy in *P. pastoris* (Farre et al., 2008). PpAtg30 interacts with peroxisomes via two peroxisomal membrane proteins, Pex3 and Pex14, and with autophagy machinery via PpAtg11 and PpAtg17, which organize the PAS. Several residues on PpAtg30 are phosphorylated under pexophagy conditions and, similarly to Atg32, such phosphorylation, especially that of Ser-112, is required for PpAtg11 interaction. The isolation membrane then expands and surrounds the PpAtg30-localizing peroxisomes, in order to selectively degrade surplus peroxisomes. Unlike other receptor proteins, PpAtg30 has no AIMs, so that PpAtg30 is unable to interact directly with PpAtg8. It is important to understand how the isolation membrane expands around the peroxisome surface while excluding cytosolic contents, and this is speculated to involve the interaction of PpAtg30 with an unidentified protein in the isolation membrane, or the interaction of PpAtg8 with another protein on the peroxisome surface.

4. Receptor protein for selective autophagy in *C. elegans*

Germ granules are restricted to the germ cells of many higher eukaryotes and are believed to carry germ cell determinants (Strome and Lehmann, 2007). Germ granules in *Caenorhabditis elegans*, also known as P granules, are maternally contributed and dispersed throughout the cytoplasm of a newly fertilized embryo. During *C. elegans* embryogenesis, some P granules are left in the cytoplasm destined for the somatic daughter cell and these P granules are quickly disassembled and/or degraded (Hird et al., 1996). Recently, Zhang *et al.* provided evidence that the P granule components PGL-1 and PGL-3 that remain in the cytoplasm destined for somatic daughters are selectively removed by autophagy through the receptor protein SEPA-1 (Zhang et al., 2009). In autophagy-deficient somatic cells, PGL-1 and PGL-3 extensively accumulate in the P granules, and SEPA-1 mediates the accumulation of these P granules into aggregates, termed PGL granules, through its self-oligomerization and direct interaction with PGL-3. SEPA-1 can also directly interact with LGG-1, an Atg8 homolog. Thus, PGL granules associated with SEPA-1 could be incorporated into autophagosomes through the interaction of SEPA-1 with LGG-1. Because the expression of SEPA-1 is restricted to somatic cells, the selective exclusion of P granules is ensured only in these cells. An *in vitro* pull-down assay showed that the SEPA-1 fragment containing amino acids 39 to 160 is required for both self-oligomerization and interaction with PGL-3 and that the SEPA-1 fragment containing amino acids 289 to 575 is required for interaction with LGG-1. The SEPA-1 fragment that interacts with LGG-1 contains a canonical AIM sequence, 469-YQEL-472 (Noda et al., 2010). Thus, the YQEL sequence in SEPA-1 is a potential candidate for a functional AIM.

5. Recognition of ubiquitinated cargoes by p62 and NBR1

Autophagic degradation of ubiquitinated protein aggregates is important for cell survival. Defects in autophagy cause the accumulation of ubiquitin-positive protein inclusions, leading to severe liver injury (Komatsu et al., 2005) and neurodegeneration (Hara et al., 2006; Komatsu et al., 2006). The polyubiquitin-binding protein p62, also called sequestosome 1 (SQSTM1), is a common component of protein aggregates found in both the brain and the liver of patients suffering from protein aggregation diseases. These include Lewy bodies in Parkinson's disease, neurofibrillary tangles in Alzheimer's disease, and huntingtin aggregates (Kuusisto et al., 2001, 2002; Nagaoka et al., 2004; Zatloukal et al., 2002). In the liver, Mallory bodies, hyaline bodies in hepatocellular carcinoma, and α1 antitrypsin aggregate contain p62 (Zatloukal et al., 2002); all of these aggregates contain polyubiquitinated proteins. p62 interacts with ubiquitin via its C-terminal UBA domain (Vadlamudi et al., 1996) and self-assembles via its N-terminal PB1 domain (Ponting et al., 2002), thereby forming large aggregates containing ubiquitinated proteins. p62 further interacts with LC3, a mammalian homolog of Atg8, via the LC3 interacting region (LIR;

residues 321-342), so that ubiquitinated protein aggregates containing p62 are selectively degraded by autophagy (Bjorkoy et al., 2005; Komatsu et al., 2007; Pankiv et al., 2007). Therefore, it is implied that p62 functions as a receptor protein for ubiquitinated proteins to be degraded in lysosomes. It is also hypothesized that p62 functions as a receptor for organelles such as peroxisomes and mitochondria (Kim et al., 2008; Kirkin et al., 2009), and for intracellular bacteria (Dupont et al., 2009; Yoshikawa et al., 2009; Zheng et al., 2009). Recently, neighbor of BRCA1 gene 1 (NBR1) has been identified as another autophagy receptor (Kirkin et al., 2009). The structure of NBR1 is similar to that of p62, and NBR1 can bind both LC3 via the LIR and ubiquitinated proteins via the UBA domain. Like p62, NBR1 is sequestered into the autophagosome via LC3-interaction and/or p62-interaction and markedly accumulates in autophagy-deficient tissues.

To clarify the molecular mechanism of ubiquitinated cargo recognition by p62 and NBR1, it is necessary to elucidate which proteins conjugated with either K48-linked or K63-linked polyubiquitin chains are targeted to the autophagy/lysosomal degradation pathway. Classically, proteins conjugated with K48-linked polyubiquitin chains are recognized as the proteolytic substrate by the UBD-containing proteasomal receptors. Recently, K63-linked chains have been implicated in proteolytic degradation of misfolded and aggregated proteins (Olzmann et al., 2007; Tan et al., 2008; Wooten et al., 2008). Given the reported preference of the known ubiquitin-binding autophagy receptors for K63-linked ubiquitin chains, cargoes conjugated with K63-linked ubiquitin chains may be preferentially targeted to the autophagy/lysosomal degradation pathway. However, p62 has been shown to compete for ubiquitinated cargo with the classical proteasomal receptors. Accumulation of p62 resulting from inhibition of autophagy compromised degradation of proteasomal substrates, most likely due to the excessive interaction between p62 and substrates conjugated with K48-linked polyubiquitin chains (Korolchuk et al., 2009). It remains to be clarified how p62 distinguishes between K48-linked and K63-linked polyubiquitin chains.

To date, several structural studies have been performed on p62. Isogai *et al.* determined the crystal structure of the UBA domain of mouse p62 and the solution structure of its ubiquitin-bound form (Isogai et al., 2011). In crystals, the p62 UBA domain adopts a dimeric structure, which is distinct from that of other UBA domains. In solution, the domain exists in equilibrium between the dimer and monomer forms, and ubiquitin-binding shifts the equilibrium toward the monomer to form a 1:1 complex between the UBA domain and ubiquitin. The extreme C-terminal end of the p62 UBA domain is responsible for dimerization of the domain. Mutations that inhibit dimerization of the p62 UBA domain increase the affinity of p62 for ubiquitin. These results suggest an autoinhibitory mechanism in the p62 UBA domain to avoid self-degradation by the ubiquitin-proteasomal system. The interaction between the p62 LIR and LC3 is structurally and functionally well characterized (Ichimura et al., 2008; Noda et al., 2008). In the structure of the p62 LIR in complex with LC3, the tryptophan and leucine residues in the p62 LIR, DDDWTHL, interact with two hydrophobic pockets, the W-site and the L-site, where tryptophan and leucine residues respectively interact (Noda et al., 2010) on the surface of LC3. In agreement with the structure of the p62 LIR in complex with LC3, the tryptophan and leucine residues are involved in the turnover of p62 via autophagy. Recently, the structure of the NBR1 LIR in complex with GABARAP, a LC3 paralog, has also been determined (Rozenknop et al., 2011). Similar to the interaction between p62 LIR and LC3, the tyrosine and the third isoleucine residues in the NBR1 LIR, EDYIII, interact with GABARAP in a manner typical of the interaction of AIM with Atg8 homologs.

6. Receptor proteins required for the restriction of infectious bacterial growth by autophagy

Autophagy also serves as a cell-autonomous effector mechanism of innate immunity in the cytosol. It does this through restricting bacterial proliferation by separating bacteria from the nutrient-rich cytosol and delivering them into bactericidal autolysosomes. Several examples showing that these types of autophagy are mediated by cytosolic bacteria-recognizing receptor proteins have been recently reported. In *Drosophila melanogaster*, PGRP-LE, a receptor protein for bacterial peptidoglycans, induces autophagy of wild-type but not listeriolysin-deficient *Listeria monocytogenes*, suggesting that this pathway specifically selects cytosolic bacteria for autophagy (Yano et al., 2008). However, it is unknown how PGRP-LE induces the autophagic degradation of *L. monocytogenes* and whether PGRP-LE binds the *D. melanogaster* Atg8 orthologs.

Salmonella enterica Typhimurium (*S.* Typhimurium) typically occupies a membrane bound compartment, the Salmonella-containing vacuole (SCV), in host cells. In mammalian cells, *S.* Typhimurium and other bacteria enter the cytosol and are released from SCVs, then become coated with a dense layer of ubiquitin (Perrin et al., 2004) and are delivered to lysosomes via autophagy (Birmingham et al., 2006). It was reported that p62 functions as a receptor protein for delivering such ubiquitin-coated bacteria into autophagosomes (Dupont et al., 2009; Zheng et al., 2009). In addition to p62, two other receptor proteins were identified, NDP52 (Thurston et al., 2009) and Optineurin (Wild et al., 2011), both of which recognize ubiquitin-coated bacteria. NDP52 is recruited by ubiquitin-coated *S.* Typhimurium and binds both ubiquitin via a zinc-finger domain (residues 420-446) and LC3. Although NDP52 interacts with LC3, it has not been clarified whether NDP52 has an AIM. NDP52 also coordinates a signaling complex including Tank-binding kinase (TBK1), Sintbad and Nap1. *In vitro* binding studies revealed that a SKICH domain (residues 1-127) in NDP52 is required for direct Nap1 binding. Thereby, NDP52 recruits TBK1 to ubiquitin-coated *S.* Typhimurium. The ability of NDP52 to serve as an adaptor for TBK1 also seems to be critical in the cell-autonomous response, but it remains to be determined what role TBK1 plays in association with NDP52. OPTN is another autophagy receptor protein that binds and co-localizes with LC3 via an AIM (178-FVEI-181) and ubiquitin via its ubiquitin binding in ABIN and NEMO (UBAN) domains. The N-terminal region of OPTN (residues 1-127) also interacts with TBK1 (Morton et al., 2008). When TBK1 is recruited into *S.* Typhimurium via OPTN, it becomes activated and phosphorylates OPTN at Ser-177, one residue N-terminal to the OPTN AIM. The phosphorylation of Ser-177 increases the affinity between OPTN and LC3, which is consistent with the previous review showing that acidic residues are preferred at the N-terminal side of the AIM for higher affinity with Atg8 homologs (Noda et al., 2010). Although p62, NDP52 and OPTN target the same bacteria, NDP52 and OPTN localize to microdomains on the surface of ubiquitinated bacteria where p62 does not co-localize (Cemma et al., 2011; Wild et al., 2011). Because these autophagy receptors have their respective ubiquitin-binding domains, the distinct specificities for different ubiquitin chains may result in partitioning of the receptors to different subdomains on the bacterium. However, it is still unknown which type of ubiquitin chains are conjugated to the bacterial surface components.

7. Concluding remarks

Autophagy receptor proteins have two main functions: recognizing autophagic cargoes, and interacting with Atg8 homologs. Because of these functions, autophagy receptor proteins can tether autophagic cargoes to the isolation membrane (e.g., Atg19; Figure 2) so that the cargoes are selectively and efficiently engulfed by an autophagosome and transported into the vacuole/lysosome. Although the mechanism of direct interaction with Atg8 homologs

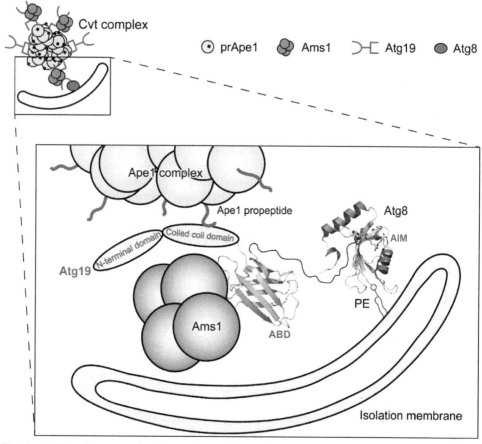

Fig. 2. Atg19 tethers autophagic cargoes to the isolation membrane. Atg19 interacts with prApe1 and Ams1 via the coiled coil domain and ABD, respectively. Atg19 also interacts with Atg8, which is conjugated to phosphatidylethanolamine (PE) and associates with the isolation membrane, via the AIM. Therefore, autophagic cargoes are tethered to the isolation membrane.

via the AIM is common among most autophagy receptor proteins, the recognition mechanisms of autophagic cargoes by autophagy receptor proteins are too divergent to be elucidated. We reported that Atg19 and Atg34 ABDs, similarly to the camelid antibody and monobody, recognize Ams1, an autophagic cargo, using the loops clustered at one side of their immunoglobulin fold. Recent proteomics analysis has identified proteins that are selectively degraded by autophagy (Onodera and Ohsumi, 2004). Although the recognition mechanism of these target proteins by autophagy has not been established, autophagy-specific receptor proteins possessing an ABD-like fold might be responsible. Identification and structural analysis of other autophagy receptor proteins are required for further clarification of the molecular mechanism of specific cargo recognition during autophagy.

8. Acknowledgment

This work was supported by Grants-in-Aids for Young Scientists (A) from the Ministry of Education, Culture, Sports, Science and Technology of Japan and for JSPS Fellows from Japan Society for the Promotion of Science.

9. References

Aoki, Y., Kanki, T., Hirota, Y., Kurihara, Y., Saigusa, T., Uchiumi, T. & Kang, D. (2011). Phosphorylation of Serine 114 on Atg32 mediates mitophagy. *Mol Biol Cell*, Vol.22, No.17, Sep, pp. 3206-3217, 1059-1524

Baba, M., Osumi, M., Scott, S.V., Klionsky, D.J. & Ohsumi, Y. (1997). Two distinct pathways for targeting proteins from the cytoplasm to the vacuole/lysosome. *J Cell Biol*, Vol.139, No.7, Dec 29, pp. 1687-1695, 0021-9525

Baba, M., Takeshige, K., Baba, N. & Ohsumi, Y. (1994). Ultrastructural analysis of the autophagic process in yeast: detection of autophagosomes and their characterization. *J Cell Biol*, Vol.124, No.6, Mar, pp. 903-913, 0021-9525

Birmingham, C.L., Smith, A.C., Bakowski, M.A., Yoshimori, T. & Brumell, J.H. (2006). Autophagy controls Salmonella infection in response to damage to the Salmonella-containing vacuole. *J Biol Chem*, Vol.281, No.16, Apr 21, pp. 11374-11383, 0021-9258

Bjorkoy, G., Lamark, T., Brech, A., Outzen, H., Perander, M., Overvatn, A., Stenmark, H. & Johansen, T. (2005). p62/SQSTM1 forms protein aggregates degraded by autophagy and has a protective effect on huntingtin-induced cell death. *J Cell Biol*, Vol.171, No.4, Nov 21, pp. 603-614, 0021-9525

Cemma, M., Kim, P.K. & Brumell, J.H. (2011). The ubiquitin-binding adaptor proteins p62/SQSTM1 and NDP52 are recruited independently to bacteria-associated microdomains to target Salmonella to the autophagy pathway. *Autophagy*, Vol.7, No.3, Mar, pp. 341-345, 1554-8627

Dupont, N., Lacas-Gervais, S., Bertout, J., Paz, I., Freche, B., Van Nhieu, G.T., van der Goot, F.G., Sansonetti, P.J. & Lafont, F. (2009). Shigella phagocytic vacuolar membrane remnants participate in the cellular response to pathogen invasion and are regulated by autophagy. *Cell Host Microbe*, Vol.6, No.2, Aug 20, pp. 137-149, 1931-3128

Farre, J.C., Manjithaya, R., Mathewson, R.D. & Subramani, S. (2008). PpAtg30 tags peroxisomes for turnover by selective autophagy. *Dev Cell*, Vol.14, No.3, Mar, pp. 365-376, 1534-5807

Goldman, S.J., Taylor, R., Zhang, Y. & Jin, S. (2010). Autophagy and the degradation of mitochondria. *Mitochondrion*, Vol.10, No.4, Jun, pp. 309-315, 1567-7249

Gutierrez, M.G., Master, S.S., Singh, S.B., Taylor, G.A., Colombo, M.I. & Deretic, V. (2004). Autophagy is a defense mechanism inhibiting BCG and Mycobacterium tuberculosis survival in infected macrophages. *Cell*, Vol.119, No.6, Dec 17, pp. 753-766, 0092-8674

Hara, T., Nakamura, K., Matsui, M., Yamamoto, A., Nakahara, Y., Suzuki-Migishima, R., Yokoyama, M., Mishima, K., Saito, I., Okano, H., *et al.* (2006). Suppression of basal autophagy in neural cells causes neurodegenerative disease in mice. *Nature*, Vol.441, No.7095, Jun 15, pp. 885-889, 0028-0836

Hird, S.N., Paulsen, J.E. & Strome, S. (1996). Segregation of germ granules in living Caenorhabditis elegans embryos: cell-type-specific mechanisms for cytoplasmic localisation. *Development*, Vol.122, No.4, Apr, pp. 1303-1312, 0950-1991

Holm, L. & Park, J. (2000). DaliLite workbench for protein structure comparison. *Bioinformatics*, Vol.16, No.6, Jun, pp. 566-567, 1367-4803

Hutchins, M.U. & Klionsky, D.J. (2001). Vacuolar localization of oligomeric alpha-mannosidase requires the cytoplasm to vacuole targeting and autophagy pathway components in Saccharomyces cerevisiae. *J Biol Chem*, Vol.276, No.23, Jun 8, pp. 20491-20498, 0021-9258

Ichimura, Y., Kirisako, T., Takao, T., Satomi, Y., Shimonishi, Y., Ishihara, N., Mizushima, N., Tanida, I., Kominami, E., Ohsumi, M., *et al.* (2000). A ubiquitin-like system mediates protein lipidation. *Nature*, Vol.408, No.6811, Nov 23, pp. 488-492, 0028-0836

Ichimura, Y., Kumanomidou, T., Sou, Y.S., Mizushima, T., Ezaki, J., Ueno, T., Kominami, E., Yamane, T., Tanaka, K. & Komatsu, M. (2008). Structural basis for sorting mechanism of p62 in selective autophagy. *J Biol Chem*, Vol.283, No.33, Jun 4, pp. 22847-22857, 0021-9258

Isogai, S., Morimoto, D., Arita, K., Unzai, S., Tenno, T., Hasegawa, J., Sou, Y.S., Komatsu, M., Tanaka, K., Shirakawa, M., *et al.* (2011). Crystal Structure of the Ubiquitin-associated (UBA) Domain of p62 and Its Interaction with Ubiquitin. *J Biol Chem*, Vol.286, No.36, Sep 9, pp. 31864-31874, 0021-9258

Kageyama, T., Suzuki, K. & Ohsumi, Y. (2009). Lap3 is a selective target of autophagy in yeast, Saccharomyces cerevisiae. *Biochem Biophys Res Commun*, Vol.378, No.3, Jan 16, pp. 551-557, 1090-2104

Kanki, T., Wang, K., Cao, Y., Baba, M. & Klionsky, D.J. (2009). Atg32 is a mitochondrial protein that confers selectivity during mitophagy. *Dev Cell*, Vol.17, No.1, Jul, pp. 98-109, 1534-5807

Kim, P.K., Hailey, D.W., Mullen, R.T. & Lippincott-Schwartz, J. (2008). Ubiquitin signals autophagic degradation of cytosolic proteins and peroxisomes. *Proc Natl Acad Sci U S A*, Vol.105, No.52, Dec 30, pp. 20567-20574, 0027-8424

Kirkin, V., Lamark, T., Sou, Y.S., Bjorkoy, G., Nunn, J.L., Bruun, J.A., Shvets, E., McEwan, D.G., Clausen, T.H., Wild, P., *et al.* (2009). A role for NBR1 in autophagosomal degradation of ubiquitinated substrates. *Mol Cell*, Vol.33, No.4, Feb 27, pp. 505-516, 1097-4164

Klionsky, D.J., Cueva, R. & Yaver, D.S. (1992). Aminopeptidase I of Saccharomyces cerevisiae is localized to the vacuole independent of the secretory pathway. *J Cell Biol*, Vol.119, No.2, Oct, pp. 287-299, 0021-9525

Komatsu, M., Waguri, S., Chiba, T., Murata, S., Iwata, J., Tanida, I., Ueno, T., Koike, M., Uchiyama, Y., Kominami, E., *et al.* (2006). Loss of autophagy in the central nervous system causes neurodegeneration in mice. *Nature*, Vol.441, No.7095, Jun 15, pp. 880-884, 0028-0836

Komatsu, M., Waguri, S., Koike, M., Sou, Y.S., Ueno, T., Hara, T., Mizushima, N., Iwata, J., Ezaki, J., Murata, S., *et al.* (2007). Homeostatic levels of p62 control cytoplasmic inclusion body formation in autophagy-deficient mice. *Cell*, Vol.131, No.6, Dec 14, pp. 1149-1163, 0092-8674

Komatsu, M., Waguri, S., Ueno, T., Iwata, J., Murata, S., Tanida, I., Ezaki, J., Mizushima, N., Ohsumi, Y., Uchiyama, Y., *et al.* (2005). Impairment of starvation-induced and constitutive autophagy in Atg7-deficient mice. *J Cell Biol*, Vol.169, No.3, May 9, pp. 425-434, 0021-9525

Korolchuk, V.I., Mansilla, A., Menzies, F.M. & Rubinsztein, D.C. (2009). Autophagy inhibition compromises degradation of ubiquitin-proteasome pathway substrates. *Mol Cell*, Vol.33, No.4, Feb 27, pp. 517-527, 1097-2765

Kuusisto, E., Salminen, A. & Alafuzoff, I. (2001). Ubiquitin-binding protein p62 is present in neuronal and glial inclusions in human tauopathies and synucleinopathies. *Neuroreport*, Vol.12, No.10, Jul 20, pp. 2085-2090, 0959-4965

Kuusisto, E., Salminen, A. & Alafuzoff, I. (2002). Early accumulation of p62 in neurofibrillary tangles in Alzheimer's disease: possible role in tangle formation. *Neuropathol Appl Neurobiol*, Vol.28, No.3, Jun, pp. 228-237, 0305-1846

Leber, R., Silles, E., Sandoval, I.V. & Mazon, M.J. (2001). Yol082p, a novel CVT protein involved in the selective targeting of aminopeptidase I to the yeast vacuole. *J Biol Chem*, Vol.276, No.31, Aug 3, pp. 29210-29217, 0021-9258

Lynch-Day, M.A. & Klionsky, D.J. (2010). The Cvt pathway as a model for selective autophagy. *FEBS Lett*, Vol.584, No.7, Apr 2, pp. 1359-1366, 0014-5793

Manjithaya, R., Nazarko, T.Y., Farre, J.C. & Subramani, S. (2010). Molecular mechanism and physiological role of pexophagy. *FEBS Lett*, Vol.584, No.7, Apr 2, pp. 1367-1373, 0014-5793

Martinez, E., Jimenez, M.A., Segui-Real, B., Vandekerckhove, J. & Sandoval, I.V. (1997). Folding of the presequence of yeast pAPI into an amphipathic helix determines transport of the protein from the cytosol to the vacuole. *J Mol Biol*, Vol.267, No.5, Apr 18, pp. 1124-1138, 0022-2836

Mijaljica, D., Prescott, M. & Devenish, R.J. (2011). Microautophagy in mammalian cells: revisiting a 40-year-old conundrum. *Autophagy*, Vol.7, No.7, Jul, pp. 673-682, 1554-8627

Mizushima, N. (2007). Autophagy: process and function. *Genes Dev*, Vol.21, No.22, Nov 15, pp. 2861-2873, 0890-9369

Mizushima, N., Yoshimori, T. & Ohsumi, Y. (2010). The Role of Atg Proteins in Autophagosome Formation. *Annu Rev Cell Dev Biol*, Oct 29, pp., 1081-0706

Morton, S., Hesson, L., Peggie, M. & Cohen, P. (2008). Enhanced binding of TBK1 by an optineurin mutant that causes a familial form of primary open angle glaucoma. *FEBS Lett*, Vol.582, No.6, Mar 19, pp. 997-1002, 0014-5793

Muyldermans, S. (2001). Single domain camel antibodies: current status. *J Biotechnol*, Vol.74, No.4, Jun, pp. 277-302, 0168-1656

Nagaoka, U., Kim, K., Jana, N.R., Doi, H., Maruyama, M., Mitsui, K., Oyama, F. & Nukina, N. (2004). Increased expression of p62 in expanded polyglutamine-expressing cells and its association with polyglutamine inclusions. *J Neurochem*, Vol.91, No.1, Oct, pp. 57-68, 0022-3042

Nakagawa, I., Amano, A., Mizushima, N., Yamamoto, A., Yamaguchi, H., Kamimoto, T., Nara, A., Funao, J., Nakata, M., Tsuda, K., et al. (2004). Autophagy defends cells against invading group A Streptococcus. Science, Vol.306, No.5698, Nov 5, pp. 1037-1040, 0036-8075

Nakatogawa, H., Suzuki, K., Kamada, Y. & Ohsumi, Y. (2009). Dynamics and diversity in autophagy mechanisms: lessons from yeast. Nat Rev Mol Cell Biol, Vol.10, No.7, Jul, pp. 458-467, 1471-0072

Noda, N.N., Kumeta, H., Nakatogawa, H., Satoo, K., Adachi, W., Ishii, J., Fujioka, Y., Ohsumi, Y. & Inagaki, F. (2008). Structural basis of target recognition by Atg8/LC3 during selective autophagy. Genes Cells, Vol.13, No.12, Dec, pp. 1211-1218, 1356-9597

Noda, N.N., Ohsumi, Y. & Inagaki, F. (2010). Atg8-family interacting motif crucial for selective autophagy. FEBS Lett, Vol.584, No.7, Apr 2, pp. 1379-1385, 0014-5793

Noda, T. & Yoshimori, T. (2009). Molecular basis of canonical and bactericidal autophagy. Int Immunol, Vol.21, No.11, Nov, pp. 1199-1204, 0953-8178

Novak, I., Kirkin, V., McEwan, D.G., Zhang, J., Wild, P., Rozenknop, A., Rogov, V., Lohr, F., Popovic, D., Occhipinti, A., et al. (2010). Nix is a selective autophagy receptor for mitochondrial clearance. EMBO Rep, Vol.11, No.1, Jan, pp. 45-51, 1469-3178 (Electronic)

Oda, M.N., Scott, S.V., Hefner-Gravink, A., Caffarelli, A.D. & Klionsky, D.J. (1996). Identification of a cytoplasm to vacuole targeting determinant in aminopeptidase I. J Cell Biol, Vol.132, No.6, Mar, pp. 999-1010, 0021-9525

Okamoto, K., Kondo-Okamoto, N. & Ohsumi, Y. (2009). Mitochondria-anchored receptor Atg32 mediates degradation of mitochondria via selective autophagy. Dev Cell, Vol.17, No.1, Jul, pp. 87-97, 1534-5807

Olzmann, J.A., Li, L., Chudaev, M.V., Chen, J., Perez, F.A., Palmiter, R.D. & Chin, L.S. (2007). Parkin-mediated K63-linked polyubiquitination targets misfolded DJ-1 to aggresomes via binding to HDAC6. J Cell Biol, Vol.178, No.6, Sep 10, pp. 1025-1038, 0021-9525

Onodera, J. & Ohsumi, Y. (2004). Ald6p is a preferred target for autophagy in yeast, Saccharomyces cerevisiae. J Biol Chem, Vol.279, No.16, Apr 16, pp. 16071-16076, 0021-9258

Orenstein, S.J. & Cuervo, A.M. (2010). Chaperone-mediated autophagy: molecular mechanisms and physiological relevance. Semin Cell Dev Biol, Vol.21, No.7, Sep, pp. 719-726, 1084-9521

Pankiv, S., Clausen, T.H., Lamark, T., Brech, A., Bruun, J.A., Outzen, H., Overvatn, A., Bjorkoy, G. & Johansen, T. (2007). p62/SQSTM1 binds directly to Atg8/LC3 to facilitate degradation of ubiquitinated protein aggregates by autophagy. J Biol Chem, Vol.282, No.33, Aug 17, pp. 24131-24145, 0021-9258

Perrin, A.J., Jiang, X., Birmingham, C.L., So, N.S. & Brumell, J.H. (2004). Recognition of bacteria in the cytosol of Mammalian cells by the ubiquitin system. Curr Biol, Vol.14, No.9, May 4, pp. 806-811, 0960-9822

Ponting, C.P., Ito, T., Moscat, J., Diaz-Meco, M.T., Inagaki, F. & Sumimoto, H. (2002). OPR, PC and AID: all in the PB1 family. Trends Biochem Sci, Vol.27, No.1, Jan, pp. 10, 0968-0004

Rozenknop, A., Rogov, V.V., Rogova, N.Y., Lohr, F., Guntert, P., Dikic, I. & Dotsch, V. (2011). Characterization of the interaction of GABARAPL-1 with the LIR motif of NBR1. *J Mol Biol*, Vol.410, No.3, Jul 15, pp. 477-487, 0022-2836

Scott, S.V., Guan, J., Hutchins, M.U., Kim, J. & Klionsky, D.J. (2001). Cvt19 is a receptor for the cytoplasm-to-vacuole targeting pathway. *Mol Cell*, Vol.7, No.6, Jun, pp. 1131-1141, 1097-2765

Shintani, T., Huang, W.P., Stromhaug, P.E. & Klionsky, D.J. (2002). Mechanism of cargo selection in the cytoplasm to vacuole targeting pathway. *Dev Cell*, Vol.3, No.6, Dec, pp. 825-837, 1534-5807

Strome, S. & Lehmann, R. (2007). Germ versus soma decisions: lessons from flies and worms. *Science*, Vol.316, No.5823, Apr 20, pp. 392-393, 1095-9203 (Electronic) 0036-8075 (Linking)

Suzuki, K., Kamada, Y. & Ohsumi, Y. (2002). Studies of cargo delivery to the vacuole mediated by autophagosomes in Saccharomyces cerevisiae. *Dev Cell*, Vol.3, No.6, Dec, pp. 815-824, 1534-5807

Suzuki, K., Kondo, C., Morimoto, M. & Ohsumi, Y. (2010). Selective transport of alpha-mannosidase by autophagic pathways: identification of a novel receptor, Atg34p. *J Biol Chem*, Vol.285, No.39, Sep 24, pp. 30019-30025, 0021-9258

Suzuki, K., Morimoto, M., Kondo, C. & Ohsumi, Y. (2011). Selective Autophagy Regulates Insertional Mutagenesis by the Ty1 Retrotransposon in Saccharomyces cerevisiae. *Dev Cell*, Vol.21, No.2, Aug 16, pp. 358-365, 1534-5807

Suzuki, K. & Ohsumi, Y. (2010). Current knowledge of the pre-autophagosomal structure (PAS). *FEBS Lett*, Vol.584, No.7, Apr 2, pp. 1280-1286, 0014-5793

Tan, J.M., Wong, E.S., Kirkpatrick, D.S., Pletnikova, O., Ko, H.S., Tay, S.P., Ho, M.W., Troncoso, J., Gygi, S.P., Lee, M.K., et al. (2008). Lysine 63-linked ubiquitination promotes the formation and autophagic clearance of protein inclusions associated with neurodegenerative diseases. *Hum Mol Genet*, Vol.17, No.3, Feb 1, pp. 431-439, 0964-6906

Thurston, T.L., Ryzhakov, G., Bloor, S., von Muhlinen, N. & Randow, F. (2009). The TBK1 adaptor and autophagy receptor NDP52 restricts the proliferation of ubiquitin-coated bacteria. *Nat Immunol*, Vol.10, No.11, Nov, pp. 1215-1221, 1529-2908

Uttenweiler, A. & Mayer, A. (2008). Microautophagy in the yeast Saccharomyces cerevisiae. *Methods Mol Biol*, Vol.445, pp. 245-259, 1064-3745

Vadlamudi, R.K., Joung, I., Strominger, J.L. & Shin, J. (1996). p62, a phosphotyrosine-independent ligand of the SH2 domain of p56lck, belongs to a new class of ubiquitin-binding proteins. *J Biol Chem*, Vol.271, No.34, Aug 23, pp. 20235-20237, 0021-9258

Watanabe, Y., Noda, N.N., Honbou, K., Suzuki, K., Sakai, Y., Ohsumi, Y. & Inagaki, F. (2009). Crystallization of Saccharomyces cerevisiae alpha-mannosidase, a cargo protein of the Cvt pathway. *Acta crystallographica Section F, Structural biology and crystallization communications*, Vol.65, No.Pt 6, Jun 1, pp. 571-573, 1744-3091

Watanabe, Y., Noda, N.N., Kumeta, H., Suzuki, K., Ohsumi, Y. & Inagaki, F. (2010). Selective transport of alpha-mannosidase by autophagic pathways: structural basis for cargo recognition by Atg19 and Atg34. *J Biol Chem*, Vol.285, No.39, Sep 24, pp. 30026-30033, 0021-9258

Wild, P., Farhan, H., McEwan, D.G., Wagner, S., Rogov, V.V., Brady, N.R., Richter, B., Korac, J., Waidmann, O., Choudhary, C., et al. (2011). Phosphorylation of the autophagy receptor optineurin restricts Salmonella growth. Science, Vol.333, No.6039, Jul 8, pp. 228-233, 0036-8075

Wooten, M.W., Geetha, T., Babu, J.R., Seibenhener, M.L., Peng, J., Cox, N., Diaz-Meco, M.T. & Moscat, J. (2008). Essential role of sequestosome 1/p62 in regulating accumulation of Lys63-ubiquitinated proteins. J Biol Chem, Vol.283, No.11, Mar 14, pp. 6783-6789, 0021-9258

Yano, T., Mita, S., Ohmori, H., Oshima, Y., Fujimoto, Y., Ueda, R., Takada, H., Goldman, W.E., Fukase, K., Silverman, N., et al. (2008). Autophagic control of listeria through intracellular innate immune recognition in drosophila. Nat Immunol, Vol.9, No.8, Aug, pp. 908-916, 1529-2908

Yoshikawa, Y., Ogawa, M., Hain, T., Yoshida, M., Fukumatsu, M., Kim, M., Mimuro, H., Nakagawa, I., Yanagawa, T., Ishii, T., et al. (2009). Listeria monocytogenes ActA-mediated escape from autophagic recognition. Nat Cell Biol, Vol.11, No.10, Oct, pp. 1233-1240, 1465-7392

Yuga, M., Gomi, K., Klionsky, D.J. & Shintani, T. (2011). Aspartyl aminopeptidase is imported from the cytoplasm to the vacuole by selective autophagy in Saccharomyces cerevisiae. J Biol Chem, Vol.286, No.15, Apr 15, pp. 13704-13713, 0021-9258

Zatloukal, K., Stumptner, C., Fuchsbichler, A., Heid, H., Schnoelzer, M., Kenner, L., Kleinert, R., Prinz, M., Aguzzi, A. & Denk, H. (2002). p62 Is a common component of cytoplasmic inclusions in protein aggregation diseases. Am J Pathol, Vol.160, No.1, Jan, pp. 255-263, 0002-9440

Zhang, Y., Yan, L., Zhou, Z., Yang, P., Tian, E., Zhang, K., Zhao, Y., Li, Z., Song, B., Han, J., et al. (2009). SEPA-1 mediates the specific recognition and degradation of P granule components by autophagy in C. elegans. Cell, Vol.136, No.2, Jan 23, pp. 308-321, 0092-8674

Zheng, Y.T., Shahnazari, S., Brech, A., Lamark, T., Johansen, T. & Brumell, J.H. (2009). The adaptor protein p62/SQSTM1 targets invading bacteria to the autophagy pathway. J Immunol, Vol.183, No.9, Nov 1, pp. 5909-5916, 0022-1767

Part 2

Regulatory Molecules

9

Role of Ceramide 1-Phosphate in the Regulation of Cell Survival and Inflammation

Alberto Ouro, Lide Arana, Patricia Gangoiti and Antonio Gomez-Muñoz
Department of Biochemistry and Molecular Biology
Faculty of Science and Technology, University of the Basque Country, Bilbao
Spain

1. Introduction

Cell and tissue homeostasis is essential for normal development of an organism. When this is altered, metabolic dysfunctions and disease are prone to occur. Therefore, the maintenance of an appropriate balance in the activation / inhibition of the different metabolic pathways and cell signaling systems is simply vital.

Many lipids, including simple sphingolipids, are known to regulate cell activation and metabolism (Gomez-Munoz et al., 1992; Gomez-Munoz, 1998; Gomez-Munoz, 2004; Gomez-Munoz, 2006; Hannun & Obeid, 2008; Chen et al., 2011; Hannun & Obeid, 2011) . Some of them, including sphingosine, ceramides and their phosphorylated forms, sphingosine 1-phosphate (S1P) and ceramide 1-phosphate (C1P) have been described as crucial regulators of key processes that are essential for normal development, and have also been involved in the establishment and progression of different diseases (Gangoiti et al., 2008a; Arana et al., 2010). In particular, ceramides can induce cell growth arrest and cause apoptosis, when they are generated (Hannun et al., 1986; Kolesnick, 1987; Kolesnick & Hemer, 1990; Merrill & Jones, 1990; Merrill, 1991; Hannun, 1994; Kolesnick & Golde, 1994; Hannun & Obeid, 1995; Hannun, 1996; Spiegel & Merrill, 1996; Merrill et al., 1997; Kolesnick et al., 2000; Hannun & Obeid, 2002; Merrill, 2002). Nonetheless, although in general, ceramides are negative signals for cell survival, in neurons they can induce cell growth (Goodman & Mattson, 1996; Ping & Barrett, 1998; Brann et al., 1999; Song & Posse de Chaves, 2003; Plummer et al., 2005). Also, ceramides play important roles in the regulation of cell differentiation, inflammation, tumor development (Okazaki et al., 1990; Mathias et al., 1991; Dressler et al., 1992; Hannun, 1994; Kolesnick & Golde, 1994; Hannun & Obeid, 1995; Gomez-Munoz, 1998; Menaldino et al., 2003), bacterial and viral infections, and ischemia-reperfusion injury(Gulbins & Kolesnick, 2003). More recently, ceramides have been associated with insulin resistance through activation of protein phosphatase 2A and the subsequent dephosphorylation and inactivation of protein kinase B (PKB) (Schmitz-Peiffer, 2002; Adams et al., 2004; Stratford et al., 2004), and toll-like receptor 4 (TLR4)-dependent induction of inflammatory cytokines, a fact essential for TLR4-dependent insulin resistance (Holland et al., 2011).

Concerning ceramide generation, there are three different mechanisms by which these molecules can be synthesized in cells. Ceramides can be generated by i) *de novo* synthesis, which takes place in the endoplasmic reticulum (ER), ii) by the action of different sphingomyelinases (SMases) in the plasma membrane, lysosomes, or mitochondria, and iii) by reacylation of sphingosine, a pathway known as the salvage or recycled pathway (Hannun & Obeid, 2011). The biosynthetic and degradative pathways of ceramide are shown in figure 1, where further products of ceramide metabolism are also indicated.

Natural ceramides tipically have long N-acyl chains ranging from 16 to 26 carbons in length (Merrill, 2002; Pettus et al., 2003a; Merrill et al., 2005), and some times longer in tissues such as skin. Many studies have used a short-chain analog (N-acetylsphingosine, or C_2-ceramide) in experiments with cells in culture because it can be incorporated into cells more easily and rapidly than long-chain ceramides. Of note, although C_2-ceramide was suggested not to occur *in vivo*, recent studies demonstrated that C_2-ceramide does exist in mammalian tissues. In particular, C_2-ceramide was found in rat liver cells (Merrill et al., 2001; Van Overloop et al., 2007), and brain tissue (Van Overloop et al., 2007). Ceramide generation is also relevant because this sphingolipid is the precursor of important bioactive molecules that can also regulate cellular functions. For instance, stimulation of ceramidases results in generation of sphingosine (Fig. 1), which was first described as a physiological inhibitor of protein kinase C (PKC) (Hannun et al., 1986). There are numerous reports in the scientific literature showing that PKC is inhibited by exogenous addition of sphingosine to cells in culture. Moreover, Merrill and co-workers demonstrated that addition of the ceramide synthase inhibitor fumonisin B1 to J774.A1 macrophages to increase the levels of endogenous sphingoid bases, also inhibited protein kinase C (Smith et al., 1997). Further work showed that sphingosine can affect the activity of other important enzymes that are involved in the regulation of metabolic or cell signaling pathways such as the Mg^{2+} dependent form of phosphatidate phosphohydrolase (Jamal et al., 1991; Gomez-Munoz et al., 1992), phospholipase D (PLD) (Natarajan et al., 1994), or diacylglycerol kinase (DAGK) (Sakane et al., 1989; Yamada et al., 1993). Sphingosine, in turn, can be phosphorylated by the action of sphingosine kinases to generate S1P, which is a potent mitogenic agent and can also inhibit apoptosis in many cell types (Olivera & Spiegel, 1993; Wu et al., 1995; Spiegel et al., 1996; Spiegel & Merrill, 1996; Spiegel & Milstien, 2002; Spiegel & Milstien, 2003). More recently, we demonstrated that S1P stimulates cortisol (Rabano et al., 2003) and aldosterone secretion (Brizuela et al., 2006) in cells of the zona fasciculata or zona glomerulosa, respectively, of bovine adrenal glands, suggesting that S1P plays an important role in the regulation of steroidogenesis.

A major metabolite of ceramide in cells is ceramide-1-phosphate (C1P), which is formed directly through phosphorylation of ceramide by the action of ceramide kinase (CerK) (Fig. 1). There is increasing evidence suggesting that C1P can regulate cell proliferation and apoptosis (Reviewed in (Gomez-Munoz, 1998; Gomez-Munoz, 2004)), and Chalfant and co-workers have implicated C1P in inflammatory responses (Reviewed in (Chalfant & Spiegel, 2005; Lamour & Chalfant, 2005)). In addition, Shayman's group demonstrated that C1P plays a key role in phagocytosis (Hinkovska-Galcheva & Shayman; Hinkovska-Galcheva et al., 1998; Hinkovska-Galcheva et al., 2005).

The aim of the present chapter is to review and update recent progress on the regulation of cell survival and inflammation by C1P.

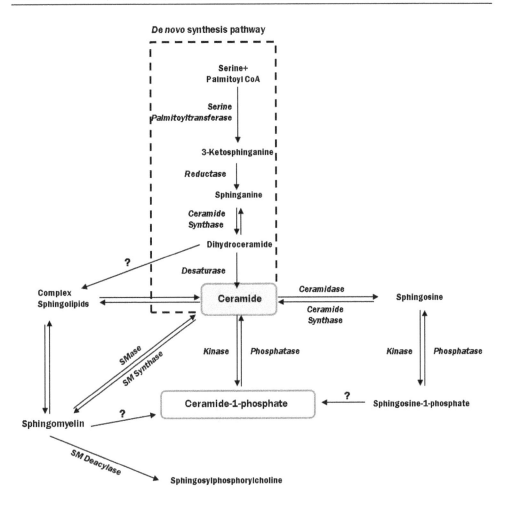

Fig. 1. Biosynthesis of simple sphingolipids in mammalian cells. Ceramide is the central core of sphingolipid metabolism. It can be produced by *de novo* synthesis through the concerted action of serine palmitoyltransferase and dihydroceramide synthase or by degradation of sphingomyelin (SM) through sphingomyelinase (SMase) activation. Ceramides can also be generated through metabolism of more complex sphingolipids. Phosphorylation of ceramide by ceramide kinase gives rise to ceramide-1-phosphate. The reverse reaction is catalyzed by ceramide-1-phosphate phosphatase, or by lipid phosphate phosphatases. Alternatively, ceramide can be degraded by ceramidases to form sphingosine, which can, in turn, be phosphorylated to sphingosine-1-phosphate by sphingosine kinases. The reverse reaction is catalyzed by sphingosine-1-phosphate phosphatases, or by lipid phosphate phosphatases. Sphingosine-1-phosphate lyase breaks down Sphingosine-1-phosphate to hexadecenal and ethanolamine phosphate, both of which can be recycled back to generate phosphatidylethanolamine. Sphingomyelin *N*-deacylase generates sphingosylphosphorylcholine, also known as lysosphingomyelin.

2. Biosynthesis of ceramide 1-phosphate. The essential role of ceramide kinase

At present, the only enzyme known to produce C1P in mammalian cells is ceramide kinase (CerK). This enzyme was first observed in brain synaptic vesicles (Bajjalieh et al., 1989), and was later found in human leukemia HL-60 cells (Kolesnick & Hemer, 1990). Cerk was first reported to be confined to the microsomal membrane fraction, but more recent studies indicate that it is mainly located in the cytosol (Mitsutake et al., 2004). These contradictory observations may arise from the different degrees of enzyme expression in different cell types, and it may also be possible that subcellular localization of this enzyme varies depending on cell metabolism. In this connection, Van Veldhoven and co-workers found that tagged forms of human CerK (FLAG-HsCerK and EGFP-HsCerK fusions), upon expression in Chinese Hamster Ovary (CHO) cells, were mainly localized to the plasma membrane, whereas no evidence for association with the ER was observed (Van Overloop et al., 2006). These findings are in agreement with those of Boath et al. (Boath et al., 2008) who showed that ceramides are not phosphorylated at the ER but must be transported to the Golgi apparatus for phosphorylation by CerK. When C1P is synthesized, it traffics from the Golgi network along the secretory pathway to the plasma membrane, where it can be back-exchanged into the extracellular environment and then bind to acceptor proteins such as albumin or lipoproteins (Boath et al., 2008). These observations are consistent with published work by Chalfant's group (Lamour et al., 2007), and it was demonstrated that CerK utilizes ceramide transported to the trans-Golgi apparatus by ceramide transport protein (CERT). In fact, downregulation of CERT by RNA interference resulted in strong inhibition of newly synthesized C1P, suggesting that CERT plays a critical role in C1P formation. However, Boat et al (Boath et al., 2008) reported that the transport of ceramides to the vicinity of CerK is not dependent upon CERT intervention. The reason for such discrepancy is unknown at the present time, but it is possible that the different experimental approaches used in those studies rendered different results. Specifically, whilst Lamour and co-workers used siRNA technology to inhibit CERT (Lamour et al., 2007), Boath and co-workers utilized pharmacological inhibitors (Boath et al., 2008). Also, it might be possible that different cell types may have different subcellular distribution of CerK, and / or that expression of this enzyme activity is not the same in all cell types.

With regards to the regulation of CerK, its ability to move intracellularly from one compartment to another and the dependency on cations (mainly Ca^{2+} ions) for activity seem to be well established. More recently, CerK has been proposed to be regulated by phosphorylation/dephosphorylation processes (Baumruker et al., 2005), and that it can be myristoylated at its N-terminus, a feature that is related to targeting proteins to membranes. Nonetheless, cleavage of the myristoylated moiety did not affect the intracellular localization of the enzyme. In addition, both CerK location and activity seem to require the integrity of its PH domain, which actually includes the myristoylation site, as deletion of this domain abolishes both the specific subcellular localization of the enzyme, as well as its activity (Baumruker et al., 2005).

Although CerK is thought to be the only enzyme for production of C1P, it was reported that bone marrow-derived macrophages (BMDM) from CerK-null mice (CerK-/-) still had significant levels of C1P (Boath et al., 2008). This observation suggests that there are other metabolic pathways, at least in mammals, capable of generating C1P independently of CerK.

Specifically, formation of C_{16}-C1P, which is a major species of C1P in cells, was not abolished in CerK-/- BMDM. Two alternative pathways for generation of C1P in cells might be: i) acylation of S1P by a putative acyl transferase that would catalyze the formation of a N-linked fatty acid in the S1P moiety to form C1P, and ii) cleavage of sphingomyelin (SM) by the action of a D-type SMase (SMase D), which would generate choline and C1P in an analogous manner to that of phospholipase D acting on phosphatidylcholine to produce choline and phosphatidic acid (PA). However, work from our own lab (Gomez-Munoz et al., 1995a) and that of others (Boath et al., 2008) demonstrated that acylation of S1P to form C1P does not occur in mammalian cells. Also, formation of C1P by the action of a putative SMase D has not yet been reported for mammalian cells. SMase D is a major component of the venom of a variety of arthropods including spiders of the gender *Loxosceles* (the brown recluse spider), such as *L. reclusa*. SMase D is also present in the toxins of some bacteria including *Corynebacterium pseudotuberculosis*, or *Vibrio damsela* (Truett & King, 1993). The bites of this spider result in strong inflammatory responses and may lead to renal failure, and occasionally lead to death (Lee & Lynch, 2005). Although we found no evidence for an analogous activity of SMase D in rat fibroblasts (Gomez-Munoz et al., 1995a), this possibility should be explored in more detail using different types of cells; so it is possible that SMase D may still be the cause for C1P generation in selective tissues.

Concerning regulation, mammalian CerK was demonstrated to be highly dependent on Ca^{2+} ions for activity (Van Overloop et al., 2006). More recently, it has been shown that treatment of human lung adenocarcinoma A549 cells and Chinese hamster ovary cells (CHO) with orthovanadate, a potent inhibitor of tyrosine phosphatases, increased CerK expression potently (Tada et al., 2010), suggesting a possible regulation of CerK by phosphorylation/dephosphorylation processes on tyrosine residues. Also, it has been suggested that CerK expression can be regulated through activation of Toll-like receptor 4 (TLR-4) by agonists such as the bacterial toxin lipopolysaccharide (Rovina et al., 2010).

The cloning of CerK (Sugiura et al., 2002) opened a new avenue of research that led to determination of important structural properties of this enzyme. The protein sequence has 537 amino acids with two protein sequence motifs, an N-terminus pleckstrin homology (PH) domain, and a C-terminal region containing a Ca^{2+}/calmodulin binding domain. Using site-directed mutagenesis, it was found that leucine 10 in the PH domain is essential for the catalytic activity of CerK (Kim et al., 2005). In addition, it was reported that the interaction between the PH domain of CERK and phosphatidylinositol 4,5-bisphosphate regulates the plasma membrane targeting and the levels of C1P (Kim et al., 2006). CERK also contains the five conserved sequence stretches (C1-C5) that are specific for lipid kinases (Reviewed in (Baumruker et al., 2005)).

With regards to substrate specificity, it was reported that phosphorylation of ceramide by CERK is stereospecific (Wijesinghe et al., 2005). The latter report also showed that a minimum of a 12-carbon acyl chain was required for normal CERK activity, whereas the short-chain ceramide analogues C_8-ceramide, C_4-ceramide, or C_2-ceramide were poor substrates for CERK. It was concluded that CERK phosphorylates only the naturally occurring D-erythro-ceramides (Wijesinghe et al., 2005). However, C_2-ceramide has been shown to also be a good substrate for CerK, especially when albumin is used as a carrier, and that C_2-ceramide can be converted to C_2-C1P within cells (Van Overloop et al., 2007). This raises the possibility that C_2-C1P is also a natural sphingolipid, capable of eliciting important biologic effects, as previously demonstrated (i.e. stimulation of cell proliferation (Gomez-Munoz et al., 1995a)). These observations suggested that substrate presentation is an important factor when testing CerK activity and that the use of different vehicles may result in different outcomes. Also, it should be borne in mind that CerK expression may not be the same in all cell types. The importance of CERK in cell signaling was emphasized in experiments using specific small interfering RNA (siRNA) to silence the gene encoding for CerK. Downregulation of CerK blocked the response of the enzyme to treatment with ATP, the calcium ionophore A23187, or interleukin 1-betta (Pettus et al., 2003b; Chalfant & Spiegel, 2005), and led to a potent inhibition of arachidonic acid release and PGE_2 formation in A549 lung adenocarcinoma cells. The relevance of CerK in cell biology was also highlighted in studies using CerK null mice; specifically, a potent reduction in the amount of neutrophils in the blood and spleen of these animals compared to their wild type counterparts was observed, whereas de amount of leukocytes, other than neutrophils, was increased in those mice. These observations suggested an important role of CerK in neutrophil biology (Graf et al., 2008). In addition to CerK, a ceramide kinase-like (CERKL) protein was identified in human retina (Tuson et al., 2004), and this was subsequently cloned (Bornancin et al., 2005). However, CERKL failed to phosphorylate ceramide or other related lipids, under conditions commonly used to measure CERK activity. Therefore, the role of this protein in cell biology is unclear at the present time.

CerK has also been reported to exist in dicotyledonous plants, where it was associated to the regulation of cell survival (Bi et al., 2011). Also, it has been recently found that a conserved cystein motif is critical for rice CerK activity and function (Bi et al., 2011). However, no reports on the possible existence of Cerk in monocot plants are available at the present time.

3. Catabolism of ceramide 1-phosphate

From the above discussion, it should be apparent that C1P is a bioactive metabolite, capable of altering cell metabolism rapidly and potently. So, the existence of enzymes capable of degrading C1P seemed to be feasible for regulation of C1P levels. The identification of a specific C1P phosphatase in rat brain (Shinghal et al., 1993), and hepatocytes (Boudker & Futerman, 1993), together with the existence of CerK suggested that ceramide and C1P are interconvertible in cells. C1P phosphatase is enriched in brain synaptosomes and liver plasma membrane fractions, and appeared to be distinct from PA phosphohydrolase, the phosphatase that hydrolyzes PA. Nonetheless, C1P can also be converted to ceramide by the action of a PA phosphohydrolase that is specifically located in the plasma membrane of cells (Waggoner et al., 1996). The latter enzyme belongs to a family of at least three mammalian lipid phosphate phosphatases (LPPs) (Brindley & Waggoner, 1998). LPPs have recently been shown to regulate cell survival by controlling the levels of intracellular PA and S1P pools (Long et al., 2005), and also to regulate leukocyte infiltration and airway inflammation (Zhao et al., 2005). Dephosphorylation of C1P might be a way of terminating its regulatory effects, although the resulting formation of ceramide could potentially be detrimental for cells. Controlling the levels of ceramide and C1P by the coordinated action of CERK and C1P phosphatases, may be of crucial importance for the metabolic or signaling pathways that are regulated by these two sphingolipids. It could be speculated that another possibility for degradation of C1P might be its deacylation to S1P, which could then be cleaved by lyase activity to render a fatty aldehyde and ethanolamine phosphate (Merrill & Jones, 1990), or to sphingosine by the action of S1P phosphatases (Fig. 1). However, no C1P deacylases or lyases have so far been identified in mammalian tissues, suggesting that the only pathway for degradation of C1P in mammals is through phosphatase activity.

4. Ceramide 1-phosphate and the control of cell growth and death

The first report showing that C1P was biologically active was published in 1995 (Gomez-Munoz et al., 1995a). C1P was found to have mitogenic properties as it stimulated DNA synthesis and cell division in rat or mouse fibroblasts (Gomez-Munoz et al., 1995a; Gomez-Munoz et al., 1997). Subsequent studies using primary macrophages, demonstrated that like for most growth factors, the mechanisms whereby C1P exerted its mitogenic effects implicated stimulation of the mitogen-activated protein kinase kinase (MEK)/Extracellularly regulated kinases 1-2 (ERK1-2), phosphatidylinositol 3-kinase (PI3-K)/protein kinase B (PKB, also known as Akt), and c-Jun terminal kinase (JNK) pathways (Gangoiti et al., 2008b). In addition, C1P caused stimulation of the DNA binding activity of the transcription factor NF-κB, and the selective inhibitors of MEK, PI3-K, and JNK (PD98059, LY290042, and SP600125), respectively) completely blocked NF-κB activation. Another major target of PKB is glycogen synthase kinase-3β (GSK-3 β), which expression was increased in the presence of C1P. This led to up-regulation of cyclin D1, and c-Myc, two important markers of cell proliferation that are targets of GSK-3β.

In addition, we found that C1P-stimulated macrophage proliferation, involved activation of sphingomyelin synthase (SMS), an enzyme that catalyzes the transfer of phosphocholine from phosphatidylcholine (PC) to ceramide to synthesize sphingomyelin (SM). The other by-product of this reaction is diacylglycerol (DAG), which is a well-established activator of protein kinase C (PKC). Conventional and novel PKC isoforms respond to DAG by translocating to the plasma membrane so that these enzymes can then express their activity and act on signaling events. In this connection, C1P stimulated the translocation and activation of the alpha isoform of PKC (PKC-α) in macrophages, and this resulted to be essential for stimulation of cell growth by C1P (Gangoiti et al., 2010c).

In a more recent report, it has been demonstrated that another essential kinase involved in the regulation of cell proliferation by C1P is the mammalian target of rapamycin (mTOR) (Gangoiti et al., 2010a). Activation of this kinase was tested my measuring the phosphorylation state of its downstream target p70S6K after treatment with C1P. Activation of mTOR/ p70S6K was dependent upon prior activation of PI3-K, as selective inhibition of this kinase blocked mTOR phosphorylation and activation. In addition, C1P caused phosphorylation of PRAS40, a component of the mTOR complex 1 (mTORC1) that is absent in mTORC2, and inhibition of the small G protein Ras homolog enriched in brain (Rheb), which is also a specific component of mTORC1, completely blocked C1P-stimulated mTOR phosphorylation, DNA synthesis and macrophage growth. C1P also caused phosphorylation of another Ras homolog gene family member, RhoA, and inhibition of its downstream effector RhoA-associated kinase (ROCK) also blocked C1P-stimulated mTOR and cell proliferation. It was concluded that mTORC1, and RhoA/ROCK are essential components of the mechanism whereby C1P stimulates macrophage proliferation. However, phospholipase D (PLD), and cAMP are not involved in the mitogenic effect of C1P (Gomez-Munoz et al., 1995a; Gomez-Munoz et al., 1997). Concerning intracellular calcium levels, which have also been implicated in the regulation of cell proliferation, the situation is controversial.

Although short-chain C1Ps failed to induce Ca^{2+} mobilization in fibroblasts (Gomez-Munoz et al., 1995a; Gomez-Munoz et al., 1997) or neutrophils (Rile et al., 2003), and natural C_{16}-C1P did not alter intracellular Ca^{2+} concentrations in A549 cells (Pettus et al., 2004), C_2-C1P- or C_8-C1P, caused intracellular Ca^{2+} mobilization in calf pulmonary artery endothelial (CAPE) cells (Gijsbers et al., 1999), thyroid FRTL-5 (Hogback et al., 2003), or Jurkat T-cells (Colina et al., 2005), suggesting that regulation of Ca^{2+} homeostasis may be cell type specific.

Finally, it should be pointed out that C1P has been recently shown to be a key mediator in the development and survival of retina photoreceptors, and to also play a critical role in photoreceptor differentiation (Miranda et al., 2011)

Apart from its mitogenic effect, another mechanism by which C1P controls cell homeostasis is by prevention of apoptosis (reviewed in (Gangoiti et al., 2010b)). We previously demonstrated that natural C1P blocked apoptosis in bone marrow-derived macrophages (Gomez-Munoz et al., 2004; Gomez-Munoz et al., 2005), and this was confirmed by Mitra and co-workers (Mitra et al., 2007) who found that down-regulation of CerK in mammalian cells reduced growth, and promoted apoptosis. Also, downregulation of CerK blocked epithelial growth factor-induced cell proliferation. However, in contrast to these observations, it was reported that addition of the cell-permeable C_2-ceramide to cells overexpressing CerK led to C_2-C1P formation and stimulation of apoptosis (Graf et al., 2007). This controversy can be explained by the fact that overexpression of CerK would substantially increase the intracellular levels of C1P, especially when cells are supplied with high concentrations of exogenous cell permeable C_2-ceramide; this action would cause overproduction of C_2-C1P inside the cells, which is toxic at high concentrations (Gomez-Munoz et al., 1995a; Gomez-Munoz et al., 2004).

When cells become apoptotic, their metabolism undergoes important changes from early stages. For example, apoptotic bone marrow-derived macrophages express high acid sphingomyelinase (A-SMase) activity and show high levels of ceramides compared to non-apoptotic cells (Gomez-Munoz et al., 2003; Hundal et al., 2003). Of interest, inhibition of A-SMase activation resulted to be one of the mechanisms by which C1P blocks apoptosis (Gomez-Munoz et al., 2004). C1P also blocked the activity of A-SMase in cell-free systems (in vitro), suggesting that inhibition of this enzyme takes place by direct physical interaction of C1P with the enzyme.

Recent work by our group (Granado et al., 2009a) showed that ceramide levels are also increased in alveolar NR8383 macrophages when they become apoptotic. However, A-SMase activity was only slightly enhanced in these cells under apoptotic conditions. This suggested the intervention of a different pathway for ceramide generation in these cells. In subsequent work we demonstrated that the mechanism whereby ceramide levels increased in apoptotic alveolar macrophages involved activation of serine palmitoyltransferase (SPT), the key regulatory enzyme of the de novo pathway of ceramide synthesis. Like for A-SMase, inhibition of SPT activation by treatment with C1P prevented the alveolar macrophages from entering apoptosis. These findings led to conclude that C1P promotes macrophage survival by blocking ceramide accumulation, and action that can be brought about through inhibition of either A-SMase activity, or SPT, depending on cell type.

The prosurvival effect of C1P was highlighted by the demonstration that intracellular levels of C1P were substantially decreased when the cells became apoptotic. It was hypothesized that depletion of intracellular C1P could result in the release of A-SMase from inhibition, thereby triggering ceramide generation an apoptotic cell death (Gomez-Munoz et al., 2004). Once generated, ceramides act on different intracellular targets to induce apoptosis. One of these targets is protein kinase B (or Akt), a kinase that lies downstream of PI3-K, a major signaling pathway through which growth factors promote cell survival. Using two different experimental approaches, it was demonstrated that PI3-K was also a target of C1P (Gomez-Munoz et al., 2005). On one hand, PI3-K activation was demonstrated by immunoprecipitation of the enzyme from whole cell lysates and assayed in vitro using ^{32}P-phosphatidylinositol. On the other hand, an in vivo approach provided evidence of phosphatidylinositol (3,4,5)-trisphosphate (PIP3) formation in intact cells that were prelabeled with ^{32}P-orthophosphate (Gomez-Munoz et al., 2005). PIP3 is a major product of PI3-K, and was shown to directly inhibit A-SMase (Testai et al., 2004). Therefore, it could be speculated that PI3-K activation might potentiate the inhibitory effect of C1P on A-SMase through generation of PIP3. C1P stimulated the phosphorylation of PKB, which was sensitive to inhibition by wortmannin or LY294002, thereby confirming that PI3-K was the enzyme responsible for its phosphorylation. These two PI3-K inhibitors also blocked the prosurvival effect of C1P, as expected (Gomez-Munoz et al., 2005). Another relevant finding was that C1P caused IkB phosphorylation and stimulation of the DNA binding activity of NF-kB in primary cultures of mouse macrophages (Gomez-Munoz et al., 2005). Of note, C1P up-regulated the expression of anti-apoptotic Bcl-XL, which is a downstream target of NF-kB. The latter results provided the first evidence for a novel biological role of natural C1P in the regulation of cell survival by the PI3-K/PKB/NF-kB pathway in mammalian cells (Gomez-Munoz et al., 2005).

As mentioned above, C1P can be metabolized to ceramide by different phosphatases, and then further converted to sphingosine and S1P by the coordinated actions of ceramidases and sphingosine kinases. Therefore, it could be speculated that the effects of C1P might be mediated through C1P-derived metabolites. However, usually ceramides and C1P exert opposing effects, (i.e. on PLD activation, adenylyl cyclase inhibition, or Ca^{2+} mobilization), and C1P is not able to reproduce the effects of S1P (Gomez-Munoz et al., 1995a; Gomez-Munoz et al., 1995b; Gomez-Munoz et al., 1997; Gomez-Munoz, 1998). Also, ceramides can decrease the expression of Bcl-XL (Chalfant & Spiegel, 2005), whereas C1P causes its up-regulation (Gomez-Munoz et al., 2005). Finally, no ceramidases capable of converting C1P into S1P have so far been reported to exist in mammalian cells, and S1P and C1P inhibit A-SMase through different mechanisms (Gomez-Munoz et al., 2003; Gomez-Munoz et al., 2004). Therefore, it can be concluded that C1P acts on its own right to regulate cell homeostasis. The above observations suggest that regulation of the enzyme activities involved in ceramide and C1P metabolism is essential for cell fate. Elucidation of the mechanisms controlling ceramide and C1P levels may help develop new molecular strategies for preventing metabolic disorders, or designing novel therapeutic agents for treatment of disease.

5. Ceramide 1-phosphate and the control of inflammation

Inflammation is, in principle, a beneficial process for protecting the organism against infection or injury. However, it can be detrimental when it becomes out of control. Apart from the classical signaling pathways and metabolites that are involved in the regulation of inflammation, it is now well accepted that ceramides are key elements in the inflammatory response (Lamour & Chalfant, 2005; Wijesinghe et al., 2008; Gomez-Munoz et al., 2010). For instance, it was reported that activation of A-SMase and the subsequent formation of ceramides play an important role in pulmonary infections as it facilitates internalization of bacteria into lung epithelial cells (Gulbins & Kolesnick, 2003). In this context, inhibition of A-SMase by C1P could be important to reduce or prevent infection in the lung.

Inflammatory mediators include chemokines, cytokines, vasoactive amines, products of proteolytic cascades, phospholipases, or lipids such as eicosanoids and sphingolipids. A major mediator of inflammation is PLA_2 activity. In particular, group IV cytosolic $cPLA_2$ (or $cPLA_2$-alpha) has been involved in receptor-dependent and independent production of eicosanoids, which are major components of inflammatory responses. Sphingolipids, including ceramides, have also been described as key mediators of inflammation (Hayakawa et al., 1996; Serhan et al., 1996; Manna & Aggarwal, 1998; Newton et al., 2000). More recently a role for ceramide in the development of allergic asthmatic responses and airway inflammation was established (Masini et al., 2008), and exogenous addition of C_2-ceramide to cultured astrocytes induced 12-lipoxigenase leading to generation of reactive oxygen species (ROS) and inflammation (Prasad et al., 2008). Also, A-SMase-derived ceramide was involved in platelet activating factor (PAF)-mediated pulmonary edema (Goggel et al., 2004). Subsequently, it was proposed that at least some of the pro-inflammatory effects of ceramides might in fact be mediated by its further metabolite C1P. The first report on the regulation of arachidonic acid (AA) release and the production of prostaglandins by C1P was from the laboratory of Charles Chalfant (Pettus et al., 2003b). This group demonstrated that C1P was able to stimulate AA release and prostanoid synthesis in A549 lung adenocarcinoma cells. In a follow up report, the same group showed that the mechanism whereby C1P stimulates AA release occurs through direct activation of $cPLA_2$ (Pettus et al., 2004). Subsequently, it was found that C1P is a positive allosteric activator of $cPLA_2$-alpha, and that it enhances the interaction of the enzyme with PC (Subramanian et al., 2005). In further work, the same group demonstrated that activation of $cPLA_2$-alpha by C1P is chain length-specific; in particular, C1P bearing acyl chains equal or higher than six carbons were able to efficiently activate $cPLA_2$-alpha in vitro, whereas shorter acyl chains (in particular C_2-C1P) were unable to activate the enzyme. It was concluded that the biological activity of C_2-C1P does not occur via eicosanoid synthesis (Wijesinghe et al., 2008). Also, C1P was shown to act in coordination with S1P to ensure maximal production of prostaglandins. Specifically, S1P was shown to induce cyclooxigenase-2 (COX-2) activity, which then uses $cPLA_2$-derived AA as substrate to synthesize prostaglandins (Pettus et al., 2005). Further details on the role of C1P in inflammatory response can be found in different reviews (Chalfant & Spiegel, 2005; Lamour et al., 2007; Wijesinghe et al., 2007), Wijesinghe et al., and recent work by Murayama and co-workers (Nakamura et al., 2011).

6. Ceramide 1-phosphate and the control of cell migration

Macrophage populations in tissues are determined by the rates of recruitment of monocytes from the bloodstream into the tissue, the rates of macrophage proliferation and apoptosis, and the rate of macrophage migration or efflux. Recently, our group demonstrated that exogenous addition of C1P to cultured Raw 264.7 macrophages stimulated cell migration

(Granado et al., 2009b). Interestingly, this action could only be observed when C1P was applied to the cells exogenously, and not by increasing the intracellular levels of C1P (i.e. through agonist stimulation of CerK, or by using the "caging" strategy to deliver C1P intracellularly (Lankalapalli et al., 2009)). This observation led us to identify a specific receptor through which C1P stimulates chemotaxis. This putative receptor seems to be located in the plasma membrane, has low affinity for C1P and has an apparent Kd of approximately 7.8 μM. The receptor is specific for C1P and is coupled to Gi proteins. Ligation of this receptor with C1P caused phosphorylation of ERK1–2, and PKB, and inhibition of either of these pathways completely abolished C1P-stimulated macrophage migration. Moreover, C1P stimulated the DNA binding activity of NF-kB, and blockade of this transcription factor resulted in full inhibition of macrophage migration. These observations suggest that MEK/ERK1-2, PI3-K/PKB (or Akt) and NF-kB are crucial signaling pathways for regulation of cell migration by C1P. It was concluded that this newly identified receptor could be an important drug target for treatment of illnesses in which cell migration is a major cause of pathology, as it occurs in atherosclerosis or in the metastasis of tumors.

7. Other relevant biological actions of C1P

In a previous report, Hinkovska-Galcheva et al (Hinkovska-Galcheva et al., 1998) showed that endogenous C1P can be generated during the phagocytosis of antibody-coated erythrocytes in human neutrophils that were primed with formylmethionylleucylphenylalanine. More recently, the same group demonstrated that C1P is a key mediator of neuthophil phagocytosis (Hinkovska-Galcheva et al., 2005). In addition, it was reported that C1P can be formed in neutrophils upon incubation with cell-permeable [^{3}H]N-hexanoylsphingosine (C_6-ceramide) (Rile et al., 2003), and Riboni and co-workers (Riboni et al., 2002) found that C1P can be generated in cerebellar granule cells both from SM-derived ceramide and through the recycling of sphingosine produced by ganglioside catabolism. C1P can be also generated by the action of interleukin 1-betta on A549 lung adenocarcinoma cells (Pettus et al., 2003b), or by stimulation of bone marrow-derived macrophages with macrophage-colony stimulating factor (M-CSF) (Gangoiti et al., 2008b). We found that C1P is present in normal bone marrow-derived macrophages isolated from healthy mice (Gomez-Munoz et al., 2004), and that C1P levels are substantially decreased in apoptotic macrophages. These observations are consistent with recent findings showing that CerK plays a key role in the stimulation of cell proliferation in A549 human lung adenocarcinoma cells (Mitra et al., 2007), and the induction of neointimal formation via cell proliferation and cell cycle progression in vascular smooth muscle cells by C1P (Kim et al., 2011).

8. Conclusion

The implication of simple sphingolipids in the regulation of cell activation and metabolism has acquired special relevance in the last two decades. Most attention was first paid to the effects elicited by ceramide because this sphingolipid turned out to be essential in the regulation of cell death, differentiation, senescence, and various metabolic disorders and diseases. However, its phosphorylated form, C1P, was thought not to be so important. However, C1P has emerged as a crucial bioactive sphingolipid, and this chapter highlights the relevance of C1P in cell biology. Specifically, C1P has now been established as key regulator of cell growth and survival, and its relevance in the regulation of cell migration is beginning to emerge. Also importantly, the discovery that C1P can act both intracellularly or as receptor ligand opens a broad avenue to investigate its implication in controlling cell metabolism. In addition to this, C1P has been postulated to be a potent proinflammatory agent, acting directly on cPLA$_2$ to trigger eicosanoid production. Therefore, C1P and CerK, the major enzyme responsible for its biosynthesis, may be key targets for developing new pharmacological strategies for treatment of illnesses associated to cell growth and death, and cell migration , such as chronic inflammation, cardiovascular diseases, neurodegeneration, or cancer.

9. Acknowledgement

Work in AGM lab is supported by Ministerio de Ciencia e Innovación (Madrid, Spain), Departamento de Educación, Universidades e Investigación del Gobierno Vasco (Gazteiz-Vitoria, Basque Country), and Departamento de Industria, Comercio y Turismo del Gobierno Vasco (Gazteiz-Vitoria, Basque Country).

10. References

Adams, J. M., 2nd, Pratipanawatr, T., Berria, R., Wang, E., DeFronzo, R. A., Sullards, M. C. and Mandarino, L. J. (2004). Ceramide content is increased in skeletal muscle from obese insulin-resistant humans. *Diabetes* 53, 1,(Jan, 2004) 25-31, 0012-1797

Arana, L., Gangoiti, P., Ouro, A., Trueba, M. and Gomez-Munoz, A. (2010). Ceramide and ceramide 1-phosphate in health and disease. *Lipids Health Dis* 9, 2010) 15, 1476-511X

Bajjalieh, S. M., Martin, T. F. and Floor, E. (1989). Synaptic vesicle ceramide kinase. A calcium-stimulated lipid kinase that co-purifies with brain synaptic vesicles. *J Biol Chem* 264, 24,(Aug 25, 1989) 14354-60, 0021-9258

Baumruker, T., Bornancin, F. and Billich, A. (2005). The role of sphingosine and ceramide kinases in inflammatory responses. *Immunol Lett* 96, 2,(Jan 31, 2005) 175-85, 0165-2478

Bi, F. C., Zhang, Q. F., Liu, Z., Fang, C., Li, J., Su, J. B., Greenberg, J. T., Wang, H. B. and Yao, N. (2011). A conserved cysteine motif is critical for rice ceramide kinase activity and function. *PLoS One* 6, 3,(2011) e18079, 1932-6203

Boath, A., Graf, C., Lidome, E., Ullrich, T., Nussbaumer, P. and Bornancin, F. (2008). Regulation and traffic of ceramide 1-phosphate produced by ceramide kinase: comparative analysis to glucosylceramide and sphingomyelin. *J Biol Chem* 283, 13,(Mar 28, 2008) 8517-26, 0021-9258

Bornancin, F., Mechtcheriakova, D., Stora, S., Graf, C., Wlachos, A., Devay, P., Urtz, N., Baumruker, T. and Billich, A. (2005). Characterization of a ceramide kinase-like protein. *Biochim Biophys Acta* 1687, 1-3,(Feb 21, 2005) 31-43, 0005-2736

Boudker, O. and Futerman, A. H. (1993). Detection and characterization of ceramide-1-phosphate phosphatase activity in rat liver plasma membrane. *J Biol Chem* 268, 29,(Oct 15, 1993) 22150-5, 0021-9258

Brann, A. B., Scott, R., Neuberger, Y., Abulafia, D., Boldin, S., Fainzilber, M. and Futerman, A. H. (1999). Ceramide signaling downstream of the p75 neurotrophin receptor mediates the effects of nerve growth factor on outgrowth of cultured hippocampal neurons. *J Neurosci* 19, 19,(Oct 1, 1999) 8199-206, 0270-6474

Brindley, D. N. and Waggoner, D. W. (1998). Mammalian lipid phosphate phosphohydrolases. *J Biol Chem* 273, 38,(Sep 18, 1998) 24281-4, 0021-9258

Brizuela, L., Rabano, M., Pena, A., Gangoiti, P., Macarulla, J. M., Trueba, M. and Gomez-Munoz, A. (2006). Sphingosine 1-phosphate: a novel stimulator of aldosterone secretion. *J Lipid Res* 47, 6,(Jun, 2006) 1238-49, 0022-2275

Colina, C., Flores, A., Castillo, C., Garrido Mdel, R., Israel, A., DiPolo, R. and Benaim, G. (2005). Ceramide-1-P induces Ca2+ mobilization in Jurkat T-cells by elevation of Ins(1,4,5)-P3 and activation of a store-operated calcium channel. *Biochem Biophys Res Commun* 336, 1,(Oct 14, 2005) 54-60, 0006-291X

Chalfant, C. E. and Spiegel, S. (2005). Sphingosine 1-phosphate and ceramide 1-phosphate: expanding roles in cell signaling. *J Cell Sci* 118, Pt 20,(Oct 15, 2005) 4605-12, 0021-9533

Chen, Y., Liu, Y., Sullards, M. C. and Merrill, A. H., Jr. (2011). An introduction to sphingolipid metabolism and analysis by new technologies. *Neuromolecular Med* 12, 4,(Dec, 2011) 306-19, 1535-1084

Dressler, K. A., Mathias, S. and Kolesnick, R. N. (1992). Tumor necrosis factor-alpha activates the sphingomyelin signal transduction pathway in a cell-free system. *Science* 255, 5052,(Mar 27, 1992) 1715-8, 0036-8075

Gangoiti, P., Arana, L., Ouro, A., Granado, M. H., Trueba, M. and Gomez-Munoz, A. (2010a). Activation of mTOR and RhoA is a major mechanism by which Ceramide 1-phosphate stimulates macrophage proliferation. *Cell Signal* 23, 1,(Aug 18, 2010a) 27-34,

Gangoiti, P., Camacho, L., Arana, L., Ouro, A., Granado, M. H., Brizuela, L., Casas, J., Fabrias, G., Abad, J. L., Delgado, A. and Gomez-Munoz, A. (2010b). Control of metabolism and signaling of simple bioactive sphingolipids: Implications in disease. *Prog Lipid Res* 49, 4,(Oct, 2010b) 316-34, 0163-7827

Gangoiti, P., Granado, M. H., Alonso, A., Goñi, F. M. and Gómez-Muñoz, A. (2008a). Implication of Ceramide, Ceramide 1-Phosphate and Sphingosine 1-Phosphate in Tumorigenesis. *Translational Oncogenomics* 3 2008a) 67-79, 1177-2727

Gangoiti, P., Granado, M. H., Arana, L., Ouro, A. and Gomez-Munoz, A. (2010c). Activation of protein kinase C-alpha is essential for stimulation of cell proliferation by ceramide 1-phosphate. *FEBS Lett* 584, 3,(Feb 5, 2010c) 517-24, 0014-5793

Gangoiti, P., Granado, M. H., Wang, S. W., Kong, J. Y., Steinbrecher, U. P. and Gomez-Munoz, A. (2008b). Ceramide 1-phosphate stimulates macrophage proliferation through activation of the PI3-kinase/PKB, JNK and ERK1/2 pathways. *Cell Signal* 20, 4,(Apr, 2008b) 726-36, 0898-6568

Gijsbers, S., Mannaerts, G. P., Himpens, B. and Van Veldhoven, P. P. (1999). N-acetyl-sphingenine-1-phosphate is a potent calcium mobilizing agent. *FEBS Lett* 453, 3,(Jun 25, 1999) 269-72, 0014-5793

Goggel, R., Winoto-Morbach, S., Vielhaber, G., Imai, Y., Lindner, K., Brade, L., Brade, H., Ehlers, S., Slutsky, A. S., Schutze, S., Gulbins, E. and Uhlig, S. (2004). PAF-mediated pulmonary edema: a new role for acid sphingomyelinase and ceramide. *Nat Med* 10, 2,(Feb, 2004) 155-60, 1078-8956

Gomez-Munoz, A. (1998). Modulation of cell signalling by ceramides. *Biochim Biophys Acta* 1391, 1,(Mar 6, 1998) 92-109, 0005-2736

Gomez-Munoz, A. (2004). Ceramide-1-phosphate: a novel regulator of cell activation. *FEBS Lett* 562, 1-3,(Mar 26, 2004) 5-10, 0014-5793

Gomez-Munoz, A. (2006). Ceramide 1-phosphate/ceramide, a switch between life and death. *Biochim Biophys Acta* 1758, 12,(Dec, 2006) 2049-56, 0005-2736

Gomez-Munoz, A., Duffy, P. A., Martin, A., O'Brien, L., Byun, H. S., Bittman, R. and Brindley, D. N. (1995a). Short-chain ceramide-1-phosphates are novel stimulators of DNA synthesis and cell division: antagonism by cell-permeable ceramides. *Mol Pharmacol* 47, 5,(May, 1995a) 833-9, 0026-895X

Gomez-Munoz, A., Frago, L. M., Alvarez, L. and Varela-Nieto, I. (1997). Stimulation of DNA synthesis by natural ceramide 1-phosphate. *Biochem J* 325 (Pt 2), Jul 15, 1997) 435-40, 0264-6021

Gomez-Munoz, A., Gangoiti, P., Granado, M. H., Arana, L. and Ouro, A. (2010). Ceramide 1-Phosphate in Cell Survival and Inflammatory Signaling. Sphingolipids as Signaling and Regulatory Molecules. C. Chalfant and M. D. Poeta. Austin (Tx), Landes Bioscience and Springer Science+Business Media: 118-130.

Gomez-Munoz, A., Hamza, E. H. and Brindley, D. N. (1992). Effects of sphingosine, albumin and unsaturated fatty acids on the activation and translocation of phosphatidate phosphohydrolases in rat hepatocytes. *Biochim Biophys Acta* 1127, 1,(Jul 9, 1992) 49-56, 0005-2736

Gomez-Munoz, A., Kong, J., Salh, B. and Steinbrecher, U. P. (2003). Sphingosine-1-phosphate inhibits acid sphingomyelinase and blocks apoptosis in macrophages. *FEBS Lett* 539, 1-3,(Mar 27, 2003) 56-60, 0014-5793

Gomez-Munoz, A., Kong, J. Y., Parhar, K., Wang, S. W., Gangoiti, P., Gonzalez, M., Eivemark, S., Salh, B., Duronio, V. and Steinbrecher, U. P. (2005). Ceramide-1-phosphate promotes cell survival through activation of the phosphatidylinositol 3-kinase/protein kinase B pathway. *FEBS Lett* 579, 17,(Jul 4, 2005) 3744-50, 1177-5793

Gomez-Munoz, A., Kong, J. Y., Salh, B. and Steinbrecher, U. P. (2004). Ceramide-1-phosphate blocks apoptosis through inhibition of acid sphingomyelinase in macrophages. *J Lipid Res* 45, 1,(Jan, 2004) 99-105, 0022-2275

Gomez-Munoz, A., Waggoner, D. W., O'Brien, L. and Brindley, D. N. (1995b). Interaction of ceramides, sphingosine, and sphingosine 1-phosphate in regulating DNA synthesis and phospholipase D activity. *J Biol Chem* 270, 44,(Nov 3, 1995b) 26318-25, 0022-2275

Goodman, Y. and Mattson, M. P. (1996). Ceramide protects hippocampal neurons against excitotoxic and oxidative insults, and amyloid beta-peptide toxicity. *J Neurochem* 66, 2,(Feb, 1996) 869-72, 0022-3042

Graf, C., Rovina, P., Tauzin, L., Schanzer, A. and Bornancin, F. (2007). Enhanced ceramide-induced apoptosis in ceramide kinase overexpressing cells. *Biochem Biophys Res Commun* 354, 1,(Mar 2, 2007) 309-14, 0006-291X

Graf, C., Zemann, B., Rovina, P., Urtz, N., Schanzer, A., Reuschel, R., Mechtcheriakova, D., Muller, M., Fischer, E., Reichel, C., Huber, S., Dawson, J., Meingassner, J. G., Billich, A., Niwa, S., Badegruber, R., Van Veldhoven, P. P., Kinzel, B., Baumruker, T. and Bornancin, F. (2008). Neutropenia with Impaired Immune Response to Streptococcus pneumoniae in Ceramide Kinase-Deficient Mice. *J Immunol* 180, 5,(Mar 1, 2008) 3457-66, 0022-1767

Granado, M. H., Gangoiti, P., Ouro, A., Arana, L. and Gomez-Munoz, A. (2009a). Ceramide 1-phosphate inhibits serine palmitoyltransferase and blocks apoptosis in alveolar macrophages. *Biochim Biophys Acta* 1791, 4,(Apr, 2009a) 263-72, 0005-2736

Granado, M. H., Gangoiti, P., Ouro, A., Arana, L., Gonzalez, M., Trueba, M. and Gomez-Munoz, A. (2009b). Ceramide 1-phosphate (C1P) promotes cell migration Involvement of a specific C1P receptor. *Cell Signal* 21, 3,(Mar, 2009b) 405-12, 0898-6568

Gulbins, E. and Kolesnick, R. (2003). Raft ceramide in molecular medicine. *Oncogene* 22, 45,(Oct 13, 2003) 7070-7, 0950-9232

Hannun, Y. A. (1994). The sphingomyelin cycle and the second messenger function of ceramide. *J Biol Chem* 269, 5,(Feb 4, 1994) 3125-8, 0021-9258

Hannun, Y. A. (1996). Functions of ceramide in coordinating cellular responses to stress. *Science* 274, 5294,(Dec 13, 1996) 1855-9,

Hannun, Y. A., Loomis, C. R., Merrill, A. H., Jr. and Bell, R. M. (1986). Sphingosine inhibition of protein kinase C activity and of phorbol dibutyrate binding in vitro and in human platelets. *J Biol Chem* 261, 27,(Sep 25, 1986) 12604-9, 0021-9258

Hannun, Y. A. and Obeid, L. M. (1995). Ceramide: an intracellular signal for apoptosis. *Trends Biochem Sci* 20, 2,(Feb, 1995) 73-7, 0968-0004

Hannun, Y. A. and Obeid, L. M. (2002). The Ceramide-centric universe of lipid-mediated cell regulation: stress encounters of the lipid kind. *J Biol Chem* 277, 29,(Jul 19, 2002) 25847-50,

Hannun, Y. A. and Obeid, L. M. (2008). Principles of bioactive lipid signalling: lessons from sphingolipids. *Nat Rev Mol Cell Biol* 9, 2,(Feb, 2008) 139-50, 0028-0836

Hannun, Y. A. and Obeid, L. M. (2011). Many ceramides. *J Biol Chem* 286, 32,(Aug 12, 2011) 27855-62, 0021-9258

Hayakawa, M., Jayadev, S., Tsujimoto, M., Hannun, Y. A. and Ito, F. (1996). Role of ceramide in stimulation of the transcription of cytosolic phospholipase A2 and cyclooxygenase 2. *Biochem Biophys Res Commun* 220, 3,(Mar 27, 1996) 681-6, 0006-291X

Hinkovska-Galcheva, V., Boxer, L. A., Kindzelskii, A., Hiraoka, M., Abe, A., Goparju, S., Spiegel, S., Petty, H. R. and Shayman, J. A. (2005). Ceramide 1-phosphate, a mediator of phagocytosis. *J Biol Chem* 280, 28,(Jul 15, 2005) 26612-21, 0021-9258

Hinkovska-Galcheva, V. and Shayman, J. A. Ceramide-1-phosphate in phagocytosis and calcium homeostasis. *Adv Exp Med Biol* 688, 131-40, 0065-2598

Hinkovska-Galcheva, V. T., Boxer, L. A., Mansfield, P. J., Harsh, D., Blackwood, A. and Shayman, J. A. (1998). The formation of ceramide-1-phosphate during neutrophil phagocytosis and its role in liposome fusion. *J Biol Chem* 273, 50,(Dec 11, 1998) 33203-9, 0021-9258

Hogback, S., Leppimaki, P., Rudnas, B., Bjorklund, S., Slotte, J. P. and Tornquist, K. (2003). Ceramide 1-phosphate increases intracellular free calcium concentrations in thyroid FRTL-5 cells: evidence for an effect mediated by inositol 1,4,5-trisphosphate and intracellular sphingosine 1-phosphate. *Biochem J* 370, Pt 1,(Feb 15, 2003) 111-9, 0264-6021

Holland, W. L., Bikman, B. T., Wang, L. P., Yuguang, G., Sargent, K. M., Bulchand, S., Knotts, T. A., Shui, G., Clegg, D. J., Wenk, M. R., Pagliassotti, M. J., Scherer, P. E. and Summers, S. A. (2011). Lipid-induced insulin resistance mediated by the proinflammatory receptor TLR4 requires saturated fatty acid-induced ceramide biosynthesis in mice. *J Clin Invest* 121, 5,(May 2, 2011) 1858-70, 0021-9738

Hundal, R. S., Gomez-Munoz, A., Kong, J. Y., Salh, B. S., Marotta, A., Duronio, V. and Steinbrecher, U. P. (2003). Oxidized low density lipoprotein inhibits macrophage apoptosis by blocking ceramide generation, thereby maintaining protein kinase B activation and Bcl-XL levels. *J Biol Chem* 278, 27,(Jul 4, 2003) 24399-408, 0022-9258

Jamal, Z., Martin, A., Gomez-Munoz, A. and Brindley, D. N. (1991). Plasma membrane fractions from rat liver contain a phosphatidate phosphohydrolase distinct from that in the endoplasmic reticulum and cytosol. *J Biol Chem* 266, 5,(Feb 15, 1991) 2988-96, 0021-9258

Kim, T. J., Kang, Y. J., Lim, Y., Lee, H. W., Bae, K., Lee, Y. S., Yoo, J. M., Yoo, H. S. and Yun, Y. P. (2011). Ceramide 1-phosphate induces neointimal formation via cell proliferation and cell cycle progression upstream of ERK1/2 in vascular smooth muscle cells. *Exp Cell Res* 317, 14,(Aug 15, 2011) 2041-51, 0014-4827

Kim, T. J., Mitsutake, S. and Igarashi, Y. (2006). The interaction between the pleckstrin homology domain of ceramide kinase and phosphatidylinositol 4,5-bisphosphate regulates the plasma membrane targeting and ceramide 1-phosphate levels. *Biochem Biophys Res Commun* 342, 2,(Apr 7, 2006) 611-7, 0006-291X

Kim, T. J., Mitsutake, S., Kato, M. and Igarashi, Y. (2005). The leucine 10 residue in the pleckstrin homology domain of ceramide kinase is crucial for its catalytic activity. *FEBS Lett* 579, 20,(Aug 15, 2005) 4383-8, 0014-5793

Kolesnick, R. and Golde, D. W. (1994). The sphingomyelin pathway in tumor necrosis factor and interleukin-1 signaling. *Cell* 77, 3,(May 6, 1994) 325-8, 0272-4340

Kolesnick, R. N. (1987). 1,2-Diacylglycerols but not phorbol esters stimulate sphingomyelin hydrolysis in GH3 pituitary cells. *J Biol Chem* 262, 35,(Dec 15, 1987) 16759-62, 0021-9258

Kolesnick, R. N., Goni, F. M. and Alonso, A. (2000). Compartmentalization of ceramide signaling: physical foundations and biological effects. *J Cell Physiol* 184, 3,(Sep, 2000) 285-300, 0021-9541

Kolesnick, R. N. and Hemer, M. R. (1990). Characterization of a ceramide kinase activity from human leukemia (HL-60) cells. Separation from diacylglycerol kinase activity. *J Biol Chem* 265, 31,(Nov 5, 1990) 18803-8, 0021-9258

Lamour, N. F. and Chalfant, C. E. (2005). Ceramide-1-phosphate: the "missing" link in eicosanoid biosynthesis and inflammation. *Mol Interv* 5, 6,(Dec, 2005) 358-67, 1543-2548

Lamour, N. F., Stahelin, R. V., Wijesinghe, D. S., Maceyka, M., Wang, E., Allegood, J. C., Merrill, A. H., Jr., Cho, W. and Chalfant, C. E. (2007). Ceramide kinase uses ceramide provided by ceramide transport protein: localization to organelles of eicosanoid synthesis. *J Lipid Res* 48, 6,(Jun, 2007) 1293-304, 0022-2275

Lankalapalli, R. S., Ouro, A., Arana, L., Gomez-Munoz, A. and Bittman, R. (2009). Caged ceramide 1-phosphate analogues: synthesis and properties. *J Org Chem* 74, 22,(Nov 20, 2009) 8844-7, 5163-5166

Lee, S. and Lynch, K. R. (2005). Brown recluse spider (Loxosceles reclusa) venom phospholipase D (PLD) generates lysophosphatidic acid (LPA). *Biochem J* 391, Pt 2,(Oct 15, 2005) 317-23, 0264-6021

Long, J., Darroch, P., Wan, K. F., Kong, K. C., Ktistakis, N., Pyne, N. J. and Pyne, S. (2005). Regulation of cell survival by lipid phosphate phosphatases involves the modulation of intracellular phosphatidic acid and sphingosine 1-phosphate pools. *Biochem J* 391, Pt 1,(Oct 1, 2005) 25-32, 0264-6021

Manna, S. K. and Aggarwal, B. B. (1998). IL-13 suppresses TNF-induced activation of nuclear factor-kappa B, activation protein-1, and apoptosis. *J Immunol* 161, 6,(Sep 15, 1998) 2863-72, 0022-1767

Masini, E., Giannini, L., Nistri, S., Cinci, L., Mastroianni, R., Xu, W., Comhair, S. A., Li, D., Cuzzocrea, S., Matuschak, G. M. and Salvemini, D. (2008). Ceramide: a key signaling molecule in a Guinea pig model of allergic asthmatic response and airway inflammation. *J Pharmacol Exp Ther* 324, 2,(Feb, 2008) 548-57., 0022-3565

Mathias, S., Dressler, K. A. and Kolesnick, R. N. (1991). Characterization of a ceramide-activated protein kinase: stimulation by tumor necrosis factor alpha. *Proc Natl Acad Sci U S A* 88, 22,(Nov 15, 1991) 10009-13, 1091-6490

Menaldino, D. S., Bushnev, A., Sun, A., Liotta, D. C., Symolon, H., Desai, K., Dillehay, D. L., Peng, Q., Wang, E., Allegood, J., Trotman-Pruett, S., Sullards, M. C. and Merrill, A. H., Jr. (2003). Sphingoid bases and de novo ceramide synthesis: enzymes involved, pharmacology and mechanisms of action. *Pharmacol Res* 47, 5,(May, 2003) 373-81, 1043-6618

Merrill, A. H., Jr. (1991). Cell regulation by sphingosine and more complex sphingolipids. *J Bioenerg Biomembr* 23, 1,(Feb, 1991) 83-104, 0145-479X

Merrill, A. H., Jr. (2002). De novo sphingolipid biosynthesis: a necessary, but dangerous, pathway. *J Biol Chem* 277, 29,(Jul 19, 2002) 25843-6, 0021-9258

Merrill, A. H., Jr. and Jones, D. D. (1990). An update of the enzymology and regulation of sphingomyelin metabolism. *Biochim Biophys Acta* 1044, 1,(May 1, 1990) 1-12, 0005-2736

Merrill, A. H., Jr., Schmelz, E. M., Dillehay, D. L., Spiegel, S., Shayman, J. A., Schroeder, J. J., Riley, R. T., Voss, K. A. and Wang, E. (1997). Sphingolipids--the enigmatic lipid class: biochemistry, physiology, and pathophysiology. *Toxicol Appl Pharmacol* 142, 1,(Jan, 1997) 208-25, 0041-008X

Merrill, A. H., Jr., Sullards, M. C., Allegood, J. C., Kelly, S. and Wang, E. (2005). Sphingolipidomics: high-throughput, structure-specific, and quantitative analysis of sphingolipids by liquid chromatography tandem mass spectrometry. *Methods* 36, 2,(Jun, 2005) 207-24, 1548-7091

Merrill, A. H., Jr., Sullards, M. C., Wang, E., Voss, K. A. and Riley, R. T. (2001). Sphingolipid metabolism: roles in signal transduction and disruption by fumonisins. *Environ Health Perspect* 109 Suppl 2, May, 2001) 283-9, 0091-6765

Miranda, G. E., Abrahan, C. E., Agnolazza, D. L., Politi, L. E. and Rotstein, N. P. (2011). Ceramide-1-phosphate, a new mediator of development and survival in retina photoreceptors. *Invest Ophthalmol Vis Sci* 52, 9,(2011) 6580-8, 1552-5783

Mitra, P., Maceyka, M., Payne, S. G., Lamour, N., Milstien, S., Chalfant, C. E. and Spiegel, S. (2007). Ceramide kinase regulates growth and survival of A549 human lung adenocarcinoma cells. *FEBS Lett* 581, 4,(Feb 20, 2007) 735-40, 0014-5793

Mitsutake, S., Kim, T. J., Inagaki, Y., Kato, M., Yamashita, T. and Igarashi, Y. (2004). Ceramide kinase is a mediator of calcium-dependent degranulation in mast cells. *J Biol Chem* 279, 17,(Apr 23, 2004) 17570-7, 0021-9258

Nakamura, H., Tada, E., Makiyama, T., Yasufuku, K. and Murayama, T. (2011). Role of cytosolic phospholipase A(2)alpha in cell rounding and cytotoxicity induced by ceramide-1-phosphate via ceramide kinase. *Arch Biochem Biophys* 512, 1,(Aug 1, 2011) 45-51, 0003-9861

Natarajan, V., Jayaram, H. N., Scribner, W. M. and Garcia, J. G. (1994). Activation of endothelial cell phospholipase D by sphingosine and sphingosine-1-phosphate. *Am J Respir Cell Mol Biol* 11, 2,(Aug, 1994) 221-9, 1044-1549

Newton, R., Hart, L., Chung, K. F. and Barnes, P. J. (2000). Ceramide induction of COX-2 and PGE(2) in pulmonary A549 cells does not involve activation of NF-kappaB. *Biochem Biophys Res Commun* 277, 3,(Nov 2, 2000) 675-9, 0006-291X

Okazaki, T., Bielawska, A., Bell, R. M. and Hannun, Y. A. (1990). Role of ceramide as a lipid mediator of 1 alpha,25-dihydroxyvitamin D3-induced HL-60 cell differentiation. *J Biol Chem* 265, 26,(Sep 15, 1990) 15823-31, 0021-9258

Olivera, A. and Spiegel, S. (1993). Sphingosine-1-phosphate as second messenger in cell proliferation induced by PDGF and FCS mitogens. *Nature* 365, 6446,(Oct 7, 1993) 557-60, 0028-0836

Pettus, B. J., Bielawska, A., Kroesen, B. J., Moeller, P. D., Szulc, Z. M., Hannun, Y. A. and Busman, M. (2003a). Observation of different ceramide species from crude cellular extracts by normal-phase high-performance liquid chromatography coupled to atmospheric pressure chemical ionization mass spectrometry. *Rapid Commun Mass Spectrom* 17, 11,(2003a) 1203-11, 0951-4198

Pettus, B. J., Bielawska, A., Spiegel, S., Roddy, P., Hannun, Y. A. and Chalfant, C. E. (2003b). Ceramide kinase mediates cytokine- and calcium ionophore-induced arachidonic acid release. *J Biol Chem* 278, 40,(Oct 3, 2003b) 38206-13, 0022-9258

Pettus, B. J., Bielawska, A., Subramanian, P., Wijesinghe, D. S., Maceyka, M., Leslie, C. C., Evans, J. H., Freiberg, J., Roddy, P., Hannun, Y. A. and Chalfant, C. E. (2004). Ceramide 1-phosphate is a direct activator of cytosolic phospholipase A2. *J Biol Chem* 279, 12,(Mar 19, 2004) 11320-6, 0021-9258

Pettus, B. J., Kitatani, K., Chalfant, C. E., Taha, T. A., Kawamori, T., Bielawski, J., Obeid, L. M. and Hannun, Y. A. (2005). The coordination of prostaglandin E2 production by sphingosine-1-phosphate and ceramide-1-phosphate. *Mol Pharmacol* 68, 2,(Aug, 2005) 330-5, 0026-895X

Ping, S. E. and Barrett, G. L. (1998). Ceramide can induce cell death in sensory neurons, whereas ceramide analogues and sphingosine promote survival. *J Neurosci Res* 54, 2,(Oct 15, 1998) 206-13, 1097-4547

Plummer, G., Perreault, K. R., Holmes, C. F. and Posse De Chaves, E. I. (2005). Activation of serine/threonine protein phosphatase-1 is required for ceramide-induced survival of sympathetic neurons. *Biochem J* 385, Pt 3,(Feb 1, 2005) 685-93, 0264-6021

Prasad, V. V., Nithipatikom, K. and Harder, D. R. (2008). Ceramide elevates 12-hydroxyeicosatetraenoic acid levels and upregulates 12-lipoxygenase in rat primary hippocampal cell cultures containing predominantly astrocytes. *Neurochem Int* 53, 6-8,(Nov-Dec, 2008) 220-9, 0197-0186

Rabano, M., Pena, A., Brizuela, L., Marino, A., Macarulla, J. M., Trueba, M. and Gomez-Munoz, A. (2003). Sphingosine-1-phosphate stimulates cortisol secretion. *FEBS Lett* 535, 1-3,(Jan 30, 2003) 101-5, 0014-5793

Riboni, L., Bassi, R., Anelli, V. and Viani, P. (2002). Metabolic formation of ceramide-1-phosphate in cerebellar granule cells: evidence for the phosphorylation of ceramide by different metabolic pathways. *Neurochem Res* 27, 7-8,(Aug, 2002) 711-6, 0364-3190

Rile, G., Yatomi, Y., Takafuta, T. and Ozaki, Y. (2003). Ceramide 1-phosphate formation in neutrophils. *Acta Haematol* 109, 2,(2003) 76-83, 0001-5792

Rovina, P., Graf, C. and Bornancin, F. (2010). Modulation of ceramide metabolism in mouse primary macrophages. *Biochem Biophys Res Commun* 399, 2,(Aug 20, 2010) 150-4, 0006-291X

Sakane, F., Yamada, K. and Kanoh, H. (1989). Different effects of sphingosine, R59022 and anionic amphiphiles on two diacylglycerol kinase isozymes purified from porcine thymus cytosol. *FEBS Lett* 255, 2,(Sep 25, 1989) 409-13, 0014-5793

Schmitz-Peiffer, C. (2002). Protein kinase C and lipid-induced insulin resistance in skeletal muscle. *Ann N Y Acad Sci* 967, Jun, 2002) 146-57, 0077-8923

Serhan, C. N., Haeggstrom, J. Z. and Leslie, C. C. (1996). Lipid mediator networks in cell signaling: update and impact of cytokines. *Faseb J* 10, 10,(Aug, 1996) 1147-58, 0892-6638

Shinghal, R., Scheller, R. H. and Bajjalieh, S. M. (1993). Ceramide 1-phosphate phosphatase activity in brain. *J Neurochem* 61, 6,(Dec, 1993) 2279-85, 0022-3042

Smith, E. R., Jones, P. L., Boss, J. M. and Merrill, A. H., Jr. (1997). Changing J774A.1 cells to new medium perturbs multiple signaling pathways, including the modulation of protein kinase C by endogenous sphingoid bases. *J Biol Chem* 272, 9,(Feb 28, 1997) 5640-6, 0021-9258

Song, M. S. and Posse de Chaves, E. I. (2003). Inhibition of rat sympathetic neuron apoptosis by ceramide. Role of p75NTR in ceramide generation. *Neuropharmacology* 45, 8,(Dec, 2003) 1130-50, 0028-3908

Spiegel, S., Foster, D. and Kolesnick, R. (1996). Signal transduction through lipid second messengers. *Curr Opin Cell Biol* 8, 2,(Apr, 1996) 159-67, 0955-0674

Spiegel, S. and Merrill, A. H., Jr. (1996). Sphingolipid metabolism and cell growth regulation. *Faseb J* 10, 12,(Oct, 1996) 1388-97,

Spiegel, S. and Milstien, S. (2002). Sphingosine 1-phosphate, a key cell signaling molecule. *J Biol Chem* 277, 29,(Jul 19, 2002) 25851-4, 0021-9258

Spiegel, S. and Milstien, S. (2003). Sphingosine-1-phosphate: an enigmatic signalling lipid. *Nat Rev Mol Cell Biol* 4, 5,(May, 2003) 397-407, 1471-0072

Stratford, S., Hoehn, K. L., Liu, F. and Summers, S. A. (2004). Regulation of insulin action by ceramide: dual mechanisms linking ceramide accumulation to the inhibition of Akt/protein kinase B. *J Biol Chem* 279, 35,(Aug 27, 2004) 36608-15, 0021-9258

Subramanian, P., Stahelin, R. V., Szulc, Z., Bielawska, A., Cho, W. and Chalfant, C. E. (2005). Ceramide 1-phosphate acts as a positive allosteric activator of group IVA cytosolic phospholipase A2 alpha and enhances the interaction of the enzyme with phosphatidylcholine. *J Biol Chem* 280, 18,(May 6, 2005) 17601-7, 0021-9258

Sugiura, M., Kono, K., Liu, H., Shimizugawa, T., Minekura, H., Spiegel, S. and Kohama, T. (2002). Ceramide kinase, a novel lipid kinase. Molecular cloning and functional characterization. *J Biol Chem* 277, 26,(Jun 28, 2002) 23294-300, 0021-9258

Tada, E., Toyomura, K., Nakamura, H., Sasaki, H., Saito, T., Kaneko, M., Okuma, Y. and Murayama, T. (2010). Activation of ceramidase and ceramide kinase by vanadate via a tyrosine kinase-mediated pathway. *J Pharmacol Sci* 114, 4,(2010) 420-32, 1347-8613

Testai, F. D., Landek, M. A., Goswami, R., Ahmed, M. and Dawson, G. (2004). Acid sphingomyelinase and inhibition by phosphate ion: role of inhibition by phosphatidyl-myo-inositol 3,4,5-triphosphate in oligodendrocyte cell signaling. *J Neurochem* 89, 3,(May, 2004) 636-44, 0022-3042

Truett, A. P., 3rd and King, L. E., Jr. (1993). Sphingomyelinase D: a pathogenic agent produced by bacteria and arthropods. *Adv Lipid Res* 26, 1993) 275-91, 0065-2849

Tuson, M., Marfany, G. and Gonzalez-Duarte, R. (2004). Mutation of CERKL, a novel human ceramide kinase gene, causes autosomal recessive retinitis pigmentosa (RP26). *Am J Hum Genet* 74, 1,(Jan, 2004) 128-38, 0002-9297

Van Overloop, H., Denizot, Y., Baes, M. and Van Veldhoven, P. P. (2007). On the presence of C2-ceramide in mammalian tissues: possible relationship to etherphospholipids and phosphorylation by ceramide kinase. *Biol Chem* 388, 3,(Mar, 2007) 315-24, 0264-6021

Van Overloop, H., Gijsbers, S. and Van Veldhoven, P. P. (2006). Further characterization of mammalian ceramide kinase: substrate delivery and (stereo)specificity, tissue distribution, and subcellular localization studies. *J Lipid Res* 47, 2,(Feb, 2006) 268-83, 0022-2275

Waggoner, D. W., Gomez-Munoz, A., Dewald, J. and Brindley, D. N. (1996). Phosphatidate phosphohydrolase catalyzes the hydrolysis of ceramide 1-phosphate, lysophosphatidate, and sphingosine 1-phosphate. *J Biol Chem* 271, 28,(Jul 12, 1996) 16506-9, 0021-9258

Wijesinghe, D. S., Lamour, N. F., Gomez-Munoz, A. and Chalfant, C. E. (2007). Ceramide kinase and ceramide-1-phosphate. *Methods Enzymol* 434, 2007) 265-92, 0076-6879

Wijesinghe, D. S., Massiello, A., Subramanian, P., Szulc, Z., Bielawska, A. and Chalfant, C. E. (2005). Substrate specificity of human ceramide kinase. *J Lipid Res* 46, 12,(Dec, 2005) 2706-16, 0022-2275

Wijesinghe, D. S., Subramanian, P., Lamour, N. F., Gentile, L. B., Granado, M. H., Szulc, Z., Bielawska, A., Gomez-Munoz, A. and Chalfant, C. E. (2008). The chain length specificity for the activation of group IV cytosolic phospholipase A2 by ceramide-1-phosphate. Use of the dodecane delivery system for determining lipid-specific effects. *J Lipid Res* 50, 10,(Dec 15, 2008) 1986-95, 0022-2275

Wu, J., Spiegel, S. and Sturgill, T. W. (1995). Sphingosine 1-phosphate rapidly activates the mitogen-activated protein kinase pathway by a G protein-dependent mechanism. *J Biol Chem* 270, 19,(May 12, 1995) 11484-8, 0021-9258

Yamada, K., Sakane, F., Imai, S. and Takemura, H. (1993). Sphingosine activates cellular diacylglycerol kinase in intact Jurkat cells, a human T-cell line. *Biochim Biophys Acta* 1169, 3,(Sep 8, 1993) 217-24, 0005-2736

Zhao, Y., Usatyuk, P. V., Cummings, R., Saatian, B., He, D., Watkins, T., Morris, A., Spannhake, E. W., Brindley, D. N. and Natarajan, V. (2005). Lipid phosphate phosphatase-1 regulates lysophosphatidic acid-induced calcium release, NF-kappaB activation and interleukin-8 secretion in human bronchial epithelial cells. *Biochem J* 385, Pt 2,(Jan 15, 2005) 493-502, 0264-6021

10

Stobadine – An Indole Type Alternative to the Phenolic Antioxidant Reference Trolox

Ivo Juranek, Lucia Rackova and Milan Stefek

Institute of Experimental Pharmacology and Toxicology, Slovak Academy of Sciences
Slovakia

1. Introduction

Treatment of free radical pathologies by antioxidants has been substantiated by studies in animal models of diseases. However, so far the therapy of oxidative stress-related diseases has not found satisfactory application in clinical practice. This may be due to an insufficient efficacy of the antioxidants available, their unsuitable pharmacokinetics, lack of selectivity, presence of adverse side effects, their toxicity, etc. Thus, new antioxidants have to be identified. In numerous studies searching for novel antioxidant compounds, trolox, a water soluble analogue of alpha-tocopherol, is commonly utilized as a reference antioxidant. Chemically, trolox represents a carboxylic acid chromane (Fig. 1A). Due to its good water solubility, this antioxidant has been broadly used as a standard when screening antioxidant efficacy of other prospectively active compounds in studies involving chemical, subcellular, cellular and tissue models of oxidative stress mediated injury (Aruoma, 2003; Huang et al., 2005; Prior et al., 2005). On the other hand, a fairly large and specific group of substances with beneficial antioxidant effects is derived from indole structure (Suzen, 2007). This puts a demand on the reassessment of the suitability of a phenol-type reference trolox, particularly in studies on screening of nitrogen heterocyclic antioxidants.

The pyridoindole stobadine (Fig. 1A) has been found out to be an effective chain-breaking antioxidant scavenging a variety of reactive oxygen species (Kagan et al., 1993; Steenken et al., 1992; Stefek & Benes, 1991). The input of stobadine into the literary data on indole-type antioxidants comprises more than two hundred PubMed references. Several comprehensive reviews cover stobadine action as determined in a variety of models including simple chemical systems, biological models at subcellular, cellular or organ level, followed by extensive studies in vivo in a number of free-radical disease models, and that also in comparison with other drugs (Horakova et al., 1994; Horakova & Stolc, 1998; Juranek et al., 2010; Stolc et al., 1997; Stefek et al., 2010). The main goal of the present paper is to provide an overview on the current data of the indole-type antioxidant stobadine and to compare them with those of the phenol-type antioxidant trolox.

2. Comparison of stobadine with trolox

Structural features, physicochemical properties, mechanism of action and efficiency of stobadine in various models of free radical damage are summarized and compared with those of trolox. Consequently, stobadine may be highlighted as a promising reference antioxidant, which may readily be utilized as a standard in studies testing antioxidative efficacy of other indole-type substances.

A)

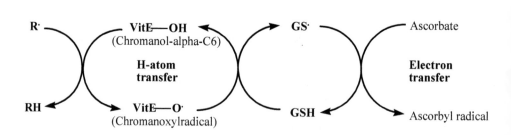

Stobadine   **Trolox**

B)

Fig. 1. Structures of stobadine and trolox (A) and possible mechanisms of free radical scavenging by stobadine and vitamin E/trolox (B).
Biologically relevant coupled reactions that might recycle stobadine (Kagan et al., 1993) and vitamin E/trolox (Davies et al., 1988) are depicted.

2.1 Physico-chemical properties

Trolox, 6-hydroxy-2,5,7,8-tetramethylchroman-2-carboxylic acid, has the character of organic acid (Nonell et al., 1995), while stobadine, (-)-cis-2,8-dimethyl-2,3,4,4a,5,9b-hexahydro-1H-pyrido[4,3b]indole, is an organic base. Despite the fact that trolox is more lipophilic than stobadine (log P values 2.83 and 1.95, respectively), at the physiological pH 7 stobadine preferentially distributes into lipid compartment, while trolox preferentially resides in the water phase. The acidobasic behaviour accounts for this apparent discrepancy. With the pKa value of the carboxyl group 3.89 (Barclay & Vinqvist, 1994), trolox undergoes virtually complete dissociation at physiological pH (99.92% of COO- form of trolox). On the other hand, stobadine with the pKa value of the tertiary nitrogen 8.5 (Stefek et al., 1989) has around 92% of the basic nitrogen in protonated form at pH 7. As a result of the acidobasic equilibrium, the corresponding distribution ratios at pH 7 of trolox and stobadine, D = 0.33 (Barclay et al., 1995) and 3.72 (Kagan et al., 1993), respectively, clearly favour partitioning of stobadine into the lipid phase, yet not that of trolox. This may explain the profound drop of the apparent antioxidant efficiency of trolox in experimental models involving membranous systems (Horakova et al., 2003; Juskova et al., 2010; Rackova et al., 2002; Rackova et al., 2004; Stefek et al., 2008).

2.2 Redox properties

Early pulse radiolysis studies indicated differences with regard both to the centre of antioxidant activity, residing in the indolic nitrogen or phenolic moiety of stobadine or trolox, respectively, and to deprotonation mechanism following the oxidation of the parent molecules (Fig. 1B). It was demonstrated that one-electron oxidation of stobadine leads to the formation of its radical cation (Steenken et al., 1992). That deprotonates from the indolic nitrogen and gives a resonance stabilised nitrogen-centred radical. With regard to the pKa value of approx. -5 of trolox-derived phenoxyl radical cation (Davies et al., 1988) and its expected extremely rapid deprotonation, no spectral evidence for generation of the trolox radical cation was obtained. However, depending on the reaction conditions, electron transfer followed by proton shift or even sequential proton loss and electron transfer (SPLET) has been suggested as a radical scavenging mechanism of phenolic antioxidants involving trolox and alpha-tocopherol (Musialik, 2005; Svanholm et al., 1974).

As shown in Table 1, stobadine and trolox are characterised by comparable rate constants of their interactions with the majority of individual reactive oxygen species tested. The major differences concern the second order rate constants of their reactions with superoxide and hydroxyl radicals. A considerably higher $k_{superoxide}$ value was found for trolox (Nishikimi & Machlim, 1975) than that of stobadine (Kagan et al., 1993) (Table 1). On the other hand, the study by Bielski (1982) showed a notably low second order rate constant of trolox for its reaction with superoxide ($k_{superoxide}$ <0.1 $M^{-1}.s^{-1}$), while Davies et al. (1988) reported an apparent absence of trolox reaction with superoxide.

Regarding the hydroxyl radical scavenging, Davies et al. (1988) reported the value $k_{.OH}$ for trolox to be comparable with that of stobadine (Table 1). Nonetheless, according to the study of Aruoma et al. (1990), the second order rate constant of trolox for scavenging HO· radicals is about one order higher than that of stobadine. These findings are in a good agreement with our data obtained in a study where the efficacy of stobadine and trolox in inhibition of hydroxyl-radical-induced cross-linking of bovine serum albumin were assessed (Kyselova et al., 2003).

Reactive oxygen species	Rate constant $(M^{-1}.s^{-1})$	
	Stobadine	Trolox
HO$\cdot$	7×10^9 15.9×10^9 (Steenken et al., 1992; Stefek & Benes, 1991)	8.5×10^{10} (Aruoma et al., 1990)
CH$_3$COO$\cdot$ Cl$_3$COO$\cdot$	$< 5 \times 10^6$ 6.6×10^8 (Steenken et al., 1992)	2.5×10^6 3.7×10^8 (Simic, 1980; Davies et al., 1988)
DPPH$\cdot$	4.9×10^2 (Rackova et al., 2004)	1.6×10^3 (Rackova et al., 2004)
C$_6$H$_6$O$\cdot$	5.1×10^8 (Steenken et al., 1992)	4.1×10^8 (Davies et al., 1988)
O$_2{}^{\cdot-}$	7.5×10^2 (Kagan et al., 1993)	1.7×10^4 <0.1 (Nishikimi & Machlin, 1975; Bielski, 1982)
^{1}O$_2$	1.3×10^8 (Steenken et al., 1992)	pH6 3.5×10^8 (Nonell et al., 1995)

Table 1. Second-order rate constants of stobadine and trolox interaction with reactive oxygen species and 1,1`-diphenyl-2-picrylhydrazyl (DPPH) stable free radical.

Redox potential of stobadine (E = 0.58 V) (Steenken et al., 1992) is more positive than that of vitamin E (E = 0.48 V) (Neta & Steenken, 1982). Hence, at pH 7, the stobadine radical, formed as a consequence of stobadine free-radical-scavenging activity, may subtract proton from the trolox molecule resulting in regeneration of the parent stobadine molecule. Indeed, Steenken et al. (1992) in their pulse radiolytic study demonstrated the ability of trolox to recycle stobadine from its one-electron oxidation product, to give a corresponding trolox phenoxyl radical. When stobadine and trolox were present simultaneously in oxidatively stressed liposomes, trolox spared stobadine in the system in a dose-dependent manner (Rackova et al., 2002). Direct interaction of trolox with the stobadinyl radical resulting in the recovery of parent stobadine molecule appears to be a plausible mechanism. Thus, under physiological conditions, the antioxidant activity of stobadine may be potentiated by vitamin E. In a good agreement with this idea, Horakova et al. (1992) showed that the antioxidant action of stobadine was profoundly diminished in tocopherol-deficient rat liver microsomes.

Analogically, in biological systems, vitamin E (E = 0.48 V) can be regenerated from its phenoxyl radical via the interaction with ascorbate (Davies et al., 1988), which possesses a more negative redox potential (E = 0.30 V) (Neta & Steenken, 1982); depicted in Figure 1B. In a similar way, the stobadinyl radical was shown to be quenched by ascorbate, as demonstrated by the increased magnitude of the ascorbyl radical ESR signal generated in the presence of stobadine in the system of lipoxygenase + arachidonate (Kagan et al., 1993). Hence, one may expect that in biological systems, the antioxidant potency of both trolox and stobadine may be modulated by their mutual interactions with other lipid- or water-soluble antioxidants.

2.3 Antioxidant efficacies in various assay systems

In a homogeneous system, antioxidant activity stems from an intrinsic chemical reactivity towards radicals. In membranes, however, the reactivity may differ as there are additional factors involved, such as a relative location of the antioxidant and radicals, ruled predominantly by their distribution ratios between water and lipid compartments. As already mentioned, a notably lower distribution ratio of trolox compared to that of stobadine may account for their different efficacies in systems involving lipid interface (membranes) in comparison to homogenous units (true solutions).

In the ethanolic solution, trolox scavenged 1,1'-diphenyl-2-picrylhydrazyl (DPPH) radical more efficiently than did stobadine, based on the initial velocity measurements (Rackova et al., 2002). The finding was corroborated by the respective rate constants (Rackova et al., 2004) as shown in Table 2.

In the models of oxidative damage comprising soluble proteins in buffer solutions, the water-soluble antioxidants stobadine and trolox have free access both to free radical initiator and to protein-derived radicals. Stobadine inhibited the process of albumin cross-linking due to the oxidative modifications induced by the Fenton reaction system of $Fe^{2+}/EDTA/H_2O_2/ascorbate$ less effectively than did trolox (Kyselova et al., 2003). The experimental IC_{50} values correlated well with the reciprocal values of the corresponding second order rate constants for scavenging OH radicals.

Trolox, in comparison with stobadine, was found to be a more potent inhibitor of 2,2'-azobis-2-amidinopropane (AAPH)-induced precipitation of the soluble eye lens proteins (Stefek et al., 2005). In contrary, production of free carbonyls due to protein oxidation was more efficiently inhibited by stobadine. Both stobadine and trolox showed comparable efficacies in an experimental glycation model in preventing glycation-related fluorescence changes of bovine serum albumin as well as in lowering the yield of 2,4-dinitrophenylhydrazine-reactive carbonyls as markers of glyco-oxidation (Table 2) (Stefek et al., 1999).

On the other hand, trolox was found to be much less effective in inhibiting AAPH-induced peroxidation of di-oleoyl-phosphatidylcholine (DOPC) liposomes with respect to stobadine (Rackova et al., 2004; Rackova et al., 2006; Stefek et al., 2008), as exemplified by the respective IC_{50} values 25.3 and 93.5 µM, shown in Table 2. Stobadine, in comparison with trolox, more effectively prolonged the lag phase of Cu^{2+}-induced low-density lipoprotein (LDL) oxidation measured by diene formation (Horakova et al., 1996). The same pattern of efficacy in prevention of the lipid oxidation boost was shown in the system of tissue homogenate. Stobadine showed a more potent inhibitory effect than trolox on lipid peroxidation in rat brain homogenates exposed to $Fe^{2+}/ascorbate$ as documented by thiobarbituric acid reactive substances (TBARS) levels (Table 2; Horakova et al., 2000). Interestingly, in the case of alloxan-induced lipid peroxidation of heat denaturized rat liver microsomes, the inhibitory efficacy of stobadine and trolox was comparable (Stefek & Trnkova, 1996). This finding may indicate that the critical competition of the scavengers with the alloxan-derived initiating reactive oxygen species takes place outside the membrane in the bulk solution.

Assay system		Parameter measured	Stobadine	Trolox
AAPH induced LPO in DOPC liposomes (Rackova et al., 2002)		IC_{50} (μmol/L)	25.3 ± 14.6	93.5 ± 8.5
BSA cross-linking induced Fe^{2+}/EDTA/H_2O_2/ascorbate (Kyselova et al., 2003)		IC_{50} (μmol/L)	0.65 ± 0.08	0.13 ± 0.02
AAPH-induced oxidative modification of soluble eye lens proteins (Stefek et al., 2005)	Inhibition of protein precipitation	IC_{50} (μmol/L)	121 ± 15	79 ± 8
	Inhibition of protein oxidation		44 ± 8	131 ± 20
Oxidative modification of BSA in an experimental glycation model (Stefek et al., 1999)	Glucose attachment into the molecule of BSA	Amadori product (with respect to 8.2 ± 0.4 nmol/mg BSA for control without inhibitor)	8.1 ± 0.5 (0.25 mmol/L)	7.4 ± 0.7 (0.25 mmol/L)
	Glycation-induced fluorescence changes of BSA	Relative fluorescence (with respect to 11.2 ± 0.7 nmol/mg BSA for control without inhibitor)	7.9 ± 0.7 (0.25 mmol/L)	6.5 ± 0.4 (0.25 mmol/L)
	Formation of DNPH-reactive carbonyl groups in BSA	Carbonyl groups (with respect to 5.6 ± 0.4 nmol/mg BSA for control without inhibitor)	3.4 ± 0.5 (0.25 mmol/L)	3.3 ± 0.2 (0.25 mmol/L)
Cu^{2+}-mediated oxidation of LDL (Horakova et al., 1996)		Δt_{lag} (min) (The increase in lag time given by one stobadine molecule per single LDL particle)	1.5	0.38
Fe^{2+}/ascorbate induced oxidative damage of rat brain homogenate (Horakova et al., 2000)	Inhibition of TBARS production	IC_{50} (μmol/L)	35	98

Assay system	Parameter measured	Stobadine	Trolox
AAPH-induced haemolysis of rat erythrocytes (Juskova et al., 2010)	t_{lag} (min) (88.6 ± 2.2 for control erythrocytes)	> 300 (100 µmol/L)	144 (100 µmol/L)

LPO, lipid peroxidation; DOPC, dioleoyl phosphatidylcholine; BSA, bovine serum albumin; AAPH, 2,2`-azobis (2-amidinopropane)hydrochloride; DNPH, dinitrophenylhydrazine; TBARS, thiobarbituric acid reactive substances.

Table 2. Summary of antioxidant and protective efficacies of stobadine and trolox in experimental models of oxidative damage.

In the cellular system of intact erythrocytes exposed to peroxyl radicals generated by thermal degradation of the azoinitiator AAPH in vitro, stobadine, in comparison to trolox, protected more powerfully erythrocytes from haemolysis, as shown (Table 2) by the respective lag phase prolongations (Juskova et al., 2010). In another cellular model, stobadine increased the viability of hydrogen-peroxide treated PC12 cells more effectively than did trolox, while both compounds reduced the content of malondialdehyde with a comparable efficiency (Horakova et al., 2003).

3. Conclusion

On balance then, the present paper, by summarizing the current data on both trolox and stobadine, underscores the structural and physicochemical differences between the two compounds as respective representatives of phenolic- and indole-type antioxidants. The structural variance explains their different mechanisms of antioxidant action and variable efficacies in the range of assay systems studied. Considering a plethora of studies reported on stobadine antioxidant action, physicochemical properties, and on a variety of its other biological activities, stobadine may represent a pertinent indole-type reference antioxidant. Hence, in studies of indole compounds, stobadine antioxidant standard may be utilized as a more relevant alternative to structurally diverse trolox.

4. Acknowledgements

Financial support by grants of the EU COST Programme (Action CM1001) and the National Agencies VEGA (2/0083/09, 2/0067/11 and 2/0149/12) and APVV (51-017905) is gratefully acknowledged.

5. References

Aruoma, O.I. (2003). Methodological considerations for characterizing potential antioxidant actions of bioactive components in plant foods, *Mutation Research* Vol. 523-524: 9–20.

Aruoma, O.I., Evans, P.J., Kaur, H., Sutcliffe, L., & Halliwell, B. (1990). An evaluation of the antioxidant and potential pro-oxidant properties of food additives and of Trolox C, vitamin E and probucol, *Free Radical Research Communication* Vol. 10: 143–157.

Barclay, L.R., Artz, J.D., & Mowat, J.J. (1995). Partitioning and antioxidant action of the water-soluble antioxidant, Trolox, between the aqueous and lipid phases of phosphatidylcholine membranes: ^{14}C tracer and product studies, *Biochimica et Biophysica Acta* Vol. 1237: 77–85.

Barclay, L.R.C., & Vinqvist, M.R. (1994). Membrane peroxidation: inhibiting effects of watersoluble antioxidants on phospholipids of different charge types, *Free Radical Biology & Medicine* Vol. 16: 779–788.

Bielski, B.H.J. (1983). Evaluation of the reactivities of $HO_2\cdot/O_2^{-\cdot}$ with compounds of biological interest. In Oxygen radicals and their scavenger systems, Vol I (ed. G. Cohen and RE Greenwald), Amsterdam: Elsevier, pp. 1–7.

Davies, M.J., Fornit, L.G., & Willson, R.L. (1988). Vitamin E Analogue Trolox C. E.S.R. and Pulse-Radiolysis Studies of Free-Radical Reactions, *The Biochemical Journal* Vol. 255:513–522.

Horakova, L., Briviba, K., & Sies, H. (1992a). Antioxidant activity of the pyridoindole stobadine in liposomal and microsomal lipid peroxidation, *Chemico–Biological Interactions* Vol. 83, 85–93.

Horakova, L., Giessauf, A., Raber, G., & Esterbauer, H. (1996). Effect of stobadine on Cu $^{(++)}$ - mediated oxidation of low-density lipoprotein, *Biochemical Pharmacology* Vol. 51(10): 1277–1282.

Horakova, L., Licht, A., Sandig, G., Jakstadt, M., Durackova, Z., & Grune, T. (2003). Standardized extracts of flavonoids increase the viability of PC12 cells treated with hydrogen peroxide: effects on oxidative injury, *Archives of Toxicology* Vol. 77(1): 22–9.

Horakova, L, Ondrejickova, O., Bachrata, K., & Vajdova, M. (2000). Preventive effect of several antioxidants after oxidative stress on rat brain homogenates, *General Physiology and Biophysics* Vol. 19(2): 195–205.

Horakova, L., Sies, H., & Steenken S. (1994). Antioxidant action of stobadine, *Methods in Enzymology* (Edited by Packer L.), Part D, Academic Press, San Diego, Vol. 234, pp. 572–580.

Horakova, L., & Stolc, S. (1998) Antioxidant and pharmacodynamic effects of pyridoindole stobadine, *General Pharmacology* Vol. 30: 627–638.

Huang, D., Ou, B., & Prior, R.L. (2005). The chemistry behind antioxidant capacity assays, *Journal of Agricultural and Food Chemistry* Vol. 53: 1841–1856.

Juranek, I., Horakova, L., Rackova, L., & Stefek, M. (2010). Antioxidants in Treating Pathologies Involving Oxidative Damage: An Update on Medicinal Chemistry and Biological Activity of Stobadine and Related Pyridoindoles, *Current Medicinal Chemistry* Vol. 17: 552–570.

Juskova, M., Snirc, V., Krizanova, L., & Stefek, M. (2010). Effect of carboxymethylated pyridoindoles on free radical-induced haemolysis of rat erythrocytes *in vitro*, *Acta Biochimica Polonica* Vol. 57(2): 153–6.

Kagan, V.E., Tsuchiya, M., Serbinova, E., Packer, L., & Sies, H. (1993). Interaction of the pyridoindole stobadine with peroxyl, superoxide and chromanoxyl radicals, *Biochemical Pharmacology* Vol. 45(2): 393–400.

Kyselova, Z., Rackova, L., & Stefek, M. (2003). Pyridoindole antioxidant stobadine protected bovine serum albumin against the hydroxyl radical mediated cross-linking in vitro, *Archives of Gerontology and Geriatrics* Vol. 36: 221–229.

Musialik, M. & Litwinienko, G. (2005). Scavenging of dpph$^\cdot$ Radicals by Vitamin E Is Accelerated by Its Partial Ionization: the Role of Sequential Proton Loss Electron Transfer, *Organic Letters* Vol. 7(22): 4951–4954.

Neta, P., & Steenken, S. (1982). One electron redox potentials of phenols, hydroxy- and aminophenols and related compounds of biological interest, *The Journal of Physical Chemistry* Vol. 86: 3661–3667.

Nishikimi, M., & Machlin, L.J. (1975). Oxidation of alpha-tocopherol model compound by superoxide anion, *Archives of Biochemistry and Biophysics* Vol. 170(2):684–689.

Nonell, S., Moncayo, L., Trull, F., Amat-Guerri, F., Lissi, E.A., Soltermann, A.T., Criado, S., & Garcia, N.A. (1995). Solvent influence on the kinetics of the photodynamic degradation of trolox, a water-soluble model compound for vitamin E, *Journal of Photochemistry and Photobiology B: Biology* Vol. 29: 157–162.

Prior, R.L., Wu, X., & Schaich, K. (2005). Standardized methods for the determination of antioxidant capacity and phenolics in foods and dietary supplements, *Journal of Agricultural and Food Chemistry* Vol. 53: 4290–4302.

Rackova, L., Majekova, M., Kostalova, D., & Stefek, M. (2004). Antiradical and antioxidant activities of alkaloids isolated from Mahonia aquifolium. Structural aspects, *Bioorganic & Medicinal Chemistry* Vol.12: 4709–4715.

Rackova, L., Snirc, V., Majekova, M., Majek, P., & Stefek, M. (2006). Free Radical Scavenging and Antioxidant Activities of Substituted Hexahydropyridoindoles.Quantitative Structure-Activity Relationships, *Journal of Medicinal Chemistry* Vol. 49; 2543–2548.

Rackova, L., Stefek, M., & Majekova, M. (2002). Structural aspects of antioxidant activity of substituted pyridoindoles, *Redox Report* Vol. 7: 207–214.

Simic MG. (1980). Peroxyl radical from oleic acid. In Simic, MG. (ed.), *Autoxidation in Food and Biological Systems*. Plenum, New York, pp. 17–26.

Steenken, S., Sunquist, A.R., Jovanovic, S.V., Crockett, R., & Sies, H. (1992). Antioxidant activity of the pyridoindole stobadine. Pulse radiolytic characterization of one-electron-oxidized stobadine and quenching of singlet molecular oxygen, *Chemical Research in Toxicology* Vol. 5: 355–360.

Stefek, M., Benes, L., & Zelnik, V. (1989). N-oxygenation of stobadine, a gamma-carboline antiarrhythmic and cardioprotective agent: the role of flavin-containing monooxygenase. *Xenobiotica* Vol. 19: 143–50.

Stefek, M., & Benes, L. (1991). Pyridoindole stobadine is a potent scavenger of hydroxyl radicals, *FEBS Letters* Vol. 294: 264–266.

Stefek, M., & Trnkova, Z. (1996). The pyridoindole antioxidant stobadine prevents alloxan-induced lipid peroxidation by inhibiting its propagation, *Pharmacology & Toxicology* Vol. 78(2):77–81.

Stefek, M., Krizanova, L., & Trnkova, Z. (1999). Oxidative modification of serum albumin in an experimental glycation model of diabetes mellitus in vitro: effect of the pyridoindole antioxidant stobadine, *Life Sciences* Vol. 65(18–19):1995–1997.

Stefek, M., Kyselova, Z., Rackova, L., & Krizanova, L. (2005). Oxidative modification of rat lens proteins by peroxyl radicals *in vitro*: Protection by the chain-breaking antioxidants Stobadine and Trolox. *Biochimica et Biophysica Acta–Molecular Basis of Disease* Vol. 1741(1-2): 183–190.

Stefek, M., Snirc, V., Djoubissie, P.-O., Majekova, M., Demopoulos, V., Rackova, L., Bezakova, Z., Karasu, C., Carbone, V., & El-Kabbani, O. (2008). Carboxymethylated pyridoindole antioxidants as aldose reductase inhibitors: Synthesis, activity, partitioning, and molecular modelling, *Bioorganic & Medicinal Chemistry* Vol. 16: 4908–4920.

Advances in Biochemistry

Stefek, M., Gajdosik, A., Gajdosikova, A., Juskova, M., Kyselova, Z., Majekova, M. Rackova, L., Snirc, V., Demopoulos, V.J. & Karasu, Ç. (2010). Stobadine and related pyridoindoles as antioxidants and aldose reductase inhibitors in prevention of diabetic complications. In Stefek, M. (ed.), *Advances in molecular mechanisms and pharmacology of diabetic complications*. Transworld Research Network, Kerala, pp. 285-307.
Stolc, S., Vlkolinsky, R., & Pavlasek, J. (1997). Neuroprotection by the pyridoindole stobadine: A minireview, *Brain Research Bulletin* Vol. 42: 335-340.
Suzen, S. (2007). Antioxidant Activities of Synthetic Indole Derivatives and Possible Activity Mechanisms, *Topics in Heterocyclic Chemistry* Vol. 11: 145-178.
Svanholm, U., Beckgaard, K., & Parker, V.D. (1974). Electrochemistry in media of intermediate activity. VIII. Reversible oxidation products of α-tocopherol model compound. Cation radical, cation, and dication, *Journal of the American Chemical Society* Vol. 96: 2409-2413.

11

Cholesterol: Biosynthesis, Functional Diversity, Homeostasis and Regulation by Natural Products

J. Thomas[1], T.P. Shentu[1] and Dev K. Singh[2*]
[1]Department of Medicine, University of Illinois, Chicago
[2]Division of Developmental Biology, Department of Pediatrics
Children's Hospital of University of Illinois, University of Illinois at Chicago
USA

1. Introduction

Most of the discussions on cardiovascular diseases include the relative level of total plasma cholesterol whereas complete lipid profile is very important as clinical diagnostics for cardiovascular risk. For example, the status of triglycerides, low density lipoprotein cholesterol (LDL cholesterol), high density lipoprotein cholesterol (HDL cholesterol), thyroid functions, insulin and lipid peroxides etc as these lipids are related to cardiovascular disease either as markers of another underlying disturbance or along the same pathways of cholesterol metabolism. According to latest World Health Organization (WHO) report about 50% of the heart attacks occur in individual with high level of cholesterol. The most abundant sterol in animal system is in the form of cholesterol whereas plants lack cholesterol but they contain structurally similar other sterol and similar biosynthetic pathway exist both in plants and animals as well as some prokaryotes also synthesize some specific sterols. In 1948, the Framingham Heart Study - under the direction of the National Heart Institute (now known as the National Heart, Lung, and Blood Institute or NHLBI) - embarked on an ambitious project in health research. At the time, little was known about the general causes of heart disease and stroke, but the death rates for CVD had been increasing steadily since the beginning of the century and had become an American epidemic. The Framingham Heart Study became a joint project of the National Heart, Lung and Blood Institute and Boston University. The concern about cholesterol was largely fueled by this study and others that provided strong evidence that when large populations are observed, persons with higher than average serum total cholesterol have a higher incidence of coronary artery disease (CAD). Laboratory reports often mention two main types of cholesterol, the HDL cholesterol (often termed the "good" cholesterol) and LDL cholesterol (often mis-named the "bad" one). Even if the total LDL is lowered, the important fraction is actually the small LDL, which is more easily oxidized into a potentially atherogenic particle than its larger, more buoyant counterpart. Bacteria or other infectious agents are being looked at as part of the culprits as causative factors in initiating injury to the arterial wall. Cholesterol is then attracted to this 'rough' site on the blood vessel

* Corresponding Author

wall in an attempt to heal the wall so that blood will flow smoothly over the injured area. Cholesterol itself is not the cause of CAD. The blood cholesterol is rather only a reflection of other metabolic imbalances in the vast majority of cases. If we assume blood lipid (fats) status are associated with some risk of cardiovascular disease. The question is which lipids are the important markers, and even more important what should we do about them.

Lowering cholesterol too aggressively or in artificial circumstances or having too low total cholesterol is also undesirable. Plasma cholesterol level below 200mg% is desirable whereas as 200-239mg% and above 240mg% is considered as borderline and high level of cholesterol, respectively. Ingested cholesterol comes from animal sources (plants and prokaryotes do not contain cholesterol (Gylling and Miettinen, 1995) such as eggs, meat, dairy products, fish, and shellfish or biosynthesized from the breakdown of carbohydrates, lipids, or proteins available in the food. One study has estimated that the complete abolition of dietary cholesterol absorption would reduce plasma cholesterol by up to 62% (Gylling and Miettinen, 1995). About 50% of dietary cholesterol is absorbed through intestinal enterocytes, while the rest is excreted through feces (Ostlund et al., 1999). It is estimated that half of ingested cholesterol enters the body while the other half excreated in the feces. Normally, the more cholesterol we absorb, the less our bodies make. There is slight increase in plasma cholesterol with increase in the amount of cholesterol ingested each day, usually is not changed more than ± 15 percent by altering the amount of cholesterol in the diet. Although the response of individuals differs markedly. Cholesterol is integeral part of membranes and perform a number of vital functions in the cell and due to this property of cholesterol, each cell has the capability to biosynthesize cholesterol if it is required. Cholesterol is a component of steroid hormones, including pregnenolone, estrogens, progesterone, testosterone, vitamin D and bile acids. Bile acids are involved in lipid digestion, absorption, and excretion.

In this chapter, mode of intracellular and extracellular cholesterol transport through acceptors-donors and thereafter cholesterol trafficking pathways will be described in detail. Furthermore, we will discuss the regulation of cholesterol at enzymatic/transcriptional level and diverse functions of cholesterol in our body. Taken together, this book chapter will address recent advances in cholesterol metabolism both *in vitro* and *in vivo* models related to absorptions, biosynthesis, transport, excretion and therapeutic targets for new drugs and natural compounds.

1.1 Cholesterol biosynthetic pathway

As early as 1926, studies by Heilbron, Kamm and Owens suggested that squalene is precursor of cholesterol biosynthesis (Garrett & Grisham, 2007). In the same year, H.J. Channon, demonstrated first time that animals fed on shark oil produced more cholesterol in the tissues. In 1940, Bloch and Rittenberg, first time demonstrated that mice fed on radiolabled acetate showed significant radiolabeled cholesterol (Bloch et al., 1945; Kresge et al., 2005). In 1952, Konard Bloch and Robert Langdon showed conclusively that squalene as well as cholesterol are synthesized from acetate for which Fyodor Lynen and Bloch were awarded the Noble Prize in Medicine/Physiology in 1964. Cholesterol is biosynthesized from 2-carbon metabolic intermediate, acetyl-CoA hooked end to end involving a number of enzymatic reactions and finally get converted into the 27-carbon molecule of cholesterol. Metabolism (catabolism) of lipids, carbohydrates and proteins lead to the formation Acetyl-CoA. Proteins are generally are not catabolized for the purpose of energy and usually broken down into amino acids for denovo protein biosynthesis, under excessive protein consumption or during certain disease states, certain proteins can be catabolized to acetyl-CoA. Non-essential fatty acids, trans-fatty acids, and saturated fats, and refined carbohydrates are general source of excessive acetyl-CoAwhich pressurize our body to biosynthesize cholesterol. In other words, cholesterol is formed from excess calories which usually are generated most often from carbohydrates and fats.

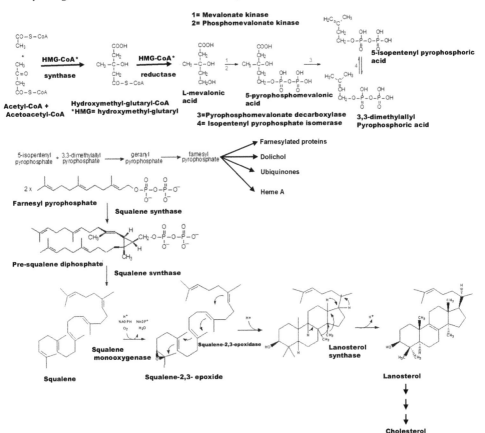

Fig. 1. Cholesterol Biosynthetic Pathway

The process of cholesterol synthesis has five major steps:

1. Acetyl-CoAs are converted to 3-hydroxy-3-methylglutaryl-CoA (HMG-CoA)
2. HMG-CoA is converted to mevalonate
3. Mevalonate is converted to the isoprene based molecule, isopentenyl pyrophosphate (IPP), with the concomitant loss of CO_2
4. IPP is converted to squalene
5. Squalene is converted to cholesterol.

Acetyl-CoA units are converted to mevalonate by a series of reactions that begins with the formation of **HMG-CoA (Figure 1).** Unlike the HMG-CoA formed during ketone body synthesis in the mitochondria, this form is synthesized in the cytoplasm. However, the pathway and the necessary enzymes are the same as those in the mitochondria. Two moles of acetyl-CoA are condensed in a reversal of the thiolase reaction, forming acetoacetyl-CoA. Acetoacetyl-CoA and a third mole of acetyl-CoA are converted to HMG-CoA by the action of HMG-CoA synthase. HMG-CoA is converted to mevalonate by HMG-CoA reductase, HMGR (this enzyme is bound in the endoplasmic reticulum, ER). HMGR absolutely requires NADPH as a cofactor and two moles of NADPH are consumed during the conversion of HMG-CoA to mevalonate. The reaction catalyzed by HMGR is the rate limiting step of cholesterol biosynthesis, and this enzyme is subject to complex regulatory controls which will be discussed in separate section of this book chapter.

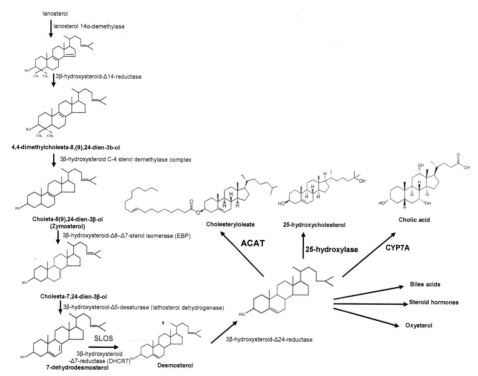

Fig. 2. Post squalene pathway of cholesterol and other sterol Biosynthesis

Mevalonate is then activated by three successive phosphorylations, yielding 5-pyrophosphomevalonate. Phosphorylation mevalonate and successive reactions maintain its solubility, since otherwise these are insoluble in water. After phosphorylation, an ATP-dependent decarboxylation yields isopentenyl pyrophosphate, IPP, an activated isoprenoid molecule. Isopentenyl pyrophosphate is in equilibrium with its isomer, dimethylallyl pyrophosphate, DMPP. One molecule of IPP condenses with one molecule of DMPP to generate geranyl pyrophosphate, GPP. GPP further condenses with another IPP molecule to yield farnesyl pyrophosphate, FPP. Finally, the NADPH-requiring enzyme, squalene synthase catalyzes the head-to-tail condensation of two molecules of FPP, yielding squalene (squalene synthase also is tightly associated with the endoplasmic reticulum). Squalene undergoes a two step cyclization to yield lanosterol catalyzed by sequalene monooxygenase and sequalene 2, 3 epoxidase enzymes. Sequalene mono oxygenase is the second committed step in cholesterol biosynthesis and lead to the formation squalene 2, 3 epoxide. This enzymatic reaction require supernatant protein factor (SPF) and NADPH as a cofactor to introduce molecular oxygen as an epoxide at the 2, 3 position of squalene. The activity of supernatant protein factor itself is regulated by phosphorylation/dephosphorylation (Singh et al., 2003). Through a series of 19 additional reactions, lanosterol is converted to cholesterol. The first sterol intermediate, lanosterol, is formed by the condensation of the 30 carbon isoprenoid squalene as explained above and figure 1., and subsequent enzymatic reactions define the 'post-squalene' half of the pathway figure 2.

The conversion of lanosterol to cholesterol involves the reduction of the C-24 double bond, removal of three methyl groups at the C-14 and C-4 positions, and 'migration' of the C-8(9) double bond (Figure 2) (for a review, see (Herman, 2003)). Some of the enzymatic reactions must occur in sequence; for example, Δ^8-Δ^7 isomerization cannot precede C-14α demethylation. The saturation of the C-24 double bond of lanosterol can occur at multiple points in the pathway, creating two immediate precursors for cholesterol, desmosterol [cholesta-5(6), 24-dien-3β-ol] and 7-dehydrocholesterol (7DHC), whose relative abundance may vary among different tissues. Desmosterol, in particular, appears to be abundant in the developing mammalian brain (Herman, 2003). Several post-squalene sterol intermediates serve additional cellular functions as well. The C-14 demethylated derivatives of lanosterol, 4,4-dimethyl-5α-cholesta-8,14,24-trien-3β-ol and 4,4-dimethyl-5α-cholesta-8,24-dien-3β-ol, have meiosis-stimulating activity and accumulate in the ovary and testis, respectively (Rozman et al., 2002). 7-Dehydrocholesterol is the immediate precursor for vitamin D synthesis. Selected human enzymes of post-squalene cholesterol biosynthesis have also been identified based on homology to sterol biosynthetic enzymes from *Arabidopsis thaliana* (Herman, 2003; Waterham et al., 2001). In addition, cholesterol and other sterol intermediates can be converted to oxysterols that can act as regulatory signaling molecules and bind orphan nuclear receptors such as LXRα (Fitzgerald et al., 2002).

Normal healthy adults synthesize cholesterol at a rate of approximately 1g/day and consume approximately 0.3g/day. A relatively constant level of cholesterol in the body (150 - 200 mg/dL) is maintained primarily by controlling the level of *de novo* synthesis which is partly regulated in part by the dietary intake of cholesterol. Cholesterol from both diet and synthesis is utilized in the formation of membranes and in the synthesis of the steroid hormones and bile acids. The greatest proportion of cholesterol is used in bile acid synthesis.

1.1.1 Regulation of cholesterol biosynthetic pathway

Regulation of the pathway includes sterol-mediated feedback of transcription of several of the genes, including the rate-limiting enzyme HMGR and HMG-CoA synthase, as well as a variety of post-transcriptional mechanisms. Recently, increased attention has been focused on the regulated degradation of HMGR in the ER. The cellular supply of cholesterol is maintained at a steady level by four distinct mechanisms:

1. Regulation of HMGR activity and levels (upstream regulation of cholesterol biosynthesis)
2. Regulation of sequalene mono-oxygenase activity by sequalene supernant protein factor (down stream biosynthesis of cholesterol). Once this step is taken place then it ends up biosynthesis of cholesterol.

3.   Regulation of excess intracellular free cholesterol through the activity of acyl-CoA: cholesterol acyltransferase, ACAT (internalization of excessive cholesterol).
4.   Regulation of plasma cholesterol levels via LDL receptor-mediated uptake and HDL-mediated reverse transport.

The first seven enzymes of cholesterol biosynthesis are soluble proteins with the exception of 3-hydroxy-3-methylglutaryl CoA reductase (HMGR), which is an integral endoplasmic reticulum (ER) membrane protein (Gaylor, 2002; Goldstein and Brown, 1990; Kovacs et al., 2002). HMG-CoA for cholesterol biosynthesis is generated within the cytosol. It is also synthesized within the mitochondria where its hydrolysis by HMG-CoA lyase generates ketones for energy during fasting (Herman, 2003). Reactions generating mevalonate can also occur within the peroxisome, and the subsequent reactions that result in the production of farnesyl-pyrophosphate are exclusively peroxisomal. The remainder of the biosynthetic reactions occurs in the ER with enzymes and substrates that are membrane-bound. Thus, cholesterol biosynthetic enzymes are compartmentalized in the cytosol, ER and/or peroxisome, adding another level of complexity to the regulation of this metabolic pathway. Particularly, direct regulation of HMGCoA reductase activity means for controlling the level of cholesterol biosynthesis. Recently, increased attention has been focused on the regulated degradation of HMGR in the ER.

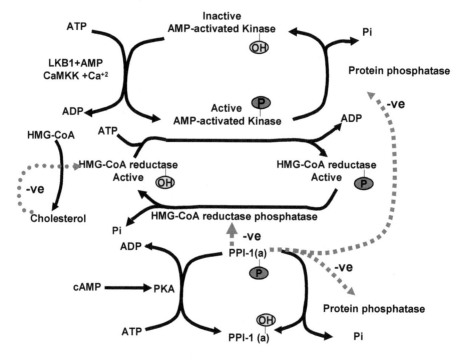

Fig. 3. Regulation of HMGCoA reductase

In vivo system, the activity of HMGCoA reductase is controlled by four distinct mechanisms: feed-back inhibition, control of gene expression, rate of enzyme degradation and phosphorylation-dephosphorylation. Cholesterol controls the first three mechanisms itself as cholesterol acts as a feed-back inhibitor of pre-existing HMGR as well as inducing rapid degradation of the enzyme. The latter is the result of cholesterol-induced polyubiquitination of HMGR and its degradation in the proteosome (explained in separate section in details). This ability of cholesterol is a consequence of the **sterol sensing domain, SSD** of HMGR. In addition, when cholesterol is in excess the amount of mRNA for HMGR is reduced as a result of decreased expression of the gene. By covalent modification of HMGreductase through the process phosphorylation/ dephosphorylation plays a very important role in regulation of HMGR. The enzyme is most active in its unmodified form. Phosphorylation of the enzyme decreases its activity (Figure 3). HMGR is phosphorylated by adenosine mono phosphate-activated protein kinase, **(AMPK)which is** itself activated via phosphorylation catalyzed by 2 enzymes. LKB1 is the primary kinase sensitive to rising AMP levels which was first identified as a gene in humans carrying an autosomal dominant mutation in Peutz-Jeghers syndrome, PJS and mutated in lung adenocarcinomas. The second AMPK phosphorylating enzyme is calmodulin-dependent protein kinase kinase-beta (CaMKKβ) which induces phosphorylation of AMPK due to increase in the level of intracellular Ca^{2+} as a result of muscle contraction. The activity of HMGR is additionally controlled by the cAMP signaling pathway (Figure 3). Enhanced level of cAMP lead to activation of cAMP-dependent protein kinase, PKA. In the context of HMGR regulation, PKA phosphorylates phosphoprotein phosphatase inhibitor-1 (PPI-1) leading to an increase in its activity. PPI-1 can inhibit the activity of numerous phosphatases including protein phosphatase 2C (PP2C) and HMG-CoA reductase phosphatase which remove phosphates from AMPK and HMGR, respectively. This maintains AMPK in the phosphorylated and active state, and HMGR in the phosphorylated and inactive state. As the stimulus leading to increased cAMP production is removed, the level of phosphorylations decreases and that of dephosphorylations increases. The net result is a return to a higher level of HMGR activity. The intracellular level of cAMP itself is regulated by hormonal stimuli, thus regulation of cholesterol biosynthesis is hormonally controlled. One of the most common hormone, insulin leads to a decrease in cAMP, which in turn activates cholesterol biosynthesis. Alternatively, glucagon and epinephrine increase the level of cAMP in turn lead to inhibition of cholesterol biosynthesis. The basic function of epinephrine and glucagon hormones is to control the availability and delivery of energy in all the cells in our body. Degradation of HMGR and inhibition of its biosynthesis, are the two long term regulatory processes for cholesterol biosynthesis as when the levels of cholesterol are high, resulted in reduction in the expression of the HMGR gene. On other hand, low levels of cholesterol activate its expression.

1.1.2 Proteolytic regulation of HMG-CoA

The stability of HMGR is regulated as the rate of flux through the mevalonate synthesis pathway changes. When the flux is high the rate of HMGR degradation is also high. When the flux is low, degradation of HMGR decreases. This phenomenon can easily be observed in the presence of the statin drugs. HMGR is localized to the ER and like SREBP contains a sterol-sensing domain, SSD. When sterol levels increase in cells there is a concomitant increase in the rate of HMGR degradation. The degradation of HMGR occurs within the proteosome, multiprotein complex dedicated for protein degradation.The primary signal directing proteins to the proteosome is ubiquitination. Ubiquitin is a 7.6kDa protein that is covalently attached to proteins targeted for degradation by ubiquitin ligases (Kimura and Tanaka, 2010). These enzymes attach multiple copies of ubiquitin allowing for recognition by the proteosome. HMGR has been shown to be ubiquitinated prior to its degradation. The primary sterol regulating HMGR degradation is cholesterol itself. As the levels of free cholesterol increase in cells, the rate of HMGR degradation increases.

1.2 Disorders of post-sequalene cholesterol biosynthesis

Seven disorders (Smith-Lemli-Opitz, desmosterolosis, X-linked dominant chondrodysplasis punctata, child syndrome, lathosterolosis and hydrops-ectopic calcification-moth-eaten skeletal dysplasia) of post-squalene cholesterol biosynthesis have been reported, only the most commonly disorder, Smith-Lemli-Opitz syndrome (SLOS) will be discussed in this section of the book as described and reviewed else where (Herman, 2003). SLOS was the first described disorder of post-squalene cholesterol biosynthesis and is by far the most common, with an incidence of approximately 1/40 000 to 1/50 000 in the USA (reviewed in (Kelley and Hennekam, 2000). It was initially described in 1964 as an autosomal recessive major malformation syndrome. In 1993, Irons et al.(Irons et al., 1993) detected decreased plasma cholesterol levels and elevated 7DHC in several patients with SLOS, suggesting an enzymatic efficiency of 7-dehydrocholesterol reductase (7DHCR).

1.3 Regulation of cholesterol biosynthesis at transcriptional level

As cells need more sterol they will induce their synthesis and uptake, conversely when the need declines synthesis and uptake are decreased. Regulation of these events is brought about primarily by sterol-regulated transcription of key rate limiting enzymes and by the regulated degradation of HMGR. Activation of transcriptional control occurs through the regulated cleavage of the membrane-bound transcription factor sterol regulated element binding protein (**SREBP**). As discussed earlier in this book chapter, degradation of HMGR is controlled by the ubiquitin-mediated pathway for proteolysis. Sterol control of transcription affects more than 30 genes involved in the biosynthesis of cholesterol, triacylglycerols, phospholipids and fatty acids. Transcriptional control requires the presence of an octamer sequence in the gene termed the sterol regulatory element-1 (SRE-1). It has been shown that SREBP is the transcription factor that binds to SRE-1 elements. It turns out that there are 2 distinct SREBP genes, SREBP-1 and SREBP-2. In addition, the SREBP-1 gene encodes 2 proteins, SREBP-1a and SREBP-1c/ADD1 (ADD1 is adipocyte differentiation-1) as a consequence of alternative exon usage. SREBP-1a regulates all SREBP-responsive genes in both the cholesterol and fatty acid biosynthetic pathways. SREBP-1c controls the expression of genes involved in fatty acid synthesis and is involved in the differentiation of adipocytes. SREBP-1c is also an essential transcription factor downstream of the actions of insulin at the level of carbohydrate and lipid metabolism. SREBP-2 is the predominant form of this transcription factor in the liver and it exhibits preference at controlling the expression of genes involved in cholesterol homeostasis, including all of the genes encoding the sterol biosynthetic enzymes. In addition SREBP-2 controls expression of the LDL receptor gene.

Regulated expression of the SREBPs is complex in that the effects of sterols are different on the SREBP-1 gene versus the SREBP-2 gene. High sterols activate expression of the SREBP-1 gene but do not exert this effect on the SREBP-2 gene. The sterol-mediated activation of the SREBP-1 gene occurs via the action of the liver X receptors (LXRs). The LXRs are members of the steroid/ thyroid hormone super family of cytosolic ligand binding receptors that migrate to the nucleus upon ligand binding and regulate gene expression by binding to specific target sequences. There are two forms of the LXRs: LXRα and LXRβ. The LXRs form heterodimers with the retinoid X receptors (RXRs) and as such can regulate gene expression either upon binding oxysterols (e.g. 22R-hydroxycholesterol) or 9-cis-retinoic acid. All 3 SREBPs are proteolytically activated and the proteolysis is controlled by the level of sterols in the cell. Full-length SREBPs have several domains and are embedded in the membrane of the endoplasmic reticulum (ER). The N-terminal domain contains a transcription factor motif of the basic helix-loop-helix (bHLH) type that is exposed to the cytoplasmic side of the ER. There are 2 transmembrane spanning domains followed by a large C-terminal domain also exposed to the cytosolic side. The C-terminal domain (CTD) interacts with a protein called SREBP cleavage-activating protein (**SCAP**). SCAP is a large protein also found in the ER membrane and contains at least 8 transmembrane spans. The C-terminal portion, which extends into the cytosol, has been shown to interact with the C-terminal domain of SREBP. This C-terminal region of SCAP contains 4 motifs called WD40 repeats. The WD40 repeats are required for interaction of SCAP with SREBP. The regulation of SREBP activity is further controlled within the ER by the interaction of SCAP with insulin regulated protein (**Insig**). When cells have sufficient sterol content SREBP and SCAP are retained in the ER via the SCAP-Insig interaction. The N-terminus of SCAP, including membrane spans 2-6, resembles HMGR which itself is subject to sterol-stimulated degradation (see above). This shared motif is called the sterol sensing domain (SSD) and as a consequence of this domain SCAP functions as the cholesterol sensor in the protein complex. When cells have sufficient levels of sterols, SCAP will bind cholesterol which promotes the interaction with Insig and the entire complex will be maintained in the ER.

The Insig proteins bind to oxysterols which in turn affects their interactions with SCAP. Insig proteins can cause ER retention of the SREBP/SCAP complex. In addition to their role in regulating sterol-dependent gene regulation, both Insig proteins activate sterol-dependent degradation of HMGR.

When sterols are scarce, SCAP does not interact with Insig. Under these conditions the SREBP-SCAP complex migrates to the Golgi where SREBP is subjected to proteolysis. The cleavage of SREBP is carried out by 2 distinct enzymes. The regulated cleavage occurs in the lumenal loop between the 2 transmembrane domains. This cleavage is catalyzed by site-1 protease, **S1P**. The function of SCAP is to positively stimulate S1P-mediated cleavage of SREBP. The second cleavage, catalyzed by site-2 protease, **S2P**, occurs in the first transmembrane span, leading to release of active SREBP. In order for S2P to act on SREBP, site-1 must already have been cleaved. The result of the S2P cleavage is the release of the N-terminal bHLH motif into the cytosol. The bHLH domain then migrates to the nucleus where it will dimerize and form complexes with transcriptional coactivators leading to the activation of genes containing the SRE motif. To control the level of SREBP-mediated transcription, the soluble bHLH domain is itself subject to rapid proteolysis.

1.4 Interaction of cholesterol with other lipids

Cholesterol plays a vital role in determining the physiochemical properties of cell membranes. However, the detailed nature of cholesterol–lipid interactions is a subject of ongoing debate. Cholesterol primarily serves as a structural component of cellular membranes. When incorporated into phospholipid bilayers, cholesterol aligns so that its polar hydroxyl group is near the interface with the aqueous environment while its hydrophobic body is buried in the bilayer (Ohvo-Rekila et al., 2002; Olsen et al., 2011a; Olsen et al., 2011b). The interaction of cholesterol with neighboring phospholipids alters membrane structure. The alignment and ordering of nearby phospholipid tails causes membrane condensation, decreasing the area of the membrane and increasing the thickness (Ohvo-Rekila et al., 2002). Cholesterol also broadens the liquid-to-solid phase transition, inducing an intermediate liquid-ordered phase that retains lateral mobility while increasing lipid order (Feigenson, 2007; Simons and Vaz, 2004; van Meer et al., 2008). These changes result in a mechanically stronger membrane with decreased permeability due to tighter packing among lipids (Ikonen, 2008; Simons and Vaz, 2004). The low activity pool consists of cholesterol that is sequestered within the phospholipids and relatively inaccessible to other molecules, while the high activity pool of cholesterol that is more accessible and mobile in the non-condensed phospholipids of the membrane (Olsen et al., 2011b). Distribution between the high and low activity pools is determined by the ability of the phospholipids to condense with cholesterol, which in turn is dependent on the phospholipid composition of the membrane. Thus, raising the plasma membrane cholesterol concentration can saturate the ability of the membrane to accommodate cholesterol in the condensed phospholipids, whereupon excess cholesterol transitions into the high activity pool where it is more available for trafficking to the ER. Sphingomyelin is other important lipid present in the biomembrane that interact with cholesterol (Garmy et al., 2005). Interaction between cholesterol and sphingomyelin has a very high biological significance as lipid-lipid interaction leads to the formation of ordered lipid domains in the plasma membrane of eukaryotic cells (Simons and Ikonen, 1997). Molecular association between cholesterol and sphingomyelin is very important as this constitute cholesterol enriched microdomains of plasma membrane and plays a very important role in cellular functions such as the control of signal transduction pathways. On the other hand, the formation of cholesterol-sphingomyelin molecular complexes in the intestinal lumen explains the mutual inhibitory effects of cholesterol and sphingomyelin on their intestinal absorption (Nyberg et al., 2000).

1.5 Cholesterol homeostasis

Every eukaryotic cell require cholesterol as its not only integral part of membrane as well as plays very important role in cell signalling pathways. That the reason, all eukaryotic cells, which have specialized methods of recruiting and synthesizing the cholesterol only when it is needed. While effectively maintaining intracellular cholesterol homeostasis, these processes leave excess circulating though the body, leading to atherosclerotic plaque development and subsequent coronary artery disease. Thus, levels of cholesterol and related lipids circulating in plasma are important predictive tools utilized clinically to diagnose the risk of a cardiovascular diseases (Daniels et al., 2009; Ikonen, 2008; Simons and Ikonen, 2000; Singh et al., 2007). Cholesterol is transported in the plasma predominantly as cholesteryl esters associated with lipoproteins and dietary cholesterol is transported from the small intestine to the liver in the form of chylomicrons. Cholesterol synthesized by the liver, as well as any dietary cholesterol in the liver that exceeds hepatic needs, is transported in the serum in the form of low density lipoproteins (LDLs). In the liver, VLDLs are biosyntheiszed and are converted to LDLs by endothelial cell-associated lipoprotein lipase. High density lipoproteins (HDLs)can extract Cholesterol found in plasma membranes and esterified by the HDL-associated enzyme LCAT. The cholesterol acquired from peripheral tissues by HDLs can then be transferred to VLDLs and LDLs via the action of cholesteryl ester transfer protein (apo-D) which is associated with HDLs. In humans, HDL levels are a very well known measurement of cardiac health due to their strong inverse relationship with coronary artery disease. Peripheral cholesterol is returned to the liver by the process called **reverse cholesterol transport by HDLs** and ultimately, cholesterol is excreted in the bile as free cholesterol or as bile salts following conversion to bile acids in the liver (Figure 4).

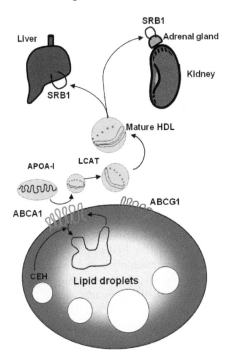

Fig. 4. Cholesterol homeostasis

1.6 Cholesterol and ion channels

A number of studies has demonstrated that the level of membrane cholesterol regulate the ion channel functions *(Levitan et al., 2010; Maguy et al., 2006; Martens et al., 2004)*. The impact of cholesterol on different types of ion channels is highly heterogeneous. In most of the cases cholesterol supresses channel activity that may include decrease in the open probability, unitary conductance and/or the number of active channels on the membrane. This effect was observed in several types of K^+ channels, voltage-gated Na^+ and Ca^{+2} channels, as well as in volume-regulated anion channels. However, there are also several types of ion channels, such as epithelial Na^+ channels (eNaC) and transient receptor potential (Trp) channels that are inhibited by the removal of membrane cholesterol. Finally, in some cases changes in membrane cholesterol affect biophysical properties of the channel such as the voltage dependence of channel activation or inactivation. Clearly, therefore, more than one mechanism has to be involved in cholesterol-induced regulation of different ion channels.

Studies from our laboratory have shown that many major and important ion channels are regulated by changes in the level of membrane cholesterol (Epshtein et al., 2009; Levitan, 2009; Levitan et al., 2010; Rosenhouse-Dantsker et al., 2011; Singh et al., 2009b; Singh et al., 2011). In general three different mechanisms may be involved in regulation of ion channels by cholesterol: (i) specific interactions (ii) changes in the physical properties of the membrane bilayer and (iii) maintaining the scaffolds for protein-protein interactions. Furthermore, our recent data for the first time demonstrate a specific cholesterol-channel binding (Singh et al., 2011). One possibility is that cholesterol may interact directly and specifically with the transmembrane domains of the channels protein. Direct interaction between channels and cholesterol as a boundary lipid was first proposed in a "lipid belt" model by (Marsh and Barrantes, 1978) suggesting that cholesterol may be a part of a lipid belt or a "shell" constituting the immediate perimeter of the channel protein. Moreover, studies from our laboratory demonstrated that inwardly-rectifying K^+ channel are sensitive to the chiral nature of the sterol analogue providing further support for the hypothesis that sensitivity of these channels to cholesterol can be due to specific sterol-protein interactions (Romanenko et al., 2002). An alternative mechanism proposed by Lundbaek and colleagues (Lundbaek and Andersen, 1999) suggested that cholesterol may regulate ion channels by hydrophobic mismatch between the transmembrane domains and the lipid bilayer. More specifically, it was proposed that when a channel goes through a change in conformation state within the viscous medium of the lipid membrane it may induce deformation of the lipid bilayer surrounding the channel. If this is the case, then a stiffer less deformable membrane will increase the energy that is required for the transition (Levitan et al., 2010). It is important to note that these mechanisms are not mutually exclusive. A lipid shell surrounding a channel may also affect the hydrophobic interactions between the channels and the lipids and increase the deformation energy required for the transitions between closed and open states. Finally, obviously, cholesterol may also affect the channels indirectly through interactions with different signalling cascades.

1.7 Cholesterol and oxysterol as major oxidized component in oxidized LDL

Oxidation of LDL is considered as the major risk factors for the development of coronary artery disease (CAD) and plaque formation (reviewed in (Berliner et al., 2001; Levitan and Shentu, 2011). Indeed, elevated levels of oxLDL are associated with an increased risk of CAD (Toshima et al., 2000) and correlate with plasma hypercholesterolemia both in humans (van Tits et al., 2006) and in the animal models of atherosclerosis (Hodis et al., 1994; Holvoet et al., 1998). It is also well-known that exposure to oxLDL induces an array of proinflammatory and proatherogenic effects but the mechanisms that underlie oxLDL-induced effects remain controversial. Most of studies involving oxLDL are based on ex-vivo oxidation of LDL. The term oxidized LDL is used to describe LDL preparations which have been oxidatively modified *ex vivo* under defined conditions, or isolated from biological sources.

The most typical procedure of LDL oxidation *ex vivo* is incubation of LDL with metal ions, Cu^{2+} in particular, that leads to the generation of multiple oxidized products in the LDL particle, including oxysterols, oxidized phospholipids, and modified apolipoprotein B (reviewed in (Burkitt, 2001; Levitan and Shentu, 2011). The oxidized LDL preparations described in the literature are broadly divided into two main categories: "minimally modified LDL" (MM-LDL) and (fully or extensively) oxidized LDL (oxLDL) based on the degree of LDL oxidation. Cu^{2+} oxidation of LDL can generate both minimally modified and fully oxidized LDLs depending on the duration of the exposure and ion concentration. Two other procedures that are also used to generate oxLDL *ex vivo* are enzymatic oxidation by 15-lypoxygenase or myeloperoxidase or by incubating LDL with 15-lypoxygenase expressing cells (Boullier et al., 2006). It is important to note that while it is controversial whether Cu^{2+} oxidation occurs *in vivo*, it was shown that there are significant similarities between Cu^{2+} oxidized LDL and oxLDL found in atherosclerotic lesions (Yla-Herttuala et al., 1989). **However, oxidation of LDL is a complex process that yields an array of bioactive compounds with different biological properties and its composition depends on the degree of LDL oxidation** It is known that oxLDL includes various oxidized products that are divided into neutral lipids, a group that includes cholesterol and cholesterol ester, and phospholipids that include SM, PE, LPC and PC (Subbanagounder et al., 2000). Our recent studies have identififed five major oxysterols in strongly oxidized LDL (16 hours of oxidation data submitted for publication) by the GC analysis: 7α-hydroxycholesterol , 7β-hydroxycholesterol, cholesterol 5α,6α- and 5β,6β-epoxides, and 7-ketocholesterol similar to earlier studies (Brown et al., 1996). Our recent unpublished data showed that the most abundant oxysterol is 7β-hydroxycholesterol, followed by 7-ketocholesterol with 7α-hydroxycholesterol and cholesterol 5β,6β-epoxide representing minor oxysterols in oxLDL. Two oxysterols, 25-hydroxycholesterol and 27-hydroxycholesterol do not constitute significant components of oxLDL complex, but were found in human atherosclerotic lesions (Bjorkhem et al., 1994; Zidovetzki and Levitan, 2007). 27-hydroxycholesterols, specially is abundant in human atherosclerotic lesions and macrophage-derived foam cells (Brown and Jessup, 1999).

1.8 Impact of oxLDL on cholesterol-rich membrane rafts

Membrane rafts were originally described as cholesterol- and sphingolipid-rich microdomains that provide platforms for protein-protein interactions in multiple signaling cascades (Brown and London, 1998; Simons and Gerl, 2010; Simons and Ikonen, 1997). Membrane rafts are small (10–200 nm), heterogeneous, highly dynamic, sterol- and sphingolipid-enriched domains that compartmentalize cellular processes" (Pike, 2006). Most recently, Simons and Gerl (Simons and Gerl, 2010) further defined membrane rafts as "dynamic, nonoscale, sterol-sphingolipid-enriched, ordered assembles of proteins and lipids" that are regulated by specific lipid-lipid, protein-lipid, and protein-protein interactions (Simons and Gerl, 2010). The goal of this section of book chapter is to discuss the recent advances in our understanding of the impact of oxLDL on membrane rafts. Oxysterols are found in abundance in Cu^{2+}-oxidized LDL, in which cholesterol is oxidized preferably at 7 positions resulting in the generation of 7-ketocholesterol, 7β-hydroxycholesterol, and 7α-hydroxycholesterol (Brown and Jessup, 1999). In addition, 27-hydroxycholesterol, that is, generated *in vivo*, has been shown to accumulate in foam cells in atherosclerotic lesions (Brown and Jessup, 1999). Several studies have shown that oxysterols result in inhibition of cholesterol efflux in the mouse and it was suggested that impairment of cholesterol homeostasis by the inhibition of cholesterol efflux may be mechanism by which oxysterols affect cellular function (Kilsdonk et al., 1995; Terasaka et al., 2008). Interestingly, 7-ketocholesterol was shown to deplete cholesterol specifically from the raft domains in human macrophages (Gaus et al., 2004) and disrupt lipid packing of the immunological synapses in sterol-enriched T lymphocytes (Rentero et al., 2008) and in cholesterol-rich membrane domains in endothelial cells (Shentu et al., 2010). These observations suggest that incorporation of oxysterols may also play an important role in oxLDL-induced disruption of cholesterol-rich membrane domains.

1.9 Manipulation of cholesterol in cellular system (*in vitro* and *in vivo* models)

The physiological importance of cholesterol in the cell plasma membrane has attracted increased attention in recent years. Consequently, the use of methods of controlled manipulation of membrane cholesterol content has also increased sharply, especially as a method of studying putative cholesterol-enriched cell membrane domains (rafts). The most common means of modifying the cholesterol content of cell membranes is the incubation of cells or model membranes with cyclodextrins, a family of compounds, which, due to the presence of relatively hydrophobic cavity, can be used to extract cholesterol from cell membranes (Zidovetzki and Levitan, 2007). Under conditions commonly used for cholesterol extraction, cyclodextrins may remove cholesterol from both raft and non-raft domains of the membrane as well as alter the distribution of cholesterol between plasma and intracellular membranes. In addition, other hydrophobic molecules such as phospholipids may also be extracted from the membranes by cyclodextrins. Here, we discuss useful control strategies that may help to verify that the observed effects are due specifically to cyclodextrin-induced changes in cellular cholesterol.

The high affinity of methyl-β-cyclodextrin (MβCDs) for cholesterol can be used not only to remove cholesterol from the biological membranes but also to generate cholesterol inclusion complexes that donate cholesterol to the membrane and increase membrane cholesterol level. MβCD-cholesterol inclusion complexes are typically generated by mixing cholesterol suspension with a cyclodextrin solution as described earlier (Christian et al., 1997; Klein et al., 1995; Levitan et al., 2000; Zidovetzki and Levitan, 2007). The ratio between the amounts of cholesterol and cyclodextrin in the complex determines whether it will act as cholesterol acceptor or as cholesterol donor (Zidovetzki and Levitan, 2007). The efficiency of cholesterol transfer from MβCD inclusion complex to biological membranes depends on MβCD:cholesterol molar ratio, MβCD-cholesterol concentration, and duration of the exposure (Zidovetzki and Levitan, 2007). Thus, it is important to note that exposing cells to MβCD-cholesterol complexes that contain saturating amounts of cholesterol typically results not just in replenishing cholesterol to control levels but in significant cholesterol enrichment. Cholesterol enrichment is observed even if the cells were first depleted of cholesterol and then exposed to MβCD-cholesterol complexes. ApoE -/- mice cholesterol ester transfer protein (CETP) knock out mice are ideal *in vivo* animal model to study cholesterol biosynthetic pathway and its regulation. Our recent study revealed that WT and ApoE -/- mice fed on normal chow diet for 4 weeks, showing a significant increase in the level of cholesterol in ApoE -/- mice as compared to WT (Shentu et al., 2011).

1.10 Treatment of hypercholesterolemia

Reductions in circulating cholesterol levels can have profound positive impacts on cardiovascular disease, particularly on atherosclerosis, as well as other metabolic disruptions of the vasculature. Control of dietary intake is one of the easiest and least cost intensive means to achieve reductions in cholesterol. But in most of the alleviated levels of cholesterol cannot be controlled merely with exercise. Drug therapy is very important to avoid the cardiovascular effects of high cholesterol levels in such patients. In this section of the book chapter, we have discussed the regulation cholesterol by drugs as well as will review recent work from our laboratory for the regulation of cholesterol by natural products such as tea/green tea, policosanol and garlic compounds.

1.10.1 Regulation of cholesterol by natural products

In this section of book the regulation of cholesterol by natural products, tea, policosanol and garlic is discussed.

1.10.1.1 Tea and tea compounds

Epidemiological studies have indicated that tea consumption is associated with a lower risk of cardiovascular disease. This decreased risk is attributed to the ability of tea to lower serum cholesterol levels, and several clinical studies have demonstrated that black tea can lower serum total- and LDL-cholesterol (Davies et al., 2003; Maron et al., 2003). Green tea has been shown to be hypocholesterolemic in animal studies, with the bulk of evidence indicating that tea polyphenols reduce the absorption of dietary and biliary cholesterol and promote its fecal excretion (Koo and Noh, 2007)..

Feeding studies have been equivocal on the ability of green tea extract to inhibit cholesterol synthesis. Although a recent study by Bursill and colleagues (Bursill et al., 2007) showed a decrease in serum lathosterol (an indicator of whole body cholesterol synthesis) in rabbits fed a green tea extract, a similar study with rats by these investigators (Bursill and Roach, 2007) was unable to demonstrate a decrease in this serum sterol, despite significant reductions in hepatic cholesterol levels and an increase in LDL receptor expression. A feeding study by Chan et al. (Chan et al., 1999) was similarly unable to demonstrate an effect of green tea extract on hepatic HMG-CoA reductase activity. Measuring cholesterol synthesis in vivo is difficult, whereas in vitro studies are more tractable. In this regard, Gebhardt and colleagues reported that several common polyphenols (luteolin, quercetin) were able to decrease cholesterol synthesis when added to cultured hepatocytes or hepatoma cell cultures (Gebhardt, 2003). This inhibition appeared to occur at the level of HMG-CoA reductase. Tea polyphenols (Abe et al., 2000), as well as the simple polyphenol resveratrol have been shown to directly inhibit squalene monooxygenase, a rate-limiting downstream enzyme in cholesterol synthesis. Two studies by Bursill and colleagues (Bursill et al., 2001; Bursill and Roach, 2006) demonstrated an increase in HMG-CoA reductase and LDL-receptor mRNA in HepG2 cells incubated with green tea extract or its principal component, epigallocatechin gallate (EGCG), and a decrease in cellular lathosterol, indicating that cholesterol synthesis was inhibited in treated cells (Singh et al., 2009a). Together, these studies suggest that green tea polyphenols are inhibitory to cholesterol synthesis by inhibiting HMGCoA reductase (Singh et al., 2009a). Moreover, the effect of black tea extract, which consists predominantly of a diverse mixture of polymerized polyphenols termed theaflavins and thearubigins, has not been examined, despite the recent clinical evidence that black tea can modestly reduce serum cholesterol levels.

1.10.1.2 Decrease in cholesterol level by policosanol

Policosanol, a mixture of very long-chain alcohols isolated from sugarcane, at doses of 10 to 20 mg/day has been shown to lower total and LDL cholesterol by up to 30%, equivalent to low-dose statin therapy (Gouni-Berthold and Berthold, 2002). In both short-term (≤12-week) and long-term (up to 2-year) randomized, placebo-controlled, double-blind studies, policosanol lowered LDL-cholesterol in normocholesterolemic patients by an average of 33%, and in hypercholesterolemic patients by 24% (for review, see (Gouni-Berthold and Berthold, 2002; Varady et al., 2003). In normocholesterolemic patients, policosanol caused a small and generally insignificant increase in high-density lipoprotein-cholesterol, whereas in seven clinical studies of dyslipidemic patients high-density lipoprotein-cholesterol was increased by an average of 17%. Policosanol is also effective in rabbits and monkeys, where it lowers blood cholesterol and reduces the development of atherosclerotic plaques (Wang et al., 2003), but it was found not to be effective in hamsters (Wang et al., 2003).

The major components of policosanol are the primary alcohols octacosanol (C28; ~60%), triacontanol (C30; 12–14%), and hexacosanol (C26; 6–12%), with lesser amounts of other alcohols with chain lengths of 24 to 34 carbons (Singh et al., 2006) . The product has no evident toxicity and is available over-the-counter in many outlets. The active component(s) has not been established, but it has been shown that very long-chain alcohols can undergo oxidation to fatty acids with subsequent peroxisomal β-oxidation, which also yields chain-shortened metabolites (Singh et al., 1987). D-003, a mixture of very long-chain saturated fatty acids, also purified from sugarcane, similarly lowers LDL and total cholesterol in normocholesterolemic patients (Castano et al., 2005) and in normocholesterolemic and casein-induced hypercholesterolemic rabbits, and a more rapid onset of effects suggests that oxidation of policosanols to very long-chain fatty acids may be necessary for their hypocholesterolemic actions (Menendez et al., 2001; Menendez et al., 2004). Several studies have demonstrated that policosanol inhibits cholesterol synthesis in laboratory animals and cultured cells, and it is thought that this is the principal mechanism by which it lowers blood cholesterol levels. In the latter study, policosanol did not affect the incorporation of [14C]mevalonate into cholesterol, indicating that policosanol was acting at or above mevalonate synthesis. However, policosanol did not inhibit HMG-CoA reductase (mevalonate synthase) when added to cell lysates, arguing against a direct interaction with this enzyme. The ability of policosanol to prevent the up-regulation of HMG-CoA reductase activity in these cells in response to lipid-depleted media suggested that policosanol suppresses HMG-CoA reductase synthesis or enhances enzyme degradation. Similar results were obtained with D-003 (Menendez et al., 2001), although neither study measured HMG-CoA reductase protein levels. Our studies explored that policosanol and identify the active component(s) of this natural product inhibits cholesterol synthesis by inhibiting HMGCoA reductase enzyme (Singh et al., 2006).

1.10.1.3 Inhibition of cholesterol biosynthesis by garlic

Garlic is rich in sulfur-containing compounds, principally S-allylcysteine and alliin, the latter of which is rapidly metabolized when garlic is crushed and alliinase is released. The highly reactive sulfenic acid that is formed from alliin condenses to allicin, which then rapidly recombines to various di- and tri-sulfides, depending on conditions. Ultimately these compounds are believed to yield allyl mercaptan and allyl methyl sulfide, which can react with cellular components or be eliminated on the breath. The organosulfur compounds formed in garlic are highly reactive with other sulfhydryl compounds, including cysteines found in proteins, and it is likely that the chemical modification of enzyme-sulfhydryls is responsible for the purported therapeutic effects of garlic. The question of which compounds are most important to the therapeutic effects of garlic remains unresolved, although several studies have shown that the diallyl disulfides, allyl mercaptan, and Salk(en)yl cysteines are effective inhibitors of cholesterol synthesis in cells (Gebhardt and Beck, 1996; Liu and Yeh, 2000; Singh and Porter, 2006). Similarly, the enzyme targets that mediate the effects of garlic have not been identified.

Our studies with hepatoma cells in which cholesterol and intermediates are radiolabeled and identified by coupled gas chromatography–mass spectrometry reveal that garlic causes the accumulation of sterol 4α-methyl oxidase substrates and that an allyl disulfide or allyl sulfhydryl group is necessary for inhibition by garlic-derived compounds (Singh and Porter, 2006).

1.10.2 Treatment of Hypercholesterolemia by available drug therapy

Drug treatment to lower plasma lipoprotein /or cholesterol is primarily aimed at reducing the risk of athersclerosis and subsequent coronary artery disease that exists in patients with elevated circulating lipids. Drug therapy usually is considered as an option only if non-pharmacologic interventions (altered diet and exercise) have failed to lower plasma lipids.

1.10.2.1 Members of statins family

These are fungal HMG-CoA reductase (HMGR) inhibitors **from members of statins family. Atorvastin (Lipitor), simvastatin (Zocor) and lovastatin (Mevacor) belongs to this family, are widely used for lowering the plasma cholesterol. During the course of treatment, cellular uptake of LDL from plama is significantly increased**, since the intracellular synthesis of cholesterol is inhibited as cells are dependent on extracellular sources of cholesterol. Important isoprenoid compounds require mevalonate as the precursor as a result long term treatment carry some risk of toxicity. A component of the natural cholesterol lowering supplement, red yeast rice, is in fact a statin-like compound. Other beneficial effects of statins other than lowering blood cholesterol levels via their actions on HMGR are ability to reduce the prenylation of numerous pro-inflammatory modulators. Thus, inhibition of this post-translational modification by the statins interferes with the important functions of many signaling proteins which is manifest by inhibition of inflammatory responses.

1.10.2.2 Fibrates compounds

Second group drugs belongs to a series of compounds are derivatives of fibric acid and although known since 1930 but identified as cholesterol lowering drugs very recently. Fibrates are activators of the peroxisome proliferator activated receptor-α (PPARα) class of proteins and are classified as nuclear receptor co-activators. Fibrates result in activation of β-oxidation and thereby decreasing the level of triacyl glycerol and cholesterol rich VLDL in liver as well as enhances the clearance of chylomicrons remnants, and increase in the level of HDLs. These drugs are also known to increase the lipase activity which in turn promotes rapid VLDL turnover. Gemifibrozil (Lopid) and Fenofibrate (Tricor) are two therapeutic drugs available in the market.

1.10.2.3 Cholestyramine or colestipol

Cholestyramine or colestipol (resins) compounds are nonabsorbable resins that bind bile acids which are then not reabsorbed by the liver but excreted. The drop in hepatic reabsorption of bile acids releases a feedback inhibitory mechanism that had been inhibiting bile acid synthesis. As a result, a greater amount of cholesterol is converted to bile acids to maintain a steady level in circulation. Additionally, the synthesis of LDL receptors increases to allow increased cholesterol uptake for bile acid synthesis, and the overall effect is a reduction in plasma cholesterol. This treatment is ineffective in homozygous FH patients since they are completely deficient in LDL receptors.

1.10.2.4 Ezetimibe

This drug is sold under the trade names Zetia® or Ezetrol® and is also combined with the statin drug simvastatin and sold as Vytorin® or Inegy®. Ezetimibe functions to reduce intestinal absorption of cholesterol, thus effecting a reduction in circulating cholesterol. The drug functions by inhibiting the intestinal brush border transporter involved in absorption of cholesterol. This transporter is known as Niemann-Pick type C1-like 1 (NPC1L1). NPC1L1 is also highly expressed in human liver. The hepatic function of NPC1L1 is presumed to limit excessive biliary cholesterol loss. NPC1L1-dependent sterol uptake is regulated by cellular cholesterol content. In addition to the cholesterol lowering effects that result from inhibition of NPC1L1, its' inhibition has been shown to have beneficial effects on components of the metabolic syndrome, such as obesity, insulin resistance, and fatty liver, in addition to atherosclerosis. Ezetimibe is usually prescribed for patients who cannot tolerate a statin drug or a high dose statin regimen. There is some controversy as to the efficacy of ezetimibe at lowering serum cholesterol and reducing the production of fatty plaques on arterial walls. The combination drug of ezetimibe and simvastatin has shown efficacy equal to or slightly greater than atorvastatin (Lipitor®) alone at reducing circulating cholesterol levels.

1.10.2.5 New approaches for the treatment of hypercholesterolemia

In the last decade, a number of epidemiological and clinical studies have demonstrated a direct correlation between the circulating levels of HDL cholesterol and a reduction in the potential for atherosclerosis and coronary heart disease (CHD). Individuals with low levels of HDL (below 40mg/dL) are at higher risk of coronary heart disease CHD) then individual with level above 50mg/dL. Clinical studies have demonstrated that infusion of HDL component, apolipoprotein A-1 (apoA-1) in patients, significantly increases the level of HDL. The newest strategies are targeted to up regulate the level of HDL cholesterol instead of decreasing the level of total cholesterol. Cholesterol ester transfer protein (CETP) is secreted primarily from the liver and plays a critical role in HDL metabolism by facilitating the exchange of cholesteryl esters (CE) from HDL for triglycerides (TG) in apoB containing lipoproteins, such as LDL and VLDL. The activity of CETP directly lowers the cholesterol levels of HDLs and enhances HDL catabolism by providing HDLs with the TG substrate of hepatic lipase. Thus, CETP plays a critical role in the regulation of circulating levels of HDL, LDL, and apoA-I. It has also been shown that in mice naturally lacking CETP most of their cholesterol is found in HDL and these mice are relatively resistant to atherosclerosis. CETP inhibitors have failed in the clinical trials as their use has increased negative cardiovascular events and death rates in test subjects.

2. Conclusion

Cholesterol is an essential component in cell membrane, as a precursor for the synthesis of steroid hormones vitamin D, and bile acids that aid in digestion and cellular signal transduction. Half of the cholesterol is *de novo* synthesized in liver and is transported through various lipoprotein. Dysfunction in cholesterol metabolism can lead to hypercholesterolemia which is a major factor in the development of atherosclerosis. Mode of intracellular and extracellular cholesterol transport through acceptors-donors and thereafter cholesterol trafficking pathways are highly co-ordinated with each other, regulated at enzymatic/transcriptional level and diverse functions of cholesterol in our body. Taken together, this book chapter addressed recent advances in cholesterol metabolism related to absorptions, biosynthesis, transport, excretion and therapeutic targets for new drugs and natural compounds.

3. Acknowledgment

We thank Professor (Dr) Papasani V. Subbaiah, Associate Professor (Dr) Irena Levitan, Professor (Dr) Todd Porter and Assistant Professor (Dr) Ramachandran Ramaswamy for their multiple critical discussions. Special thanks are to Software Professional, Mr Ravi Kesavarapu, for help in making the figures for the manuscript.

4. References

Abe, I., T. Seki, K. Umehara, T. Miyase, H. Noguchi, J. Sakakibara, and T. Ono. 2000. Green tea polyphenols: novel and potent inhibitors of squalene epoxidase. *Biochemical and Biophysical Research Communications*. 268:767-771.

Berliner, J.A., G. Subbanagounder, N. Leitinger, A.D. Watson, and D. Vora. 2001. Evidence for a role of phospholipid oxidation products in atherogenesis. *Trends in Cardiovascular Medicine*. 11:142-147.

Bjorkhem, I., O. Andersson, U. Diczfalusy, B. Sevastik, R.J. Xiu, C. Duan, and E. Lund. 1994. Atherosclerosis and sterol 27-hydroxylase: evidence for a role of this enzyme in elimination of cholesterol from human macrophages. *The Proceedings of National Academy Sciences, U S A*. 91:8592-8596.

Boullier, A., Y. Li, O. Quehenberger, W. Palinski, I. Tabas, J.L. Witztum, and Y.I. Miller. 2006. Minimally oxidized LDL offsets the apoptotic effects of extensively oxidized LDL and free cholesterol in macrophages. *Arteriosclerosis, Thrombosis and Vascular Biology*. 26:1169-1176.

Brown, A.J., R.T. Dean, and W. Jessup. 1996. Free and esterified oxysterol: formation during copper-oxidation of low density lipoprotein and uptake by macrophages. *Journal of Lipid Research*. 37:320-335.

Brown, A.J., and W. Jessup. 1999. Oxysterols and atherosclerosis. *Atherosclerosis*. 142:1-28.

Brown, D.A., and E. London. 1998. Functions of lipid rafts in biological membranes. *Annual Review of Cell and Developmental Biology*. 14:111-136.

Burkitt, M.J. 2001. A critical overview of the chemistry of copper-dependent low density lipoprotein oxidation: roles of lipid hydroperoxides, alpha-tocopherol, thiols, and ceruloplasmin. *Archieves of Biochemistry and Biophysics*. 394:117-135.

Bursill, C., P.D. Roach, C.D. Bottema, and S. Pal. 2001. Green tea upregulates the low-density lipoprotein receptor through the sterol-regulated element binding Protein in HepG2 liver cells. *Journal of Agricultural Food Chemistry*. 49:5639-5645.

Bursill, C.A., M. Abbey, and P.D. Roach. 2007. A green tea extract lowers plasma cholesterol by inhibiting cholesterol synthesis and upregulating the LDL receptor in the cholesterol-fed rabbit. *Atherosclerosis*. 193:86-93.

Bursill, C.A., and P.D. Roach. 2006. Modulation of cholesterol metabolism by the green tea polyphenol (-)-epigallocatechin gallate in cultured human liver (HepG2) cells. *Journal of Agricultural Food Chemistry.* 54:1621-1626.

Bursill, C.A., and P.D. Roach. 2007. A green tea catechin extract upregulates the hepatic low-density lipoprotein receptor in rats. *Lipids.* 42:621-627.

Castano, G., R. Mas, L. Fernandez, J. Illnait, S. Mendoza, R. Gamez, J. Fernandez, and M. Mesa. 2005. A comparison of the effects of D-003 and policosanol (5 and 10 mg/day) in patients with type II hypercholesterolemia: a randomized, double-blinded study. *Drugs under Experimental and Clinical Research.* 31 Suppl:31-44.

Chan, P.T., W.P. Fong, Y.L. Cheung, Y. Huang, W.K. Ho, and Z.Y. Chen. 1999. Jasmine green tea epicatechins are hypolipidemic in hamsters (Mesocricetus auratus) fed a high fat diet. *Journal of Nutrition.* 129:1094-1101.

Christian, A.E., M.P. Haynes, M.C. Phillips, and G.H. Rothblat. 1997. Use of cyclodextrins for manipulating cellular cholesterol content. *Journal of Lipid Research.* 38:2264-2272.

Daniels, T.F., K.M. Killinger, J.J. Michal, R.W. Wright, Jr., and Z. Jiang. 2009. Lipoproteins, cholesterol homeostasis and cardiac health. *International Journal of Biological Sciences.* 5:474-488.

Davies, M.J., J.T. Judd, D.J. Baer, B.A. Clevidence, D.R. Paul, A.J. Edwards, S.A. Wiseman, R.A. Muesing, and S.C. Chen. 2003. Black tea consumption reduces total and LDL cholesterol in mildly hypercholesterolemic adults. *Journal of Nutrition.* 133:3298S-3302S.

Epshtein, Y., A.P. Chopra, A. Rosenhouse-Dantsker, G.B. Kowalsky, D.E. Logothetis, and I. Levitan. 2009. Identification of a C-terminus domain critical for the sensitivity of Kir2.1 to cholesterol. *The Proceedings of National Academy Sciences, U S A.* 106:8055-8060.

Feigenson, G.W. 2007. Phase boundaries and biological membranes. *Annual Review of Biophysics and Biomolecular Structure.* 36:63-77.

Fitzgerald, M.L., K.J. Moore, and M.W. Freeman. 2002. Nuclear hormone receptors and cholesterol trafficking: the orphans find a new home. *Journal of Molecular Medicine (Berl).* 80:271-281.

Garmy, N., N. Taieb, N. Yahi, and J. Fantini. 2005. Interaction of cholesterol with sphingosine: physicochemical characterization and impact on intestinal absorption. *Journal of Lipid Research.* 46:36-45.

Gaus, K., L. Kritharides, G. Schmitz, A. Boettcher, W. Drobnik, T. Langmann, C.M. Quinn, A. Death, R.T. Dean, and W. Jessup. 2004. Apolipoprotein A-1 interaction with plasma membrane lipid rafts controls cholesterol export from macrophages. *The FASEB Journal.* 18:574-576.

Gaylor, J.L. 2002. Membrane-bound enzymes of cholesterol synthesis from lanosterol. *Biochemical Biophysical Research Communications.* 292:1139-1146.

Gebhardt, R. 2003. Variable influence of kaempferol and myricetin on *in vitro* hepatocellular cholesterol biosynthesis. *Planta Medica.* 69:1071-1074.

Gebhardt, R., and H. Beck. 1996. Differential inhibitory effects of garlic-derived organosulfur compounds on cholesterol biosynthesis in primary rat hepatocyte cultures. *Lipids.* 31:1269-1276.

Goldstein, J.L., and M.S. Brown. 1990. Regulation of the mevalonate pathway. *Nature.* 343:425-430.

Gouni-Berthold, I., and H.K. Berthold. 2002. Policosanol: clinical pharmacology and therapeutic significance of a new lipid-lowering agent. *American Heart Journal.* 143:356-365.

Gylling, H., and T.A. Miettinen. 1995. The effect of cholesterol absorption inhibition on low density lipoprotein cholesterol level. *Atherosclerosis.* 117:305-308.

Herman, G.E. 2003. Disorders of cholesterol biosynthesis: prototypic metabolic malformation syndromes. *Human Molecular Genetics.* 12 Spec No 1:R75-88.

Hodis, H.N., D.M. Kramsch, P. Avogaro, G. Bittolo-Bon, G. Cazzolato, J. Hwang, H. Peterson, and A. Sevanian. 1994. Biochemical and cytotoxic characteristics of an *in vivo* circulating oxidized low density lipoprotein (LDL-). *Journal of Lipid Research.* 35:669-677.

Holvoet, P., G. Theilmeier, B. Shivalkar, W. Flameng, and D. Collen. 1998. LDL hypercholesterolemia is associated with accumulation of oxidized LDL, atherosclerotic plaque growth, and compensatory vessel enlargement in coronary arteries of miniature pigs. *Arteriosclerosis, Thrombosis and Vascular Biology.* 18:415-422.

Ikonen, E. 2008. Cellular cholesterol trafficking and compartmentalization. *Nature Reviews Molecular Cell Biology.* 9:125-138.

Irons, M., E.R. Elias, G. Salen, G.S. Tint, and A.K. Batta. 1993. Defective cholesterol biosynthesis in Smith-Lemli-Opitz syndrome. *Lancet.* 341:1414.

Kelley, R.I., and R.C. Hennekam. 2000. The Smith-Lemli-Opitz syndrome. *Journal of Medical Genetics.* 37:321-335.

Kilsdonk, E.P., D.W. Morel, W.J. Johnson, and G.H. Rothblat. 1995. Inhibition of cellular cholesterol efflux by 25-hydroxycholesterol. *Journal of Lipid Research.* 36:505-516.

Kimura, Y., and K. Tanaka. 2010. Regulatory mechanisms involved in the control of ubiquitin homeostasis. *The Journal of Biochemistry.* 147:793-798.

Klein, U., G. Gimpl, and F. Fahrenholz. 1995. Alteration of the myometrial plasma membrane cholesterol content with beta-cyclodextrin modulates the binding affinity of the oxytocin receptor. *Biochemistry.* 34:13784-13793.

Koo, S.I., and S.K. Noh. 2007. Green tea as inhibitor of the intestinal absorption of lipids: potential mechanism for its lipid-lowering effect. *Journal of Nutritional Biochemistry.* 18:179-183.

Kovacs, W.J., L.M. Olivier, and S.K. Krisans. 2002. Central role of peroxisomes in isoprenoid biosynthesis. *Progress in Lipid Research.* 41:369-391.

Levitan, I. 2009. Cholesterol and Kir channels. *IUBMB Life.* 61:781-790.

Levitan, I., A.E. Christian, T.N. Tulenko, and G.H. Rothblat. 2000. Membrane cholesterol content modulates activation of volume-regulated anion current in bovine endothelial cells. *Journal of General Physiology.* 115:405-416.

Levitan, I., Y. Fang, A. Rosenhouse-Dantsker, and V. Romanenko. 2010. Cholesterol and ion channels. *Subcell Biochemistry.* 51:509-549.

Levitan, I., and T.P. Shentu. 2011. Impact of oxLDL on Cholesterol-Rich Membrane Rafts. *Journal of Lipids.* 2011:730209.

Liu, L., and Y.Y. Yeh. 2000. Inhibition of cholesterol biosynthesis by organosulfur compounds derived from garlic. *Lipids.* 35:197-203.

Lundbaek, J.A., and O.S. Andersen. 1999. Spring constants for channel-induced lipid bilayer deformations. Estimates using gramicidin channels. *Biophysical Journal.* 76:889-895.

Maguy, A., T.E. Hebert, and S. Nattel. 2006. Involvement of lipid rafts and caveolae in cardiac ion channel function. *Cardiovascular Research*. 69:798-807.

Maron, D.J., G.P. Lu, N.S. Cai, Z.G. Wu, Y.H. Li, H. Chen, J.Q. Zhu, X.J. Jin, B.C. Wouters, and J. Zhao. 2003. Cholesterol-lowering effect of a theaflavin-enriched green tea extract: a randomized controlled trial. *Archieves in Internal Medicine* 163:1448-1453.

Marsh, D., and F.J. Barrantes. 1978. Immobilized lipid in acetylcholine receptor-rich membranes from Torpedo marmorata. *The Proceedings of National Academy of Scences, U S A*. 75:4329-4333.

Martens, J.R., K. O'Connell, and M. Tamkun. 2004. Targeting of ion channels to membrane microdomains: localization of KV channels to lipid rafts. *Trends in Pharmacological Sciences*. 25:16-21.

Menendez, R., A.M. Amor, I. Rodeiro, R.M. Gonzalez, P.C. Gonzalez, J.L. Alfonso, and R. Mas. 2001. Policosanol modulates HMG-CoA reductase activity in cultured fibroblasts. *Archieves of Medical Research*. 32:8-12.

Menendez, R., R. Mas, J. Perez, R.M. Gonzalez, and S. Jimenez. 2004. Oral administration of D-003, a mixture of very long chain fatty acids prevents casein-induced endogenous hypercholesterolemia in rabbits. *Canadian Journal of Physiology and Pharmacology*. 82:22-29.

Nyberg, L., R.D. Duan, and A. Nilsson. 2000. A mutual inhibitory effect on absorption of sphingomyelin and cholesterol. *Journal of Nutritional Biochemistry*. 11:244-249.

Ohvo-Rekila, H., B. Ramstedt, P. Leppimaki, and J.P. Slotte. 2002. Cholesterol interactions with phospholipids in membranes. *Progress in Lipid Research*. 41:66-97.

Olsen, B.N., P.H. Schlesinger, D.S. Ory, and N.A. Baker. 2011a. 25-Hydroxycholesterol increases the availability of cholesterol in phospholipid membranes. *Biophysical Journal*. 100:948-956.

Olsen, B.N., P.H. Schlesinger, D.S. Ory, and N.A. Baker. 2011b. Side-chain oxysterols: From cells to membranes to molecules. *Biochimica et Biophysica Acta*. (in press)

Ostlund, R.E., Jr., M.S. Bosner, and W.F. Stenson. 1999. Cholesterol absorption efficiency declines at moderate dietary doses in normal human subjects. *Journal of Lipid Research*. 40:1453-1458.

Pike, L.J. 2006. Rafts defined: a report on the Keystone Symposium on Lipid Rafts and Cell Function. *Journal of Lipid Research*. 47:1597-1598.

Rentero, C., T. Zech, C.M. Quinn, K. Engelhardt, D. Williamson, T. Grewal, W. Jessup, T. Harder, and K. Gaus. 2008. Functional implications of plasma membrane condensation for T cell activation. *PLoS One*. 3:e2262.

Romanenko, V.G., G.H. Rothblat, and I. Levitan. 2002. Modulation of endothelial inward-rectifier K+ current by optical isomers of cholesterol. *Biophysical Journal*. 83:3211-3222.

Rosenhouse-Dantsker, A., D.E. Logothetis, and I. Levitan. 2011. Cholesterol sensitivity of KIR2.1 is controlled by a belt of residues around the cytosolic pore. *Biophysical Journal*. 100:381-389.

Rozman, D., M. Cotman, and R. Frangez. 2002. Lanosterol 14alpha-demethylase and MAS sterols in mammalian gametogenesis. *Molecular and Cellular Endocrinology*. 187:179-187.

Shentu, T.P., I. Titushkin, D.K. Singh, K.J. Gooch, P.V. Subbaiah, M. Cho, and I. Levitan. 2010. oxLDL-induced decrease in lipid order of membrane domains is inversely correlated with endothelial stiffness and network formation. *American Journal Physiology: Cell Physiology.* 299:C218-229.

Simons, K., and M.J. Gerl. 2010. Revitalizing membrane rafts: new tools and insights. *Nature Reviews Molecular Cell Biology.* 11:688-699.

Simons, K., and E. Ikonen. 1997. Functional rafts in cell membranes. *Nature.* 387:569-572.

Simons, K., and E. Ikonen. 2000. How cells handle cholesterol. *Science.* 290:1721-1726.

Simons, K., and W.L. Vaz. 2004. Model systems, lipid rafts, and cell membranes. *Annual Review of Biophysics and Biomolecular Structure.* 33:269-295.

Singh, D.K., S. Banerjee, and T.D. Porter. 2009a. Green and black tea extracts inhibit HMG-CoA reductase and activate AMP kinase to decrease cholesterol synthesis in hepatoma cells. *Journal of Nutritional Biochemistry.* 20:816-822.

Singh, D.K., L.R. Gesquiere, and P.V. Subbaiah. 2007. Role of sphingomyelin and ceramide in the regulation of the activity and fatty acid specificity of group V secretory phospholipase A2. *Archieves of Biochemistry Biophysics.* 459:280-287.

Singh, D.K., L. Li, and T.D. Porter. 2006. Policosanol inhibits cholesterol synthesis in hepatoma cells by activation of AMP-kinase. *Journal of Pharmacology and Experimental Therapeutics.* 318:1020-1026.

Singh, D.K., V. Mokashi, C.L. Elmore, and T.D. Porter. 2003. Phosphorylation of supernatant protein factor enhances its ability to stimulate microsomal squalene monooxygenase. *Journal of Biological Chemistry.* 278:5646-5651.

Singh, D.K., and T.D. Porter. 2006. Inhibition of sterol 4alpha-methyl oxidase is the principal mechanism by which garlic decreases cholesterol synthesis. *Journal of Nutrition.* 136:759S-764S.

Singh, D.K., A. Rosenhouse-Dantsker, C.G. Nichols, D. Enkvetchakul, and I. Levitan. 2009b. Direct regulation of prokaryotic Kir channel by cholesterol. *Journal of Biological Chemistry.* 284:30727-30736.

Singh, D.K., T.P. Shentu, D. Enkvetchakul, and I. Levitan. 2011. Cholesterol regulates prokaryotic Kir channel by direct binding to channel protein. *Biochimica et Biophysica Acta.* 1808:2527-2533.

Singh, H., N. Derwas, and A. Poulos. 1987. Very long chain fatty acid beta-oxidation by rat liver mitochondria and peroxisomes. *Archieves of Biochemistry Biophysics.*259:382-390.

Subbanagounder, G., A.D. Watson, and J.A. Berliner. 2000. Bioactive products of phospholipid oxidation: isolation, identification, measurement and activities. *Free Radical Biology and Medicine.* 28:1751-1761.

Terasaka, N., S. Yu, L. Yvan-Charvet, N. Wang, N. Mzhavia, R. Langlois, T. Pagler, R. Li, C.L. Welch, I.J. Goldberg, and A.R. Tall. 2008. ABCG1 and HDL protect against endothelial dysfunction in mice fed a high-cholesterol diet. *The Journal of Clinical Investigation.* 118:3701-3713.

Toshima, S., A. Hasegawa, M. Kurabayashi, H. Itabe, T. Takano, J. Sugano, K. Shimamura, J. Kimura, I. Michishita, T. Suzuki, and R. Nagai. 2000. Circulating oxidized low density lipoprotein levels. A biochemical risk marker for coronary heart disease. *Arteriosclerosis, Thrombosis and Vascular Biology.* 20:2243-2247.

van Meer, G., D.R. Voelker, and G.W. Feigenson. 2008. Membrane lipids: where they are and how they behave. *Nature Reviews Molecular Cell Biology.* 9:112-124.

van Tits, L.J., T.M. van Himbergen, H.L. Lemmers, J. de Graaf, and A.F. Stalenhoef. 2006. Proportion of oxidized LDL relative to plasma apolipoprotein B does not change during statin therapy in patients with heterozygous familial hypercholesterolemia. *Atherosclerosis*. 185:307-312.

Varady, K.A., Y. Wang, and P.J. Jones. 2003. Role of policosanols in the prevention and treatment of cardiovascular disease. *Nutrition Review*. 61:376-383.

Wang, Y.W., P.J. Jones, I. Pischel, and C. Fairow. 2003. Effects of policosanols and phytosterols on lipid levels and cholesterol biosynthesis in hamsters. *Lipids*. 38:165-170.

Waterham, H.R., J. Koster, G.J. Romeijn, R.C. Hennekam, P. Vreken, H.C. Andersson, D.R. FitzPatrick, R.I. Kelley, and R.J. Wanders. 2001. Mutations in the 3beta-hydroxysterol Delta24-reductase gene cause desmosterolosis, an autosomal recessive disorder of cholesterol biosynthesis. *The American Journal of Human Genetics*. 69:685-694.

Yla-Herttuala, S., W. Palinski, M.E. Rosenfeld, S. Parthasarathy, T.E. Carew, S. Butler, J.L. Witztum, and D. Steinberg. 1989. Evidence for the presence of oxidatively modified low density lipoprotein in atherosclerotic lesions of rabbit and man. *The Journal of Clinical Investigation*. 84:1086-1095.

Zidovetzki, R., and I. Levitan. 2007. Use of cyclodextrins to manipulate plasma membrane cholesterol content: evidence, misconceptions and control strategies. *Biochimica et Biophysica Acta*. 1768:1311-1324.

Permissions

The contributors of this book come from diverse backgrounds, making this book a truly international effort. This book will bring forth new frontiers with its revolutionizing research information and detailed analysis of the nascent developments around the world.

We would like to thank Dr. Deniz Ekinci, for lending his expertise to make the book truly unique. He has played a crucial role in the development of this book. Without his invaluable contribution this book wouldn't have been possible. He has made vital efforts to compile up to date information on the varied aspects of this subject to make this book a valuable addition to the collection of many professionals and students.

This book was conceptualized with the vision of imparting up-to-date information and advanced data in this field. To ensure the same, a matchless editorial board was set up. Every individual on the board went through rigorous rounds of assessment to prove their worth. After which they invested a large part of their time researching and compiling the most relevant data for our readers. Conferences and sessions were held from time to time between the editorial board and the contributing authors to present the data in the most comprehensible form. The editorial team has worked tirelessly to provide valuable and valid information to help people across the globe.

Every chapter published in this book has been scrutinized by our experts. Their significance has been extensively debated. The topics covered herein carry significant findings which will fuel the growth of the discipline. They may even be implemented as practical applications or may be referred to as a beginning point for another development. Chapters in this book were first published by InTech; hereby published with permission under the Creative Commons Attribution License or equivalent.

The editorial board has been involved in producing this book since its inception. They have spent rigorous hours researching and exploring the diverse topics which have resulted in the successful publishing of this book. They have passed on their knowledge of decades through this book. To expedite this challenging task, the publisher supported the team at every step. A small team of assistant editors was also appointed to further simplify the editing procedure and attain best results for the readers.

Our editorial team has been hand-picked from every corner of the world. Their multi-ethnicity adds dynamic inputs to the discussions which result in innovative outcomes. These outcomes are then further discussed with the researchers and contributors who give their valuable feedback and opinion regarding the same. The feedback is then collaborated with the researches and they are edited in a comprehensive manner to aid the understanding of the subject.

Apart from the editorial board, the designing team has also invested a significant amount of their time in understanding the subject and creating the most relevant covers. They scrutinized every image to scout for the most suitable representation of the subject and create an appropriate cover for the book.

The publishing team has been involved in this book since its early stages. They were actively engaged in every process, be it collecting the data, connecting with the contributors or procuring relevant information. The team has been an ardent support to the editorial, designing and production team. Their endless efforts to recruit the best for this project, has resulted in the accomplishment of this book. They are a veteran in the field of academics and their pool of knowledge is as vast as their experience in printing. Their expertise and guidance has proved useful at every step. Their uncompromising quality standards have made this book an exceptional effort. Their encouragement from time to time has been an inspiration for everyone.

The publisher and the editorial board hope that this book will prove to be a valuable piece of knowledge for researchers, students, practitioners and scholars across the globe.

List of Contributors

Jana Viskupicova
Institute of Experimental Pharmacology and Toxicology, Slovak Academy of Sciences, Slovakia
Department of Biochemistry and Microbiology, Slovak University of Technology in Bratislava, Slovakia

Miroslav Ondrejovic
Department of Biotechnology, University of SS. Cyril and Methodius in Trnava, Slovakia
Department of Biocentrum, Food Research Institute in Bratislava, Slovakia

Tibor Maliar
Department of Biotechnology, University of SS. Cyril and Methodius in Trnava, Slovakia

Con Dogovski, Sarah. C. Atkinson, Sudhir R. Dommaraju, Lilian Hor, Jason J. Paxman, Martin G. Peverelli, Theresa W. Qiu, Jacinta M. Wubben, Tanzeela Siddiqui and Nicole L. Taylor
Department of Biochemistry and Molecular Biology, Bio21 Molecular Science and Biotechnology Institute, University of Melbourne, Parkville, Victoria, Australia

Matthew Downton, Stephen Moore, Matthias Reumann and John Wagner
IBM Research Collaboratory for Life Sciences-Melbourne, Victorian Life Sciences Computation Initiative, University of Melbourne, Parkville, Victoria, Australia

Matthew A. Perugini
Department of Biochemistry and Molecular Biology, Bio21 Molecular Science and Biotechnology Institute, University of Melbourne, Parkville, Victoria, Australia
Department of Biochemistry, La Trobe Institute for Molecular Science, La Trobe University, Melbourne, Australia

Lei Zheng
The Sidney Kimmel Comprehensive Cancer Center, The Skip Viragh Center for Pancreatic Cancer, The Sol Goldman Pancreatic Cancer Center, Department of Oncology, and Department of Surgery, The Johns Hopkins University School of Medicine, Baltimore, Maryland, USA
Department of Surgery, the Second Affiliated Hospital, Zhejiang University College of Medicine, Hangzhou, China

Jiangtao Li
Department of Surgery, the Second Affiliated Hospital, Zhejiang University College of Medicine, Hangzhou, China

Yan Luo
Section of Biochemistry and Genetics, School of Basic Medical Sciences and Cancer Institute, the Second Affiliated Hospital, Zhejiang University College of Medicine, Hangzhou, China

Renjitha Pillai and Jamie W. Joseph
School of Pharmacy, University of Waterloo, Waterloo, Canada

Tatsuaki Tsuruyama
Department of Molecular Pathology, Graduate School of Medicine, Kyoto University, Kyoto, Kyoto Prefecture, Japan

Kuo-Hsiang Tang
Carlson School of Chemistry and Biochemistry, and Department of Biology, Clark University, Worcester, USA

Mitsushi J. Ikemoto and Taku Arano
Biomedical Research Institute, National Institute of Advanced Industrial Science and Technology (AIST), Graduate School of Science, Toho University, Japan

Yasunori Watanabe
Graduate School of Life Science, Hokkaido University, Japan
Institute of Microbial Chemistry, Tokyo, Japan

Nobuo N. Noda
Institute of Microbial Chemistry, Tokyo, Japan

Alberto Ouro, Lide Arana, Patricia Gangoiti and Antonio Gomez-Muñoz
Department of Biochemistry and Molecular Biology, Faculty of Science and Technology, University of the Basque Country, Bilbao, Spain

Ivo Juranek, Lucia Rackova and Milan Stefek
Institute of Experimental Pharmacology and Toxicology, Slovak Academy of Sciences, Slovakia

J. Thomas and T.P. Shentu
Department of Medicine, University of Illinois, Chicago, USA

Dev K. Singh
Division of Developmental Biology, Department of Pediatrics, Children's Hospital of University of Illinois, University of Illinois at Chicago, USA

Printed in the USA
CPSIA information can be obtained
at www.ICGtesting.com
JSHW011812301024
72690JS00002B/52